AF676570

Anaerobic Co-Digestion of Lignocellulosic Waste

Anaerobic Co-Digestion of Lignocellulosic Waste

Editors

Luis Isidoro Romero García
Carlos José Álvarez Gallego
Luis Alberto Fernández Güelfo

MDPI • Basel • Beijing • Wuhan • Barcelona • Belgrade • Manchester • Tokyo • Cluj • Tianjin

Editors
Luis Isidoro Romero García
University of Cadiz
Spain

Carlos José Álvarez Gallego
University of Cadiz
Spain

Luis Alberto Fernández Güelfo
University of Cadiz
Spain

Editorial Office
MDPI
St. Alban-Anlage 66
4052 Basel, Switzerland

This is a reprint of articles from the Special Issue published online in the open access journal *Applied Sciences* (ISSN 2076-3417) (available at: https://www.mdpi.com/journal/applsci/special_issues/Anaerobic_Co_Digestion).

For citation purposes, cite each article independently as indicated on the article page online and as indicated below:

LastName, A.A.; LastName, B.B.; LastName, C.C. Article Title. *Journal Name* **Year**, *Volume Number*, Page Range.

ISBN 978-3-0365-1142-9 (Hbk)
ISBN 978-3-0365-1143-6 (PDF)

© 2021 by the authors. Articles in this book are Open Access and distributed under the Creative Commons Attribution (CC BY) license, which allows users to download, copy and build upon published articles, as long as the author and publisher are properly credited, which ensures maximum dissemination and a wider impact of our publications.
The book as a whole is distributed by MDPI under the terms and conditions of the Creative Commons license CC BY-NC-ND.

Contents

About the Editors

Luis Isidoro Romero García (Professor) was born in Cádiz on 18 December 1960, and he graduated in Chemical Sciences in 1982, obtaining a Ph.D. in Chemical Sciences in 1991 at the University of Cádiz (Spain). He reached a position as Senior Lecturer of Chemical Engineering at the University of Cádiz in 1993 and finally, since 2010, he has held a position as Professor of Chemical Engineering at the Department of Chemical Engineering and Food Technology of the University of Cádiz (Spain). He is the co-author of more than 100 papers, most of them published in high impact factor journals (http://orcid.org/0000-0001-5193-7600). In addition, he is the co-author of 28 books or book chapters, more than 150 communications in congresses and 7 international patents. The main topics of interest in his research career have been related to bioenergy production (bio-hydrogen and bio-methane) from waste by anaerobic processes. Currently, he is engaged in the study of polyhydroxyalkanoates (PHAs) production from volatile fatty acids (VFAs), generated by dark fermentation. Prof. Romero-García has been involved in more than 50 Research Projects, being responsible for 14 of them. He has frequently acted as a reviewer for more than 17 high impact factor journals. In addition, he is a regular evaluator of research projects for different national and international agencies and he has previously been a member of the National Project Evaluation Commission of the Spanish Plan for Research, Development and Innovation.

Carlos José Álvarez Gallego (Senior Lecturer) was born in Cádiz (Spain) on 14 April 1973, and graduated in Chemical Sciences (1996) with a Ph.D. in Chemical Engineering (2005) from the University of Cádiz (Spain). His professional career has always been linked to the University of Cádiz. First of all, he completed his predoctoral fellowship and he reached the teaching position of assistant professor in 2000, lecturer in 2006 and Senior Lecturer in 2016, a position that he still holds. His career was developed in the Chemical Engineering and Food Technology Department of the University of Cádiz, where he worked as a Seceretary. He is the author of numerous publications (http://orcid.org/0000-0002-1222-2482) focused in the field of anaerobic treatments of waste and wastewater. Additionally, he is the co-author of the volumes I.3 and II.2 of the encyclopedia: from waste to resource, all the way to sustainability (Madrid, Spain). Mundiprensa Ed. 2015 (ISBN 978-84-8476-702-2 and 978-84-8476-700-8). His past and actual research topics have covered aerobic composting, anaerobic digestion, methane and bio-hydrogen production and related ones.

Luis Alberto Fernández Güelfo (Senior Lecturer) was born in El Puerto de Santa María (Cádiz, Spain) on 25 August 1978, and graduated in Chemical Engineering (2001) and with a Ph.D. in three specialties "Chemical", "Chemical Engineering" and "Environmental Engineering" (2008) from the University of Cádiz (Spain). His academic career has also been linked to the University of Cádiz and, additionally, he was the general manager of a spin-off company for five years (2009–2014). He became an assistant professor in 2013 and a Senior Lecturer in 2018, a position which he still holds today. His academic career was developed at the Environmental Technology Department of the University of Cádiz, where he was the academic director of two international master's degrees and deputy director of an external cathedra with companies. He is the author of numerous publications (http://orcid.org/0000-0001-7158-3352) focused in the field of biological treatments of wastes and wastewaters with the application of pretreatments. His past research topics have covered aerobic composting, anaerobic digestion, methane and bio-hydrogen production and related ones and, currently, the research he is developing is focused on bioplastics production (PHAs) from waste.

applied sciences

MDPI

Editorial

Editorial of the Special Issue "Anaerobic Co-Digestion of Lignocellulosic Wastes"

Luis Isidoro Romero-García [1,*], Carlos José Álvarez-Gallego [1] and Luis Alberto Fernández-Güelfo [2]

1 Department of Chemical Engineering and Food Technology, University of Cádiz, 11510 Puerto Real, Cádiz, Spain; carlosjose.alvarez@uca.es
2 Department of Environmental Technologies, University of Cádiz, 11510 Puerto Real, Cádiz, Spain; alberto.fdezguelfo@uca.es
* Correspondence: luisisidoro.romero@uca.es

Received: 4 October 2020; Accepted: 14 October 2020; Published: 22 October 2020

1. Introduction

Carbohydrates from vegetal biomass (wood and agricultural biomass) are the focus of biorefinery strategies [1]. In the search for a cheap, widespread and available carbon source not related to the food market (second generation biomass), all efforts have been focused on lignocellulosic biomass as the main substrate. However, lignocellulosic structures are difficult to depolymerize and pretreatments are needed in order to access the readily degradable carbon.

On the other hand, anaerobic digestion is a technologically mature microbial process which has been considered as a theoretically carbon-neutral process when vegetal biomass is used as a substrate. Even a substantial reduction of greenhouse gases impact is possible if natural and decontrolled anaerobic decomposition is considered as the usual fate of some types of waste (e.g., animal manures or forestry residuals) [2]. In this issue, Muhayodin et al. [3] present a global study about the applicability of anaerobic digestion as a process for reducing the greenhouse gases from rice straw as an example of lignocellulosic material including a mass balance for nutrients. The anaerobic digestion of rice straw is a clear case of reducing the impact of greenhouse gases. In data extracted from Scival®, the *anaerobic digestion/digester/methane production* topic (T155) is ranked 7th of 1621 topics in the *Environmental engineering* subject by prominence index based on Citation count, Scopus view count and average CiteScore, with a percentile of 99.937 [4].

In the present Special Issue, a collection of research and review papers has been presented for the discussion of the possibilities of anaerobic co-digestion of lignocellulosic biomass for bioenergy production with a special emphasis on the pretreatment of biomass to enhance the biodegradability and the subsequent biogas production. Five reviews have been included to provide a full dimension of the topic.

2. Pretreatments in the Hotspot

As a reflection of the clear interest in pretreatment procedures for the improvement of the anaerobic digestion of lignocellulosic biomass, the first half of the papers in the Special Issue are related to pretreatments. The reviews by Sayara et al. [5] and Hernández-Beltrán et al. [6] contribute with an overview on the possible types of pretreatment and the subsequent conditions of the anaerobic digestion process for biogas production. In addition, the latter paper includes a techno-economic and environmental analysis. Acid pretreatment is a highly effective chemical technique used to disrupt the lignocellulosic matrix by cleavage of glucosidic bonds, where mild acid pretreatments are the most promising technology [7]. Sorghum stalks from sixteen representative pedigreed sorghum mutant lines were tested by Xu et al. [8] with a classical diluted acid pretreatment as a previous step for enzymatic saccharification. The authors obtained a glucose recovery of 80–85% from cellulose and hemicellulose with minimal degradation products, proposing a pre-washing step where water-extractive content exceeded 35% (*w/w*), before being subjected to the acid pretreatment process in order to avoid sugar losses or degradation. Furthermore, an effective option is the organosolv pretreatment, which Sulbarán-Rangel et al. [9] verified using corncob wastes. The organosolv process is a thermochemical delignification process that consists of breaking the internal lignin and hemicellulose bonds of the lignocellulosic material with solvents. In this case, ethanol/acetic acid (1:10) was applied as organosolv reagent to obtain the hydrolysates. Improvement in biogas production was four times greater in the batch test. As a common co-substrate, animal manures are particulate organic-rich materials with negative effects on the kinetics of the anaerobic process. The effect of enzymatic (cellulase) pretreatment on particulate disintegration of this waste was studied by Tassew et al. [10]. Moreover, the review by Ahmed et al. [11] offers an in-depth insight into hydrothermal pretreatment with a section dedicated to inhibitors formed by the delignification process (mainly furfural and 5-HMF) and strategies to overcome it.

3. Anaerobic Co-Digestion Process

Aboudi et al. [12] presented a study of the mesophilic anaerobic co-digestion of lignocellulosic material (exhausted sugar beet pulp) with animal manures (pig and cow) in reactors fed in semi-continuous mode. This mixture shows the huge potential of co-substrates to improve the methane yield (70 and 31% increase in methane production, respectively) by mitigating the inhibitory effect of volatile fatty acids at high loading rate conditions. The tests were analyzed from a novel perspective, evaluating the uncoupling of anaerobic phases (hydrolysis, acidogenesis/acetogenesis and methanogenesis) through the calculation of indirect carbon-related parameters such as acidogenic substrate carbon (ASC). In the same way, Gómez-Quiroga et al. [13] refer to a similar study with the same type of lignocellulosic material (exhausted sugar beet pulp) under thermophilic operating conditions, where batch tests were assayed, showing that the activity of acetoclastic methanogens was especially affected at thermophilic conditions while organic matter solubilization was more efficient. As a final colophon, the review of Sarker et al. [14] focuses on the different types of bioreactors and configurations for the anaerobic process.

Funding: This research received no external funding.

Conflicts of Interest: The authors declare no conflict of interest.

References

1. Ajao, O.; Marinova, M.; Savadogo, O.; Paris, J. Hemicellulose based integrated forest biorefineries: Implementation strategies. *Ind. Crop. Prod.* **2018**, *126*, 250–260. [CrossRef]
2. Pabón-Pereira, C.P.; Hamelers, H.V.M.; Matilla, I.; Van Lier, J.B. New insights on the estimation of the anaerobic biodegradability of plant material: Identifying valuable plants for sustainable energy production. *Processes* **2020**, *8*, 806. [CrossRef]
3. Muhayodin, F.; Fritze, A.; Rotter, V.S. A Review on the fate of nutrients and enhancement of energy recovery from rice straw through anaerobic digestion. *Appl. Sci.* **2020**, *10*, 2047. [CrossRef]
4. Scival ®. Report from Scival in the period 2015–2019 Using Topic (T155) and Topic Clusters (TC65 and TC119). Available online: www.scival.com (accessed on 10 September 2020).
5. Sayara, T.; Sánchez, A. A review on anaerobic digestion of lignocellulosic wastes: Pretreatments and operational conditions. *Appl. Sci.* **2019**, *9*, 4655. [CrossRef]
6. Hernández-Beltrán, J.U.; Hernández-De Lira, I.O.; Cruz-Santos, M.M.; Saucedo-Luevanos, A.; Hernández-Terán, F.; Balagurusamy, N. Insight into pretreatment methods of lignocellulosic biomass to increase biogas yield: Current state, challenges, and opportunities. *Appl. Sci.* **2019**, *9*, 3721. [CrossRef]

7. Woiciechowski, A.L.; Dalmas-Neto, C.J.; Porto de Souza-Vandenberghe, L.; Pedro de Carvalho-Neto, D.; Novak-Sydney, A.C.; Junior-Letti, L.A.; Grace-Karp, S.; Zevallos-Torres, L.A.; Ricardo-Soccol, C. Lignocellulosic biomass: Acid and alkaline pretreatments and their effects on biomass recalcitrance—Conventional processing and recent advances. *Bioresour. Technol.* **2020**, *304*, 122848. [CrossRef] [PubMed]
8. Xu, Y.; Li, J.; Xin, Z.; Bean, S.R.; Tilley, M.; Wang, D. Water-soluble sugars of pedigreed sorghum mutant stalks and their recovery after pretreatment. *Appl. Sci.* **2020**, *10*, 5472. [CrossRef]
9. Sulbarán-Rangel, B.; Alarcón Aguirre, J.S.; Breton-Deval, L.; del Real-Olvera, J.; Gurubel Tun, K.J. Improvement of anaerobic digestion of hydrolysed corncob waste by organosolv pretreatment for biogas production. *Appl. Sci.* **2020**, *10*, 2785. [CrossRef]
10. Tassew, F.A.; Bergland, W.H.; Dinamarca, C.; Bakke, R. Effect of Particulate Disintegration on biomethane potential of particle-rich substrates in batch anaerobic reactor. *Appl. Sci.* **2019**, *9*, 2880. [CrossRef]
11. Ahmed, B.; Aboudi, K.; Tyagi, V.K.; Álvarez-Gallego, C.J.; Fernández-Güelfo, L.A.; Romero-García, L.I.; Kazmi, A.A. Improvement of anaerobic digestion of lignocellulosic biomass by hydrothermal pretreatment. *Appl. Sci.* **2019**, *9*, 3853. [CrossRef]
12. Aboudi, K.; Gómez-Quiroga, X.; Álvarez-Gallego, C.J.; Romero-García, L.I. Insights into anaerobic co-digestion of lignocellulosic biomass (sugar beet by-products) and animal manure in long-term semi-continuous assays. *Appl. Sci.* **2020**, *10*, 5126. [CrossRef]
13. Gómez-Quiroga, X.; Aboudi, K.; Álvarez-Gallego, C.J.; Romero-García, L.I. Enhancement of methane production in thermophilic anaerobic co-digestion of exhausted sugar beet pulp and pig manure. *Appl. Sci.* **2019**, *9*, 1791. [CrossRef]
14. Sarker, S.; Lamb, J.J.; Hjelme, D.R.; Lien, K.M. A review of the role of critical parameters in the design and operation of biogas production plants. *Appl. Sci.* **2019**, *9*, 1915. [CrossRef]

Publisher's Note: MDPI stays neutral with regard to jurisdictional claims in published maps and institutional affiliations.

© 2020 by the authors. Licensee MDPI, Basel, Switzerland. This article is an open access article distributed under the terms and conditions of the Creative Commons Attribution (CC BY) license (http://creativecommons.org/licenses/by/4.0/).

applied sciences

MDPI

Review

A Review on the Fate of Nutrients and Enhancement of Energy Recovery from Rice Straw through Anaerobic Digestion

Furqan Muhayodin [1,2,*], Albrecht Fritze [1] and Vera Susanne Rotter [1]

1 Department of Environmental Technology, Chair of Circular Economy and Recycling Technology, TU Berlin, 10623 Berlin, Germany; a.fritze@tu-berlin.de (A.F.); vera.rotter@tu-berlin.de (V.S.R.)
2 Department of Farm Machinery and Power, University of Agriculture, Faisalabad 38000, Pakistan
* Correspondence: furqan.muhayodin@campus.tu-berlin.de; Tel.: +49-(0)30-314-22619

Received: 19 February 2020; Accepted: 14 March 2020; Published: 18 March 2020

Abstract: Open field burning and tilling the rice straw (RS) back into the fields causes environmental threats by contributing to the increased greenhouse gas emissions. Energy and nutrient recovery from RS through anaerobic digestion (AD) is an effective solution for its utilization. Although RS has good methane potential, its characteristics make it a difficult substrate for AD. This paper reviews the characteristics of RS, mass balance, and distribution of nutrients into liquid and solid digestate in the AD. The present review also discusses the effect of temperature, co-digestion, mixing, inoculum, organic loading rate, recycling liquid digestate, the addition of trace elements, and their bioavailability on the enhancement of biogas/methane yield in the AD of RS. In addition, the digestion of RS at various scales is also covered in the review.

Keywords: anaerobic digestion; rice straw; nutrients; recycling; digestate

1. Introduction

Rice is ranked as the third-largest crop in the world [1] and it is one of the most important staple foods of humans [2]. Rice straw (RS) is an agricultural residue obtained at the time of harvesting of the rice crop [2]. The rate of RS generated per ton of rice crop varies from 0.4 to 4.0 [3–5] due to cultivated rice variance, cultivation technique, and harvesting method [3]. According to the International Rice Research Institute (IRRI), an average of 1.1 ton of RS are generated with each ton of rice produced (range of 0.7–1.4) [6]. There has been an increasing trend in rice production and in the generation of RS in the last decade. A total of 686 million tons of rice and approximately 754 million tons of RS were produced in 2009 [7] while their production was 769 million tons and 846 million tons in 2017 globally, according to the Food and Agriculture Organization of the United Nations [7].

RS is used for feeding animals, heating homes, construction of buildings, and making paper. However, probably the largest share of straw is left unused in the fields [2]. Open-field burning and tilling the RS back into the fields are the common practices for its disposal. Both cause environmental threats by contributing to the increased greenhouse gas emissions [8,9]. Furthermore, the incorporation of RS in the soil increases foliar diseases, which resulted in reduced crop yields [10].

Conversion of RS into more valuable products like methane, ethanol, nutrient-rich soil conditioner, or biofuels are better options compared to its conventional utilization as energy challenges and environmental concerns can be addressed together [2]. The conversion can be done by thermochemical or biological processes [11]. The thermochemical processes like gasification [1], combustion [12], and pyrolysis [13] have been used as energy recovery methods but can cause a threat to the environment. Composting is a biological process that has been used for the conversion of RS into a valuable fertilizer with other substrates [14,15]. Anaerobic digestion (AD) is another biological process which converts biomass into various biofuels or platform chemicals such as methane, ethanol, hydrogen, etc. [1]. It is considered to be one of the most environmentally friendly processes [2]. Moreover, AD technology can be used economically for dry and wet biomasses on a variety of scales, and it requires less input energy as compared to thermochemical processes [16]. Moreover, RS is a bulky material because of its low density (70 kg/m^3) which makes it difficult to transport to a storage place for energy recovery by thermochemical processes [17].

During the AD process in biogas technology, organic material is degraded biologically by microorganisms in the absence of molecular oxygen and converted into biogas. The biogas contains primarily methane (CH_4) and carbon dioxide (CO_2) with small quantities of water vapor ($H_2O_{(g)}$), ammonia (NH_3), and hydrogen sulfide (H_2S) [11]. Appels et al. [18] explained the AD in four stages: hydrolysis, acidogenesis, acetogenesis, and methanogenesis. Since carbon (C) bound in organic matter is released as CH_4 and CO_2 during the AD, the C/N ratio of the remaining digestate decreases [19]. Additionally, the plant availability of N is enhanced through the AD process, as organically bound N is mineralized into ammonium (NH_4^+) which is readily available [20].

The utilization of RS for the recovery of nutrients and energy through AD technology is more desirable than its common disposal practices (open-field burning and tilling in soil). AD technology can be viable for RS because of its high generation rate and potential to reduce greenhouse gas emissions. Energy recovery through AD of food crops is controversial since the demand for food is expected to rise due to an increasing world population [21]. RS as an agricultural residue and does not affect the food security. Furthermore, the calculated and measured biogas and methane yields of RS render it an attractive substrate for AD [22].

According to Baserga [23], biogas and methane potentials of RS can be calculated based on the crude nutrients (ash, fat, fiber, N-free extracts, protein), resulting in 409–415 and 207–211 L/kg VS, respectively [24]. The CH_4 yield measured under different experimental conditions were reported from 92 to 280 L/kg VS with various inocula and temperatures [11]. Other authors reported different CH_4 yields in their studies such as 226–281 L/kg VS [25] and 100–300 L/kg VS [26].

In addition, many emerging factors advance the utilization of RS for energy recovery through AD, like the usage of appropriate inocula and pretreatment methods. Researchers have become increasingly interested in the AD of RS as the Web of Science ® shows an increasing trend of published papers with "anaerobic digestion" and "rice straw" as topics for the last 10 years as shown in Figure 1.

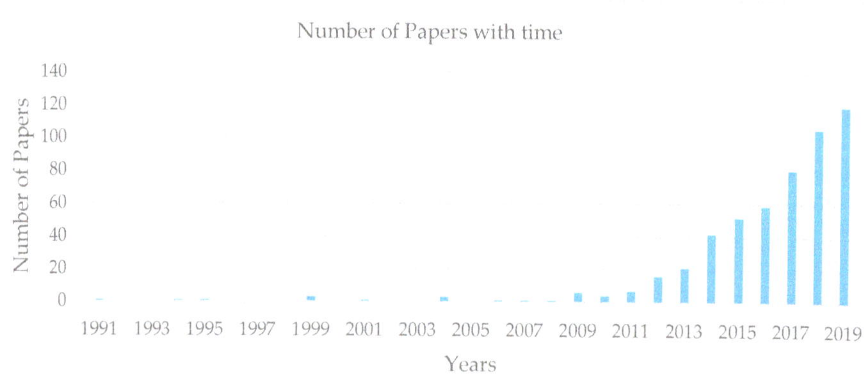

Figure 1. Web of Science ®bibliometric study with topics "anaerobic digestion" and "rice straw" from 1991 until December 2019. The number of publications from 1991 to 1999 was 6; from 2000 to 2008 was 7; and from 2009 to 2019 was 512.

Apart from its availability and CH_4 potential, RS has a complex, lignocellulosic structure and ligno-carbohydrate complexes that create a barrier for the microbial conversion, which make it a difficult substrate to decompose by AD [4,27]. Other challenges associated with mono-digestion of RS are the unbalanced nutritional composition (high C/N ratio) [28], low content of essential trace elements (TEs) [29], and its floatation behavior in wet AD systems [30]. Therefore, RS has not been used as an attractive substrate for the recovery of energy and nutrients through AD.

Significant research and development have taken place in the last 10 years in the field of AD of RS as evidenced by the number of published papers in the scientific literature, as shown in Figure 1. In 2013, Mussoline et al. [11] reviewed the AD of RS. This review provides insights only into certain aspects such as greenhouse gas emissions from rice fields, energy recovery potential of RS, various pretreatment strategies to enhance biogas from RS, optimal nutrient balance, and microbiological considerations. However, new knowledge and technologies are continuously unfolding in the field of AD of RS for energy and nutrients recovery.

The process parameters such as temperature, organic loading rate (OLR), and mixing are fundamental in the AD and here referred to as "factors". The other process techniques, such as adding co-substrates, usage of appropriate inoculum in a suitable amount, recycling liquid digestate (LD), the addition of TEs, and their bioavailability, are also relevant to the AD of RS and termed as "methods" in this review. The objectives of this paper are to provide an overview of the effect of these factors and methods on the enhancement of energy recovery (biogas/methane yield) from RS through AD. Additionally, the mass balance of nutrients and their distribution into LD and solid digestate (SD) during AD are discussed in this review. Moreover, the digestion of RS at various scales is also part of the current paper. The focus is the detailed highlighting of these aspects in this review and those studies with other similar substrates have also been included for comparison where fewer studies with RS were available.

The effects of various pretreatment techniques on the optimization of AD of RS have been discussed in the review conducted by Mussoline et al. [11]. The enhancement in biogas yield from lignocellulosic substrates including RS by using various pretreatment methods (mechanical, chemical, biological, and physiochemical) has been reviewed recently by Hernández-Beltrán et al. [31]. Moreover, a recent review conducted by Ahmed et al. [32] discusses the optimization of the AD of lignocellulosic substrates by hydrothermal pretreatment. These studies reviewed the effect of various pretreatment methods on the enhancement of energy recovery from RS and other similar substrates and, hence, have not been included in this review. Some aspects of the AD of RS have been reviewed by Mussoline et al. [11] in 2013, and some are discussed in this paper. Therefore, Table 1 indicates the part of the AD discussed by Mussoline et al. [11] and the part which is presented in the current review to show the complementarity.

RS is a difficult substrate to degrade under anaerobic conditions due to its complex structure [27]. It is critical to achieve a stable AD process to get efficient biogas production. The project "BIORIST" aimed at designing technically feasible and innovative RS digestion processes. During a 3 year project phase, the process technology was developed and successfully tested in a pilot-scale plant [33].

This paper focuses on factors supporting a stable AD process with RS as an input substrate to produce biogas. This will help to further enhance process technology which mostly exists at a laboratory or pilot-scale to full-scale application. All further post-treatments or optimizations such as biogas purifications can be regarded once a stable biogas technology for RS is established. Further processing of the biogas to biomethane is not considered here, although various post-treatments exist to increase the proportion of valuable biomethane from biogas up to the quality of natural gas. These treatments include water scrubbing, cryogenic separation, physical absorption, chemical absorption, pressure swing adsorption, membrane technology, in-situ methane enrichment, hydrate formation, biological methods, and photoautotrophic methods [34,35]. However, the detailed description of the principles of these treatments is not the scope of this review.

Considering the low number of published data on the experience of pilot- and farm-scale plants for RS digestion resulting from missing implementation of AD of RS, an economic comparison is not reasonable and therefore excluded from this review.

Table 1. Aspects of anaerobic digestion (AD) of rice straw (RS) discussed in both studies.

Categories	Mussoline et al. [11]	This Review
Environmental concerns	Emissions of greenhouse gas from rice fields	
Product		Mass balance of nutrients and their distribution into liquid digestate (LD) and solid digestate (SD) in the AD of RS
Potential	Biogas production potential of RS	AD of RS at various scales
Influence	Effect of the following parameters on the AD of RS • Temperature • pH • Nutrients balance • Total solids	Effect of the following factors and methods on the enhancement of energy recovery from the AD of RS • Temperature • Co-digestion • Inoculum • Mixing • Organic loading rate (OLR) • Recycling LD • Addition of trace elements (TEs) • Bioavailability of TEs
Pretreatment	Effect of the following pretreatment alternatives to enhance biogas potential of RS • Mechanical size reduction • Biological treatment • Acid treatment • Alkaline treatment	
Microorganisms	Effect of various microorganisms on the AD of RS	

2. Characteristics of RS

RS is produced when harvesting rice crop, and it can be collected from the field once it contains a dry matter of more than 75%.

RS looks like flat fibers having approximate dimensions of 20–60 cm in length and 0.5 cm in width [4].

The general characteristics of RS such as total solids (TS), volatile solids (VS), total carbon (TC), total nitrogen (TN), and C/N ratio are usually reported as a single value or mean values with a variation in the literature. The original data is further processed to generate median, maximum, minimum, and quartiles as shown in Figure 2. Box plots of TS and VS are smaller, and the whiskers do not extend as far, indicating that these properties do not differ much over the various analyzed studies, while the C/N ratio is different for the RS used in these studies with spread-out data points [10,26,28,36–42].

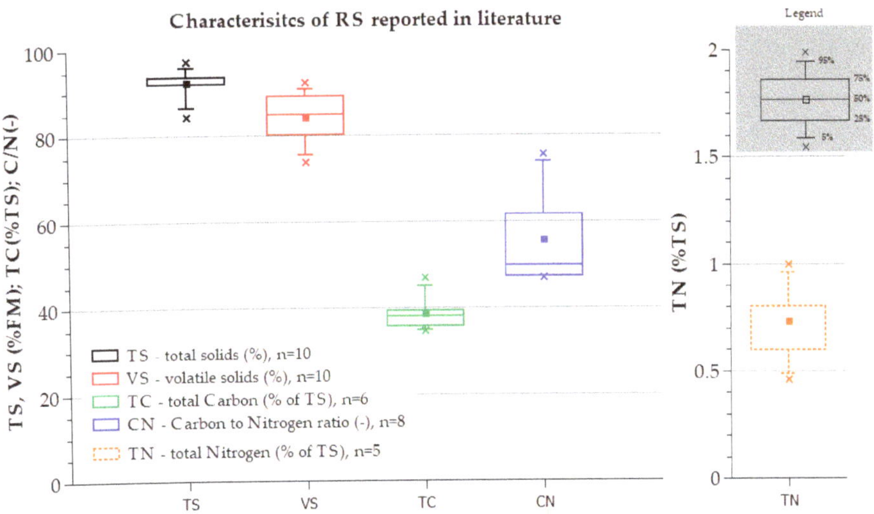

Figure 2. Statistical evaluation of RS characteristics reported in the literature. "n" is the number of values to calculate the median, maximum, minimum, and quartiles [10,26,28,36–42].

Other nutrients and TEs, like Phosphorus (P), Calcium (Ca), Magnesium (Mg), Potassium (K), Sodium (Na), Nickel (Ni), Copper (Cu), Zinc (Zn), Iron (Fe), etc., are usually not analyzed in the RS, except in two studies (Table 2).

Table 2. Nutrients and TEs in RS reported in the literature. ("± " stands for the standard deviation of the mean as reported by the authors).

References	Unit	P	Ca	Mg	K	Na	Ni	Cu	Zn	Fe
[36]	(mg/kgTS)	1082 ± 118	6669 ± 17	1572 ± 12	9676 ± 23	712 ± 11	4.4 ± 0.6	18.3 ± 1.2	54.3 ± 1.6	320 ± 24
[37]	(mg/kgTS)	-	1940 ± 23	990 ± 10	25,050 ± 13	1230 ± 80	-	-	-	-

3. Nutrients Balance and Distribution into Liquid and Solid Digestate during Anaerobic Digestion

AD is an appropriate technology to convert agricultural residues into biogas [37] and to transform the nutrients contained in the substrates into plant-available forms so that the final digestate can be used as fertilizers [43,44]. The effective use of nutrients available in the digestate is a way to improve the overall efficiency of AD technology [37]. The mass balance of nutrients in the AD is crucial for the quantification of their recovery potential [45]. It is also essential to evaluate the fate of nutrients like N, P, K, Ca, Mg, and Sulfur (S) and their distribution into LD and SD to assess the agronomic, environmental, and economic benefits of AD technology [46].

The mass balance of nutrients has been investigated in the AD of different substrates [37,46–49]. The recovery rate (RR) of nutrients is defined as the amount of nutrients available in the digestate after the AD process relative to their initial value (before the AD) and expressed as a percentage (%) [37]. The RR is usually calculated by the authors and provided in the studies.

Various substrates have been used in the studies for the investigation of the mass balance of nutrients. The categories are defined here on the basis of the nature of the substrates. The category "straws" consists of rice, corn, and wheat straw while, "manures" contains swine, cattle, and chicken manure. The RR of N from various studies using different substrates is summarized in Table 3. N showed up to 82% RR for different substrates used in these various experiments at different scales, specifically, 93–100% for manures [37,46], 98.4% for pig slurry (PS) [47], 85–91% for straws [37], 90.8% for organic fraction of municipal waste (OFMSW) [48], and 82–94% for mixed substrates [48,49]. Total N is often reported to be conserved during AD [43]. However, the primary cause of less than

100% RR of N may be gaseous NH_3 migration with the biogas flux [48,49]. The RRs of N in these studies are comparable; however, the results of one study are contradictory, as the amount of total N in the digestate after AD was 8–23% higher than before AD [45]. The authors explained that the amount of hydrolyzed microalgae and the quantity of N released to the aqueous phase might have been overestimated since all the parameters were calculated as the difference between the microalgae tests and the control test [45].

Table 3. RR of nutrients during AD of various substrates.

Substrates	RR of N	RR of P	RR of K	RR of Ca	RR of Mg	RR of S	References
Straws	85–91%	91–97%	-	-	-	-	[37]
Manures	93–100%	74.5–97%	101.9%	91.3%	81.3%	32.3%	[37,46]
PS	98.4%	64%	106.2%	56%	67.5%	83.4%	[47]
OFMSW	90.8%	94%	94%	-	-	-	[48]
Mixed substrates	82–94%	70–94%	95–98%	-	-	-	[48,49]
Microalgae	108–123%	99–100%	-	-	-	-	[45]

RR, Recovery rate; **AD**, Anaerobic digestion; **Straws** (Rice straw, corn straw, wheat straw); **Manure** (Swine manure, cattle manure, chicken manure); **PS**, Pig slurry; **OFMSW**, Organic fraction of municipal solid waste; **Mixed substrates**, (swine manure, maize silage, triticale, sorghum, barley thresh from beer industry, molasses, bakery-industry waste, and olive mill sludge; swine manure, cow manure, milk whey, and rice culture by-products; cow manure, blood, fish, paper pulp, out of date beverages, and grease trap waste).

There is also a wide range (64–100%) of RRs of P when the different substrates are used in AD (Table 3). A RR of P of up to 91% during AD is not considered far from 100% [37], while up to 75% [46] and 64% [47] of P RR have been observed in two studies. The retention of the sludge in psychrophilic anaerobic sequencing batch reactors (PASBR) could be the possible reason for such a low RR of P [46]. In another study, crystal lining was observed on the opening of the digesters, and scanning electron microscopy (SEM-EDS) clearly showed them to be composed of P, Ca, Mg, and Mn (Manganese) [47].

K is conserved in the digestate in all these studies during the AD as shown in Table 3.

There are a few studies found addressing the mass balance of Ca, Mg, and S in AD. For example, on average, 8.7% of Ca, 18.7% of Mg, and 67.7% of S were retained in the bioreactors. However, the authors did not find statistically significant Ca and Mg retention in the digesters except for S [46]. Significant retention of Ca and Mg with 44% and 32.5% has been found in another study as they crystallize as carbonates and phosphates [47].

Li et al. [37] investigated the mass balance of nutrients (N and P) during the AD of RS in batch experiments. The RRs of N and P were 85–91% and 91–97%, respectively. The mass balance of other nutrients (K, Ca, Mg, and S) were not investigated in this study. This is the only study found in which RS was used and mass balance of nutrients (N and P) was performed, thus indicating the scarcity of nutrient mass balance on AD of RS.

The ranges of all nutrients' RRs for specific substrates are shown in Table 3. The results mentioned in Table 3 are reported as a single value or as a range in the studies. The arithmetic mean is first determined based on the minimum and maximum values if the data points in the study are available as a range. Then these means are further used to calculate the overall average RR of N and P in the AD irrespective of a specific substrate. The results of 12 and 14 experiments are used to calculate the average RR of N and P, respectively, in AD, which is shown in Figure 3. It is assumed that the RR of N is not 100% due to its release over the gas phase, whereas P is suspected to crystalize with other elements and remains on the reactor walls [47]. The overall average RRs of Ca, Mg, and S was not calculated because only two studies were available.

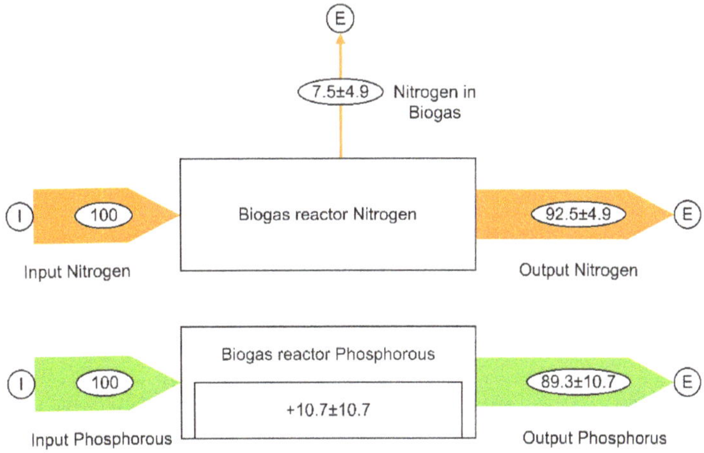

Figure 3. Calculated mass balance of N and P in percent, relative to the input amount. N is assumed to be released in the the gas phase; P is assumed to remain in the digesters as it forms crystals with other elements. The standard deviation of the mean is displayed as "±". The results of 12 [37,46–49] and 14 [37,45–49] experiments are used to calculate the average RR of N and P, respectively.

Concerning the effect of AD on the RR of nutrients, Figure 3 shows that 92.5% of N and 89.3% of P, on average, are recoverd. It can be concluded that N, P, and K are conserved in the digestate, as their RRs are close to 100%.

The separation of digestate into LD and SD with the use of separation techniques like press screws have been considered simple and low-cost practices for the management of the digestate from biogas plants and have been widely used in the European Union [43]. It has been reported that LD has a higher concentration of N in the form of ammonia, and it can be used as a fertilizer to substitute mineral fertilizer to some extent [43]. Moreover, the N/P ratio tends to increase in the LD because of separation, therefore, reducing the accumulation of P in the soil, as it offers a more balanced nutrient content for the crops [50]. The resulting SD generally used in the soil as an organic amendment [51], and it is also exported to other nearby farms to provide the solution of the excess of nutrients problem in that farm [52].

Tambone et al. [52] investigated thirteen full-scale biogas plants fed with a variety of substrates to study the distribution of nutrients into LD and SD after the usage of the screw press as a separation technique. Their results indicated that, on average, 87% of N measured as Total Kjeldahl Nitrogen (TKN) was available in the LD. In another study conducted on a full-scale biogas plant, treating the PS revealed that 85% of output N was also found in the LD [47]. Li et al. [37] carried out AD experiments in batch mode to study the distribution of N and P into LD and SD. They used three kinds of straws (rice, corn, wheat) and three kinds of manures (swine, cattle, and chicken) in their experiments. Their results indicated that on average, 34% of initial N was available in LD and 53% in SD, while the difference between initial and final N in case of straws was 9–15%. Similarly, for the manures, on average, 36% of initial N was available in LD and 59% in SD, while the difference in initial and final N after AD was 3–7% (as presented in Table 4). One possible reason for different proportions of nutrients in LD among these studies could be the usage of varying separation techniques like centrifugation [37], settling [46], and press screw [52].

Table 4. Distribution of nutrients into LD and SD. "n" is the number of experiments to calculate the standard deviation of the mean (±).

Substrate	Separation Technique	AD Mode	N		P		References
			LD (%)	SD (%)	LD (%)	SD (%)	
Straws (n = 3)	Centrifuge	Batch	33.7 ± 6.8	53.0 ± 9.7	82.5 ± 2.1	14.4 ± 3.8	[37]
Manures (n = 3)	Centrifuge	Batch	36.0 ± 39.9	59.2 ± 38.6	8.8 ± 7.7	85.7 ± 10.0	[37]
PS (n = 1)	Screening	Full-scale	85	3	90	10	[47]
Mixed Substrates (n = 13)	Screw press	Full-scale	86.8 ± 6.2	13.1 ± 6.2	71.6 ± 12.0	28.3 ± 12.0	[52]
SM (n=1)	Settling	PASBR	61.8 ± 3.3	34.5 ± 2.3	33.5 ± 2.5	62.3 ± 7.0	[46]

LD, Liquid digestate; **SD**, Solid digestate; **AD**, Anaerobic digestion; **Straws** (Rice straw, corn straw, wheat straw); **Manures** (Swine manure, cattle manure, chicken manure); **PS**, Pig slurry; **Mixed substrates** (PS, energy crops, cornmeal, milk whey, agro-industrial residues, cow slurry, molasses, poultry manure); **SM**, Swine manure; **PASBR**, psychrophilic anaerobic sequencing batch reactor.

Results obtained by Tambone et al. [52] indicated that the P available in LD accounted for 72% of the output (measured as P_2O_5). When the straws and PS used as substrates, 82% and 90% of P were available in the LD, respectively [37,47]. For manures, a significant portion of P (86%) was available in the SD [37]. One possible explanation could be that the P in the manures precipitate in solid form. This phenomenon is more likely due to a higher concentration of Mg and Ca which precipitates P [53].

A total of 14 full-scale AD plants, six batch experiments (three for straws and three for manures), and one PASBR were used in these studies to investigate the distribution of nutrients into LD and SD as presented in Table 4. The LD contained about 34–87% of N and 9–90% of P.

There is a wide range of nutrients (N and P) available in LD and SD. The explanation could be the usage of different separation techniques, various substrates, and different modes of AD. More studies need to be conducted to draw better conclusions about the distribution of nutrients into LD and SD. As data on the mass distribution of Ca, K, Mg, and S to LD and SD is scarce. Only N and P have been considered in this study. For RS, only one batch study was available on nutrients mass balance and LD/SD distribution. Future research into the mass balance of other important nutrients (Ca, K, Mg, and S) for RS co-digestion with manure or other suitable substrates, at pilot-scale or farm-scale plants, would help to evaluate the overall efficiency of AD technology and thus support establishing a mechanism to use the digestate.

4. Enhancement of Energy Recovery (Biogas/Methane Yield) from the Anaerobic Digestion of Rice Straw

The effects of the "factors" and "methods" on the improvement of energy recovery from AD of RS are discussed in the following section of this review.

4.1. Temperature

Temperature is one of the most significant parameters affecting activities, survival, and growth of microorganisms in the AD [54]. It also influences the biogas yield and digestate quality [55] in both batch and continuous mode of digestion. Temperature is a crucial parameter to maximize the efficiency of the AD but also in regards to economic input [11]. Generally, there are three different temperature ranges at which anaerobic bacteria can grow. These are: psychrophilic (10–30 °C), mesophilic (30–40 °C), and thermophilic (50–60 °C) [56].

Contreras et al. [25] conducted a study in which RS was digested for 36 days at 37 °C and 55 °C in the presence of mesophilic and thermophilic inocula, respectively. The cumulative CH_4 yield was 226 L/kg VS at 37°C (with average 56% CH_4 content) and 281 L/k VS at 55 °C (average CH_4 content: 67%). At thermophilic conditions, RS produced approximately 25% more CH_4 than at mesophilic conditions. Similar results have been obtained by Sathish et al. [54] who co-digested RS with press mud in digesters of 1 m^3 capacity. Sathish et al. [54] pointed out that cumulative biogas yield for the duration of 20 days under thermophilic conditions (45–55 °C) was 34% higher than mesophilic conditions (25–40 °C). RS mono-digestion has also been evaluated by Lianhua et al. [57] at mesophilic (35 °C) and thermophilic conditions (55 °C) at different TS (7.5% and 20%) in batch tests. Under wet AD, their results showed the CH_4 yield of 120 and 136 L/kg VS at mesophilic and thermophilic conditions, respectively. While CH_4 yield was 123 and 76 L/kg VS at mesophilic and thermophilic conditions, respectively, under dry AD. Yu et al. [42] investigated the effect of increasing temperature from 39 to 50 °C on the biogas yield from a 300 m^3 digester. The RS was added daily to the digester. The biogas yield increased by approximately 100% (from 200 L/kg TS to 402 L/kg TS) when the temperature of the system increased from 39 to 50°C. Moreover, the CH_4 content was improved from 50% to 60%.

The studies mentioned above investigated the effect of mesophilic and thermophilic conditions on the biogas/methane yield. The biogas/methane yields at both temperatures in these studies [25,42,54,57] are shown in Figure 4. About 13–100% more methane is yielded at thermophilic as compared to mesophilic conditions in the four experiments, while in one AD test, 62% more methane was produced at mesophilic conditions.

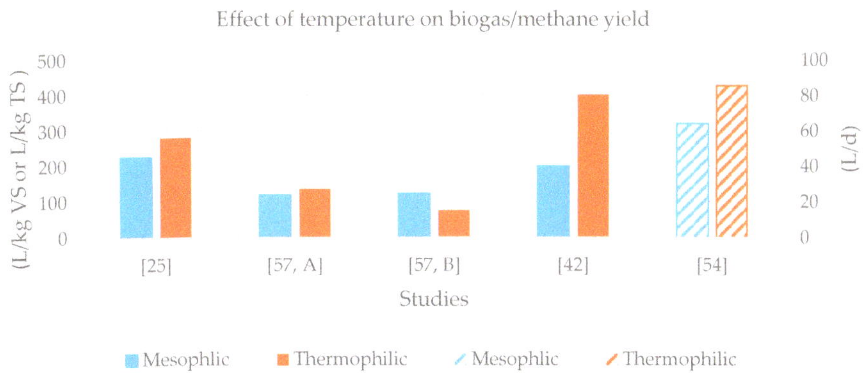

Figure 4. Effect of temperature on biogas/methane yield [25,42,54,57].

The effect of increasing temperature on the CH_4 yield from co-digestion of dairy manure (DM), chicken manure (CM), and RS has been investigated [26], and the results are shown in Figure 5. RS was co-digested with DM, CM, and also in a mixture with both manures in batch digesters at 20, 30, 40, 50, and 60 °C. CH_4 yield continuously increased by increasing the temperature. CH_4 yield in three mixtures was on average 2.5 times higher at 40 than at 20 °C, while only about 1.2 times higher at 60 than at 40 °C. No significant difference was observed in CH_4 yields between 50 and 60 °C [26].

The results of these studies demonstrate that psychrophilic AD of RS has the lowest CH_4 yield compared to mesophilic and thermophilic conditions. It illustrates the critical role of temperature in AD digester concerning energy recovery. Compared to mesophilic conditions, the AD of RS under thermophilic conditions yielded up to 100% more biogas yields. Moreover, the quality of biogas in terms of CH_4 content is also higher at higher temperatures. Thus, optimal energy recovery from the AD of RS is attained under thermophilic conditions.

Apart from the advantages of higher energy recovery, there are some disadvantages to the thermophilic AD process. For instance, variation in temperature should not exceed 0.6 °C/day to maintain a stable digestion process [18]. The mesophilic process is more robust and less sensitive to environmental changes than the thermophilic process [58].

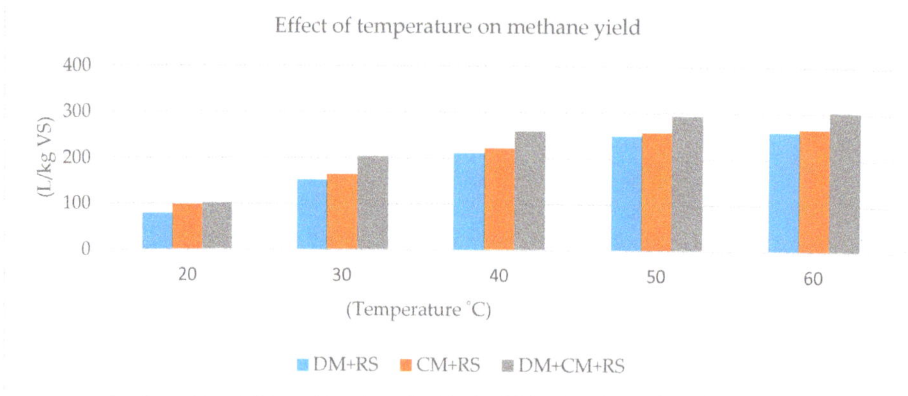

Figure 5. Effect of temperature on methane yield. Values are presented as the mean of three replicates (n = 3). **DM**, Dairy manure; **CM**, Chicken manure; **RS**, Rice straw [26].

4.2. Co-digestion of RS

RS is a promising substrate for AD because of its high biogas potential [11]. Untreated RS has a higher amount of total C (e.g., 35–47% TS) and very low content of N (e.g., ≤ 1% TS) with a typical C/N ratio of approximately 47–75 [10,39]. Optimum C/N ratio ranges from 20 to 30 [59,60]. As the optimal C/N ratio is crucial for efficient AD, an external source of N is required for the digestion of RS.

The term co-digestion describes the simultaneous digestion of a mixture of two or more substrates. Co-digestion offers various advantages like improved system stability, adjusted C/N ratio, a better nutrient balance, dilution of toxic compounds, improved buffer capacity, increased biodegradation, adjusted moisture content, a supply of TEs, and increasing OLR. It can therefore, enhance volumetric biogas recovery [61–63] and improves the fertilizer value of the digestate [64]. These advantages of co-digestion are presented in Figure 6. Co-digestion of RS with different animal manures is considered a more cost-effective method for nutrition regulation than the addition of N-containing chemical reagents like ammonium bicarbonate or urea [40].

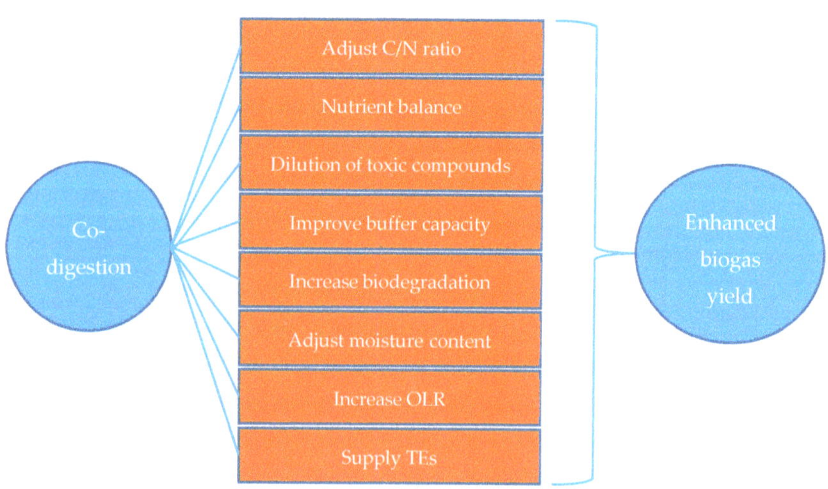

Figure 6. Advantages of anaerobic co-digestion [61–63].

Ye et al. [39] attained about 55%, 72%, and 47% higher biogas yield compared to RS mono-digestion when kitchen waste (KW) and pig manure (PM), were co-digested with RS at a mass ratio of 0:2:1, 0.4:1.6:1, and 0.8:1.2:1, respectively. Similar results were obtained by Zhan-Jiang et al. [65], where 70% more biogas was recovered in co-digestion of RS with food waste (FW) as compared to mono-digestion of RS. RS digested with goat manure (GM) performed best, with a mass mixing ratio of 50:50 (vs. 100:0, 90:10, 70:30, 30:70, 10:90 and 0:100) and yielding a 83% higher biogas yield than RS only [66].

Similarly, co-digestion of RS with various co-substrates such as cow manure (CoM), PM, CM, paper mill sludge (PMS), piggery wastewater (PWW), and municipal sewage sludge (MSS) has been reported by other researchers. The results of these studies (Table 5) show the enhancement in energy recovery from the co-digestion of RS compared to its mono-digestion [22,28,41,67,68].

Table 5. Effects of co-digestion on the methane yield of RS with other substrates.

Substrate	t_{HR} (Days)	AD Mode	Temperature (°C)	C/N	CH_4 Yield (L/kg VS)	CH_4 Enhancement (%)	References
RS RS + CoM[a] RS + CoM[a] RS + CoM[a]	40	Batch	37 ± 1	47.5 19.5 22.4 26.6	177.3 181.1 196 183.5	- 2 10.5 3.5	[28]
RS RS + PM[a] RS + PM[a] RS + PM[a]	35	Batch	37 ± 1	47.5 20.6 23.4 27.7	178 268.1 267.2 241.2	- 50 50 35.5	[41]
RS RS + CM[a] RS + CM[a] RS + CM[a]	35	Batch	37 ± 1	47.5 14.8 17.8 22.4	178 240.5 250.3 206.1	- 35 40.6 15.8	[67]
RS RS + PMS RS + PWW + PMS[a] RS + PWW + PMS[a]	90	Batch	35 ± 1	- - - -	46 340 335 302	- 640 628 556	[22]
RS RS + MSS	30	Batch	35	- -	-	26% more biogas in co-digestion	[68]

AD, Anaerobic digestion; **RS**, Rice straw; **CoM**, Cow manure; **PM**, Pig manure; **CM**, Chicken manure; **PWW**, Piggery wastewater; **PMS**, Papermill sludge; **MSS**, Municipal sewage sludge; [a], Substrates added with different ratios to achieve different C/N; t_{HR}, Hydraulic retention time.

Co-digestion of RS can be realized with several different co-substrates as the results of the above-mentioned studies illustrate. It enhances biogas yield from 26% to 83% and also the CH_4 yield from 2% to 640% relative to RS mono-digestion when optimized for the appropriate C/N ratio. Co-digestion of RS with proper C/N ratio, which can be achieved by mixing one or even more substrates in a suitable amount, is a good approach for the enhancement of energy recovery. This approach can be applied in pilot-scale or farm-scale plants.

4.3. Inoculum

The starting of the AD system is a vital and delicate stage to achieve a successful operation [69]. Inoculum contains active microorganisms and is added to the digester to initiate the AD [38,70]. Usual inoculum is from digested sludge of active biogas plants or organic animal material [71]. A suitable inoculum can improve the biogas yield by increasing the degradation rate, stabilizing the AD, and also shorten the start-up time [72]. Other positive aspects of inoculum are contained N and micronutrients [73]. The micronutrients can enhance the enzyme activity [74] while the extra N can be provided to meet the needs of the methanogens [38]. Therefore, the sources of inoculum, its treatments and the substrate to inoculum ratio (S/I) are important for the AD of RS. The term S/I ratio determines the amounts of substrate and inoculum that should be added in the AD batch digester to prevent the inhibition by avoiding the overlarge portion of the substrate [75].

Gu et al. [38] found that the sources (digested dairy manure, digested swine manure, digested chicken manure, digested municipal sludge, anaerobic granular sludge, and paper mill sludge) of the inoculum can significantly affect the AD of RS. The digested dairy manure resulted in the highest biogas yield of 325 L/kg VS as it provided suitable nutrients and the highest enzyme activities. RS inoculated with other inocula attained biogas yields of 30–280 L/kg VS. In addition, Gu et al. [38] also suggested that a suitable inoculum can also provide N, macronutrients, and micronutrients for the AD.

Haider et al. [69] compared five S/I ratios, i.e., 0.25, 0.5, 1.0, 1.5, and 2.0 for anaerobic co-digestion of FW and rice husk using the fresh cow dung as inoculum and the biogas yields were 557, 458, 267, 97 and 71 L/kg VS, respectively. The highest biogas yield was obtained at S/I ratio of 0.25. Higher S/I ratios are attributed to the presence of a higher amount of biodegradable substrate and a lower amount of inoculum. Furthermore, the authors reported an accumulation of volatile fatty acids (VFAs) at higher S/I ratios (1.5 and 2.0), indicating insufficient activity of methanogenic bacteria.

In another study, the inoculum was diluted in order to compare the effect of inoculum dilution on the CH_4 yield of pretreated RS at various S/I ratios (2, 4, 6, 8, and 10). The original inoculum (I_0) was collected from the anaerobic digester of a wastewater treatment plant. I_o was diluted twice with the recycled water to achieve the diluted inoculum (I_2). When I_0 was used as inoculum, the CH_4 yields were 238, 190, < 50, < 50, and < 50 L/kg VS at S/I ratios of 2, 4, 6, 8, and 10, respectively. While the yields were 193, 187, 176, 144 and < 50 L/kg VS when I_2 was used in the digesters. These results evidenced that I_2 had better dilution capacity for acidic compounds and mass transfer performance due to the availability of additional water in the inoculum. These characteristics of I_2 improve methanogenic activity [76].

Deng et al. [70] studied the effects of co-inoculating RS with ruminal microbiota (RM) and anaerobic sludge (AS). Five co-inoculum mixtures were prepared with RM: AS as 1:1, 1:2, 2:1, 1:0, and 0:1 based on VS content, while the S/I ratio in all mixtures was approximately 0.25. The results showed that the co-inoculation had almost threefold higher CH_4 yield than the digestion of RS in the presence of RM alone.

Kim et al. [77] pointed out that heat treatment (heating the sludge at 100 °C for 15 min and then cooling to room temperature to enrich the spore-forming bacteria of the Clostridium species) of sewage sludge, which was used as inoculum for the AD of RS, caused less biogas yield in comparison to untreated sludge.

It is demonstrated from the results of the studies that the source of inoculum affects the biogas yield. Digested dairy manure proved to be optimal inoculum for RS among the various sources of inocula. Biogas/methane yields tend to decrease with increasing S/I ratios. A high amount of substrate relative to inoculum may cause problems in AD, such as the accumulation of VFAs and lower biogas production rate. Heat treatment may cause the inoculum to be less effective. The use of co-inoculation seems to increase the biogas yield from the AD of RS.

4.4. Organic Loading Rate (OLR)

Process stability and efficiency are proved to be the basic criteria to evaluate the performance of AD [78]. OLR is an important operation parameter that greatly influences the performance of the AD, and it can be described as the amount of substrate added to the digester per day per unit volume. OLR is expressed as Equation (1):

$$OLR = \frac{C}{t_{HR}} \tag{1}$$

where C is the feeding concentration of substrate in g.VS/L, and t_{HR} is the hydraulic retention time in days (d).

Higher OLR can produce higher biogas because of the higher quantity of substrate for digestion [79]. With a higher OLR, the rate of hydrolysis/acidogenesis could be higher than methanogenesis, and the accumulation of VFAs can eventually cause process failure [80]; therefore, a suitable OLR is always necessary for stable operation.

Thermophilic co-digestion of RS with animal manures (PM, CoM, and CM) was carried out at different OLRs of 3.0, 3.6, 4.2, 4.8, 6.0, 8.0, and 12.0 g VSL^{-1} d^{-1} by Li et al. [40]. Results from co-digestion of RS and PM illustrated that a stable biogas yield, on average, 434 L/kg VS was achieved without inhibition by ammonia or VFAs while, for the co-digestion of RS with CoM, the average yield was 455 L/kg VS at all OLRs. In contrast, the AD process was inhibited by ammonia when the OLR was ≥ 6 g V S L^{-1} d^{-1} during the co-digestion of RS with CM and the specific biogas yield was 422 L/kg VS at 4.8 g VSL^{-1} d^{-1}.

Similarly, the effect of various OLRs on the anaerobic mesophilic co-digestion of RS with the PM [41], CoM [28], and CM [67] have been investigated, and at 6–8, 6, and 4.8 g VS L^{-1} d^{-1}, there was no inhibition in the digesters due to the accumulation of VFAs or ammonia, respectively. However, co-digestion of RS with PM and CoM was severely inhibited by VFAs at OLR of 12 g VS L^{-1} d^{-1}. The co-digestion of RS with CM was inhibited at OLR ≥ 6 g VS L^{-1} d^{-1} due to the accumulation of ammonia.

The effect of various OLRs on the yield of biogas from the AD of RS, and their recommended values from the literature are summarized in Table 6.

Table 6. Effects of OLR on the biogas yield in AD of RS.

Substrate	Temperature (°C)	OLR (g VS L^{-1} d^{-1})	OLR (g VS L^{-1} d^{-1}) Recommended	Biogas Yield (L/kg VS)	References
RS + PM	35 ± 0.5	0.4, 0.8, 1.3, 1.9, 2.5, and 3.1[a]	2.5[a]	460[b]	[81]
RS + PM	37 ± 2	3.0, 3.6, 4.2, 4.8, 6.0, 8.0, and 12.0	6.0–8.0 Inhibition by VFAs at 12.0	400–450	[41]
RS + CoM	37 ± 2	3.0, 3.6, 4.2, 4.8, 6.0, 8.0, and 12.0	6.0 Inhibition by VFAs at 12.0	380	[28]
RS + CM	37 ± 2	3.0, 3.6, 4.2, 4.8, 6.0, 8.0, and 12.0	4.8 Inhibition by FAN at 6.0–12.0	380	[67]
RS + PM	55 ± 2	3.0, 3.6, 4.2, 4.8, 6.0, 8.0, and 12.0	12.0 No inhibition	434	[40]
RS + CoM	55 ± 2	3.0, 3.6, 4.2, 4.8, 6.0, 8.0, and 12.0	12.0 No inhibition	455	[40]
RS + CM	55 ± 2	3.0, 3.6, 4.2, 4.8, 6.0, 8.0, and 12.0	4.8 Inhibition by FAN	422	[40]

RS, Rice straw; **PM**, Pig manure; **CoM**, Cow manure; **CM**, Chicken manure; **FAN**, Free ammonia; **VFAs**, Volatile fatty acids; [a], **OLR** measured in kg COD/ (m^3 d); [b], methane yield measured as L CH_4/ kg $COD_{removed}$.

The recommended OLRs for commercial co-digestion of RS with PM, CoM, and CM at mesophilic conditions are 6.0–8.0, 6.0, and 4.8 g VS L^{-1} d^{-1}, while recommended values for thermophilic conditions are 12.0, 12.0, and 4.8 g VS L^{-1} d^{-1}. From the results of these studies, it can be concluded that a choice of a suitable OLR is necessary for attaining a good biogas yield from the AD of the RS. The optimal OLR is influenced by the type of substrate (e.g., manure) and the operating temperature. Higher OLRs with stable biogas yield can be attained at thermophilic conditions compared to mesophilic ones. Moreover, the cause of inhibition is also different in different co-digestion systems. For CM, the process inhibits by free ammonia; VFAs inhibit co-digestion of PM and CoM at higher OLR. The co-digestion system of RS assisted with in situ removal of ammonia can be considered to avoid inhibition by ammonia at higher OLRs [40]. The methods considered for the removal of ammonia are immobilization and adaptation of microorganisms, ultrasonication, microwave, hollow fiber membranes, and microbial fuel cell applications [82].

4.5. Mixing

Mixing is an important operational factor that determines the performance of the AD. It enhances the mass transfer of organic substrates to microbial biomass due to the homogenization of the content in the digester [83]. Moreover, mixing facilitates the removal of trapped gas bubbles in the digester, preventing the sedimentation of heavy particles and establishing uniform temperature [18]. However, intense continuous mixing disrupts the structure of microbial flocks, which disturbs the syntrophic relationship between organisms and leads to a reduction in biogas yield, instability of the digester, and increased vulnerability to shock loadings [83,84]. Therefore, it is crucial to provide optimal mixing for efficient and stable AD [85].

Generally, gas circulation, mechanical pumping, and mechanical stirring are used for mixing [86]. Mechanical stirring is widely used in process industry operations involving various solid–liquid flows because of its higher homogeneity in practice [87]. Moreover, mechanical agitators are the most efficient mixing device in terms of power consumption [88].

The effect of mixing has been mostly studied for municipal solid waste, biosolids [89,90], and manures [83,91]. It is not well understood how mixing affects the AD of agricultural crop residues such as RS [85]. Hence, the optimized mixing parameters for the well-investigated traditionally used substrates could be very different for crop residues, including RS [87].

RS, like other lignocellulosic substrates, has a complex chemical structure. RS also has some specific physical characteristics like low density, high water holding capacity, poor fluidity, and heterogeneity [87,92]. RS floats on the surface in the digester and does not mix properly, especially when biogas is entrapped in the slurry. Due to poor mixing, the substrate does not get good contact with the microorganisms, thus reducing degradation and, ultimately, lowering biogas yield [92].

Kim et al. [85] conducted batch experiments using RS as a substrate to study the effect of periodic mixing (once a day, once a week, twice a week), continuous mixing (50, 150, and 300 revolutions per minute (rpm)), and no mixing on the CH_4 yield. Their results showed that the differences in CH_4 yield were not statistically significant among the periodic and no mixing conditions, while the total CH_4 yield for the 50, 150, and 300 rpm continuous mixing conditions were 13%, 25%, and 38% lower than for no mixing conditions. Kim et al. [85] also investigated the effect of three modes of continuous mixing during the AD of RS in three continuously stirred tank reactors (CSTR). The three applied regimes were: (R1) one digester with continuous mixing at 50 rpm, (R2) second digester with manually mixing of once a day, and (R3) the third digester was operated at 50 rpm continuous mixing with occasional high speed (150 rpm) mixing events. They concluded that the CH_4 content in biogas in R2 (53%) was slightly higher than in R1 (51%), while in R3, it dropped to a minimum value of 7%.

Shen et al. [87] carried out a study to improve the mixing of RS during AD to achieve higher biogas yield by computational fluid dynamics (CFD) and experimental tests. The experiments were performed at different stirring rates (40, 80, 120, and 160 rpm) to investigate their effect on biogas yield in CSTR. The highest biogas yield was achieved at a stirring rate of 80 rpm. It was higher than that of other stirring rates by about 19% (40 rpm), 10% (120 rpm), and 12% (160 rpm). Tian et al. [92] utilized the results of Shen et al. [87] for identifying proper agitation intervals to prevent floating layer formation during AD of corn stover (CS) and to improve the biogas yield. Shen et al. [87] studied the effect of various stirring rates (40, 80, 120, and 160 rpm for 5 min each time) on biogas yield of RS, while Tian et al. [92] further investigated the effect of various agitation intervals (continuous; after 2, 4, 8, and 12 h for 5 min for each time) with different OLRs (1.44, 1.78, and 2.11 g TS L^{-1} d^{-1}) on the biogas yield from AD of CS in CSTRs. The results of the study showed that, at lower OLR of 1.44 g TS L^{-1} d^{-1}, there was no floating layer observed in the digesters when the time interval for the agitation was continuous for 2 h, 4 h, and 8 h, while the performance of AD was adversely affected at 12 h. Moreover, the biogas yield for the digester agitated after 12 h was 418 L/kg TS, which was 2%, 9%, 3%, and 0% lower than those digesters stirred continuously, for 2 h, 4 h, and 8 h, respectively. The optimal agitation intervals shifted to 6 h and 2 h when the OLRs increased to higher levels of 1.78 g TS L^{-1} d^{-1} and 2.11 g TS L^{-1} d^{-1}, respectively. When the agitation time intervals were more extended than optimal, floating layers were observed in the digesters, especially at higher OLR of 2.11 g TS L^{-1} d^{-1}. The results of studies investigating the effects of various mixing conditions on biogas yield from the AD of RS and other straw have been summarized in Table 7.

Table 7. Effects of mixing on the biogas yield in AD of rice and other straw.

Substrate	Digester Type	Mixing Mode (Mixing Interval [h]/ Stirring Period [min])	Mixing Speed	OLR (g TS L^{-1} d^{-1})	Biogas Yield (L/kg TS)	References
RS	CSTR	Periodic (2 h/5 min)	40 rpm	-	299	[87]
			80 rpm		370	
			120 rpm		332	
			160 rpm		327	
MS	Batch	No mixing	-	-	414	[93]
		Once a day	-	-	491	
		Twice a week	-	-	482	
CS	CSTR	Continuous	80 rpm	1.44	427	[92]
		(2 h/5 min)			459	
		(4 h/5 min)			430	
		(8 h/5 min)			420	
		(12 h/5 min)			418	
CS	CSTR	Continuous	80 rpm	1.78	487	[92]
		(2 h/5 min)			508	
		(4 h/5 min)			505	
		(8 h/5 min)			454	
		(12 h/5 min)			416	
CS	CSTR	Continuous	80 rpm	2.11	423	[92]
		(2 h/5 min)			433	
		(4 h/5 min)			429	
		(8 h/5 min)			378	
		(12 h/5 min)			335	

RS, Rice straw; **MS**, Maize stover; **CS**, Corn stover; **CSTR**, Continuously stirred tank reactor; **rpm**, Revolutions per minute.

The results in Table 7 show that periodic mixing is better as compared to continuous mixing. The stirring rate of 80 rpm lasting for the 5 min provides proper mixing in the digesters. It has also been observed that higher OLR led to the rapid formation of a floating layer. The agitation interval with highest biogas yields has been identified as 2 h at OLRs of 1.44, 1.78, and 2.11 g TS L^{-1} d^{-1} during the AD of straw in CSTRs.

The results of these reported studies verified that the proper agitation speed, mixing time, and the mixing intervals can achieve higher biogas yield. In contrast, the very intense mixing can affect the process adversely. These results are useful for operating the digesters treating the RS or other straw in an efficient and cost-effective way.

4.6. Recycling Liquid Digestate (LD)

A separation machine (press screws) is usually used to separate the digestate from the digester into LD and SD. The SD is rather simple to handle as it is generally used as fertilizer after composting or even directly. However, the utilization or treatment of LD is a challenge. LD can be used as liquid fertilizer or can be disposed of in the water streams after treatment; however, the most promising option is its recycling to the digester [94].

The recycling of LD can help to improve the moisture content for some high solid substrates [94] and provides a range of nutrients and vitamins needed by the microorganisms [95]. However, the excessive recycling may result in process inhibition by the accumulation of VFAs, ammonia, and nonbiodegradable intermediates [94,96,97].

Hu et al. [94] investigated the effect of recycling LD during the AD of MS in the CSTR. They compared three scenarios: No recycling of LD serving as a reference, while in the other two digesters, direct recycling and recycling after aeration carried out. Their results showed that there was no significant difference in CH_4 yield, CH_4 content, substrate reduction, or pH in all three modes. This indicates that recycling did not adversely affect system stability.

Pezzolla et al. [98] evaluated the effect of the frequency of LD recycling on the biogas yield in the solid-state AD of straw with PS. They compared the impact of different recycling frequencies—once per day, twice per day, and four times per day—with a control system (no recycling). The results suggested that recycling LD had a positive effect on biogas yield when it was more than twice per day. The highest cumulative biogas yield was achieved at a recycling frequency of four times per day.

Mussoline et al. [36] aimed to define and optimize the operational parameters of a farm-scale anaerobic batch digester (13,000 m^3) using RS and PWW as substrates where recycling of LD was one of the important strategies. Increasing recycling rates of LD (from 0.04 to more than 0.14 m^3/m^3straw d^{-1}) resulted in increasing biogas yield. Higher recycling rates are recommended, but these are limited by the accumulation of VFAs which can inhibit the methanogenic process.

The effect of LD recycling in various proportions (relative to recovered amount), with different frequencies and with varying post-treatments on the biogas/methane yield in the AD of various substrates is shown in Table 8.

Table 8. Effects of recycling LD on the biogas/methane yield in AD of various substrates.

Substrate	Recycling Scenarios	Effect on AD	Effect on Biogas/Methane Yield	References
RS + PWW	Increasing the rate	Increase microbial growth	Increase	[36]
MS	No Direct Aerated	No adverse effect on system performance and stability	No significant increase	[94]
Mixture (PS and straw)	No recycling 1 x day 2 x day 4 x day	Enhanced alkalinity, buffer capacity, and methanogenic activity	Increase	[98]
WS	No After centrifugation After filtration	Higher nutrients and microbial concentration	21% increase	[99]
WS	No After centrifugation After filtration	The process could be sustained for a longer time due to dilution of nutrients	21% increase	[100]
WS + SS	No After centrifugation After filtration	Enrichment of nutrients	Increase	[100]
WS + MN	No After centrifugation After filtration	Process collapse without recycling while stable with recycling	Increase	[100]
LG	0%, 10%, 20%, 30%, 40%, 50%, and 60% of recovered LD	Provision of methanogenic bacteria and trace elements	Maximum biogas with 60% LD	[101]
DM	30%, 50%, and 80% of recovered LD	No adverse effect on performance and system stability	No significant increase	[96]
AS	0%, 50%, and 100%, of recovered LD	Inhibition at 100% while 50% provides stable operation	Increase with 50% LD	[102]
CS + OFMSW	0.3, 0.6, 1.2, 2.4, and 4.8 (daily recycling volume/total recovered volume)	Improved hydrolysis and mass transfer Accumulation of VFAs	Highest methane at 0.3	[103]

AD, Anaerobic digestion; **LD**, Liquid digestate; **RS**, Rice straw; **PWW**, Piggery wastewater; **MS**, Maize stover; **Mixture**, PS and straw; **WS**, Wheat straw; **SS**, Sewage sludge; **MN**, Macronutrients (N and P); **LG**, Lawn grass; **DM**, Dairy manure; **AS**, Alfalfa silage; **CS**, Corn straw; **OFMSW**, Organic fraction of municipal solid waste.

The studies presented in Table 8 show different results of the recycling of LD on the biogas or methane yield in the AD of various substrates. The results of the two studies show no increase or decrease in the biogas yield due to LD recycling, whereas the authors explained that there is no adverse impact of recycling on the process stability [94,96]. Some authors obtained higher biogas yield with the recycling of LD. They attributed this to increase microbial growth, enhanced buffer capacity, higher alkalinity, increased methanogenic activities, higher nutrients availability, and improved hydrolysis [36,98–100]. In some other studies, the biogas yield improved at a lower rate of LD recycling, and it reduced when more LD was recycled. The accumulation of VFAs and the dilution of essential nutrients were named as causes [100,102,103].

From these studies, it can be concluded that it is technically feasible to recycle part or all of the LD to the AD system. It is also possible that the process cannot be sustained for a longer time due to the dilution of nutrients. Co-digestion of straw with nutrients rich substrates combined with recycling LD is an excellent approach to achieve a stable AD system. In addition, an appropriate amount of LD is required because its excessive recycling may cause process failure due to the accumulation of VFAs. Therefore, LD recycling may enhance system stability and biogas yield.

4.7. Addition of Trace Elements (TEs)

Several specific TEs such as iron (Fe), cobalt (Co), nickel (Ni), copper (Cu), zinc (Zn), molybdenum (Mo), selenium (Se), and tungsten (W) are essential for enzyme cofactors involved in the biochemistry of CH_4 formation [104] and needed in a balanced AD [105]. However, the availability of TEs in higher amounts can cause inhibition to anaerobic organisms due to the disruption of an enzyme's structure and function [106]. Therefore, an appropriate amount of TEs is required to avoid the inhibition by maintaining effective growth and metabolism of microorganisms. Moreover, it has been found from the literature that the degree of inhibition depends upon several factors such as the total concentration, pH, redox potential, and chemical forms of the elements [107,108].

Recently, Mancini et al. [29] added different TEs (Co, Ni, and Se) in various mass fractions to study the enhancement of CH_4 yield in the AD of RS. The results showed a 12% improvement in the CH_4 yield with the addition of these TEs. In another similar study, Cai et al. [109] observed 18% more CH_4 yield when Fe was added to the batch AD of RS. A study conducted by Mancini et al. [110] using RS as a sole substrate showed different results as compared to these similar studies [29,109]. The authors did not observe any enhancement in CH_4 yield due to the addition of Fe and Co. They concluded that it is likely due to the presence of the elevated TEs in the inoculum used in the study. The effect of the addition of various TEs on the methane yield in the AD of straws is summarized in Table 9.

The results of the abovementioned studies show that research on several TEs has been carried out for single and multiple elements, while Fe, Ni, and Co are the most investigated. The addition of TEs has been successfully used to enhance biogas and CH_4 yield in the AD of various straw types. Specifically, the enhancement in CH_4 yield of 0–18% has been observed due to the addition of TEs (Fe, Ni, Co, and Se) to the AD of RS. Similarly, up to 62% enhancement in CH_4 yield observed with the addition of TEs to other straw (CS) (Table 9).

Literature data on the impact of TEs addition on the enhancement of CH_4 yield in the AD of RS is scarse. There were only three studies found investigating the effect of the addition of Fe, Ni, Co, and Se. Two studies show that the addition of TEs improves the CH_4 yield while the third study shows there was no improvement. Therefore, there is a need for future work in this area to understand the effect of the addition of other TEs as well. Likely, the addition of suitable elements in an appropriate combination and amount would enhance the CH_4 yield in the AD of RS as evidenced by other straws.

Table 9. Effect of addition of TEs on the methane yield in the AD of RS and other straws.

Substrate	TEs	AD Mode	Adjusted Amount of TEs	Units	Maximum Methane Enhancement (at Adjusted Amount)	References
RS	Co Ni Se	Batch	9.0–45.0 9.0–45.0 1.0–5.0	mg/kg TS	12% (Ni: 45)	[29]
RS	Fe Co	Batch	205 18	mg/kg TS	0%	[110]
RS	Fe	Batch	92,464:1, 18,492:1 9246:1, 1849:1 925:1, 185:1 92:1, 18:1, 9:1	VS:Fe	18% (925:1)	[109]
MS[a]	Fe Ni Co	Batch	0.10–0.20 0.005–0.01 0.004–0.009	mg/g VSS	35% [b] (Fe: 0.1 Ni: 0.005 Co: 0.004)	[111]
MS[a]	Ni Co Mo	Batch	0.1–0.5 0.1–0.5 0.05–0.25	mg/L	36% (Ni: 0.5 Co: 0.5 Mo: 0.25)	[112]
CS	Fe Ni Co	Batch	1–10 0.2–0.6 0.05–0.4	mg/L	62% (Fe: 1 Ni: 0.4 Co: 0.4)	[113]
MS	Fe	Batch	50.0, 200.0, 1000.0 and 2000.0	mg/L	15% (200.0–1000.0)	[114]
PS[a]	Fe	Batch	0.5, 1.0, 5.0 and 10.0	mg/L	18% [b] (10.0)	[115]
PS[a]	Ni	Batch	0.2, 0.8 and 2.0	mg/L	>18% [b] (0.8 and 2.0)	[116]
PS[a]	Cu	Batch	30.0 100.0 500.0	mg/L	44% [b] (30)	[117]

TEs, Trace elements; **AD**, Anaerobic digestion; **RS**, Rice straw; **MS[a]**, Maize silage; **CS**, Corn stover; **MS**, Maize straw; **PS[a]**, Phragmites straw, [b], Biogas enhancement.

4.8. Bioavailability of Trace Elements (TEs)

The presence of a particular TE in the AD does not necessarily imply that it can be taken up by microorganisms and incorporated into the catalytic center of the enzymes [118]. Bioavailability is defined as the degree to which a substance is available for metabolic activities [119]. The bioavailability of TEs can be grouped and approximated by sequential extraction techniques [120]. Water-soluble and exchangeable fractions of TEs are commonly considered as highly bioavailable. TEs bound in carbonates, sulfides, or organic complexes are less bioavailable as their mobility depends on the aqueous solubility of the compounds. The residual fraction is not available to microorganisms because it is non-extractable and non-dissolvable [120]. Ortner et al. [120] examined the applicability of the sequential extraction technique on slurries of AD plants. These slurries were obtained from three biogas plants: an agricultural biogas plant utilizing PM and MS[a], a plant using grass silage (GS) and MS[a], and a plant processing the slaughterhouse waste (SHW). Their results are presented in Table 10.

Table 10. Water-soluble and exchangeable fractions (relative to a total concentration of TE) of TEs in the AD of various substrates.

Substrates	AD Mode Temperature (°C)	TE	Water-Soluble and Exchangeable Fractions (%)	References
RS	Batch 37 ± 2	Fe, Co	23, 48	[110]
RS	Batch 35 ± 2	Fe	1.7–3.1	[109]
MS	Semi-continuous 37 ± 1	Fe, Ni, Co, and Mn	14, 29, 36, and 70[a]	[121]
SHW	Semi-continuous 38	Ni, Co, and Mo	62, 62, and 68	[118]
SHW	38	Fe, Ni, Co, Cu, Zn, and Mo	54, 52, 48, 57, 54, and 72	[120]
GS + MS[a]	48	Fe, Ni, Co, Cu, Zn, and Mo	22, 36, 24, 46, 43, and 0	[120]
PM + MS[a]	40	Fe, Ni, Co, Cu, Zn, and Mo	27, 62, 60, 29, 28, and 62	[120]

AD, Anaerobic digestion; **TE**, Trace element; **RS**, Rice straw; **MS**, Maize straw; **SHW**, Slaughterhouse waste; **GS**, Grass silage; **MS[a]**, Maize silage; **PM**, Pig manure; [a], also include carbonates fraction of TE.

Fe has been added alone and incombination with Co to determine whether their bioavailability could be a limiting factor for CH_4 yield using RS as a substrate in the presence of agro-zootechnical digestate as an inoculum in a study carried out by Mancini et al. [110]. The results pointed out that the amount of highly bioavailable Fe was increased from 11% to 23% when it was added alone and from 11% to 20% when it was dosed with Co. Similarly, Co addition resulted in an increase of the highly bioavailable fraction from 8% to 48%. The CH_4 yield was the same with and without the addition of TEs. Therefore, the direct relationship between the increased highly bioavailable fractions of TEs and enhanced CH_4 yield in the AD of RS was not observed. Cai et al. [109] conducted a similar study to investigate the changes in five fractions of Fe to assess its bioavailability and whether the system was deficient in Fe during the AD of RS. Their results indicated an overall very low (1.7–3.1%) bioavailability of Fe, and most of the Fe was not readily available to microorganisms in the digestion system. Thus, there was a deficiency of bioavailable Fe which may have reduced CH_4 yield.

The effect of the bioavailability of Fe, Ni, Co, and Mn on the CH_4 yield from the AD of MS in the semi-continuous CSTR was studied by González-Suárez [121]. The results indicated a decreasing trend of highly bioavailable fractions of TEs over time. At the end of the experiment (after 96 days), the highly bioavailable fractions of Fe, Ni, Co, and Mn decreased by 23%, 32%, 33%, and 20%, respectively, compared to day 1. The results confirm that fewer TEs were directly available for microbial uptake at the end of the process leading to an 11% decrease in CH_4 yield.

The results of the studies summarized in Table 10 show that the sequential extraction is a useful method to assess the bioavailability of TEs available in various AD systems. The determination of the bioavailability of TEs is a more appropriate approach compared to total available amounts concerning their requirement to monitor the process to achieve a stable and efficient process.

The results of one study show the direct relationship between the reduced amount of highly bioavailable fractions of TEs and reduction in CH_4 yield in the AD of MS. While, in another study, the CH_4 yield has not enhanced due to the increased amount of highly bioavailable fractions of TEs in the AD of RS. Therefore, the relationship between the higher amount of highly bioavailable fractions of TEs and enhanced CH_4 yield is not clear. Moreover, specifically for RS, only Fe and Co have been investigated to assess the effect of bioavailability on the CH_4 yield. A large fraction is not available for microbial uptake during the AD of RS and this aspect should be considered for an appropriate supplementation of TEs to achieve enhanced CH_4 yield.

5. Pilot-scale and Farm-scale Plants for AD of RS

Energy recovery in the form of biogas/methane from agricultural residues is highly interesting for the agricultural sector, as it is an option to produce local energy, replace fossil fuels, and thus, become self-sufficient and sustainable in terms of energy supply [122]. Introducing AD plants into the farm system, treating the animal manures along with the straws for the recovery of energy has a great potential to reduce the demand for fossil energy and reduction of greenhouse gas emissions [123].

Numerous bench-scale experiments have been carried out on RS digestion to define the optimal temperature [25,26,42], substrates for co-digestion [39,65,66], mixing intensities and duration [87,92], OLR [28,40,41], sources of inoculum [38], S/I ratio [69], recycling LD to the digester [36,98], addition of TEs and their bioavailability [110]. Bench-scale studies are important to understand the principles but it is challenging to design and operate a farm-scale plant based on these microcosms [124]. Besides, the monitoring of a farm-scale plant is necessary to obtain detailed information and a better understanding of technology [125]. It also helps to discover practical operational aspects such as the length of start-up phase and energy and co-substrates input and to check if the system is sustainable on a long-term basis [36,125]. These aspects are often missed in lab tests.

Mussoline et al. [124] conducted a batch pilot-scale study using untreated RS and PWW to define TS, straw-to-wastewater ratio, digestion temperature, and digestion time. RS was co-digested in two pilot-scale batch digesters with different amounts of PWW (Digester A: 50 kg dry straw, 150 L PWW, 20% TS; Digester B: 50 kg dry straw, 60 L PWW, 20% TS) for 189 days. The digestion temperature inside the digester A was between 30 and 40 °C. The digester B was initiated with ambient conditions (20–25 °C) and constant heating was applied later on to maintain mesophilic temperatures. The CH_4 yield from digester A was 231 L/kg VS and from the digester, B was 12 L/kg VS. The lower temperature and lesser amount of PWW were the limiting factors for lower CH_4 yield from digester B. The amount of PWW was not sufficient to establish a stable microbial community. The results of digester A were compared with the results of another study [57] of the same magnitude where the pretreated RS was digested. The pretreated (cutting into 7–8 cm pieces and treated with white-rot fungi) RS requires only 89 days for digestion while the untreated RS needs 189 days. Moreover, the pretreated RS can be adequately digested at ambient temperature (19–30 °C), while the untreated RS requires the mesophilic conditions in the presence of approximately 67% more PWW [57].

Other than RS, André et al. [126] studied the dry AD of roadside grass (RSG) with manure in a pilot-scale plant at 37 °C with three digesters. The overall average CH_4 yield (226 L/kg VS) from three digesters obtained in this study [126] was similar to the yield obtained by Mussoline et al. [124] during the co-digestion of RS with PWW. A pilot-scale plant was successfully operated with a stable AD process treating RS and manure in the BIORIST project [24]. No further studies were found to compare the performance of AD of RS with other substrates at the pilot scale.

Data from a farm-scale system treating RS is limited to one study [36]. During the initial loading event, 727 t of RS and 285 t of PWW was loaded in a batch digester. Approximately 1300 t of water was added to the digester overtime during the first year. The co-digestion of RS with PWW was completed after 422 days, and the cumulative energy production was 295 MWh. The results showed a start-up period of 200 days and the authors attributed this long phase to high straw to wastewater ratio, low ambient temperature (<15 °C), and low recycling rate of LD (< 0.04 m^3/m^3straw d^{-1}). Similarly, Wandera et al. [127] investigated the performance of a full-scale biogas plant treating the MS. The digester had a capacity of 400 m^3 and was operated at a temperature ranging from 55 to 60 °C. They obtained a CH_4 yield of 200 L/kg TS.

The abovementioned studies indicate the importance of pilot-scale experimentations to define the operational parameters such as heat input, TS, straw to wastewater ratio, digestion temperature, digestion time, loading frequency, and wastewater volume for the farm-scale plants treating RS.

Although the literature reporting data on the AD of RS in farm-scale plants is scared. The implementation of a farm-scale plant using RS co-digested with other suitable substrates (PWW, manures) is possible as evidenced by studies that co-digestion performs better than mono-digestion. The farm-scale biogas plants are good candidates for the utilization of RS to recover energy, to facilitate the development of the agriculture economy and to provide local energy. However, further investigation is needed for a better understanding of the process. Moreover, reseach about the economical viability of farm-scale AD plants for RS should be conducted and published, covering all financial cost of capital and operation, including the logistics of RS and other co-substrates.

6. Conclusions

RS generation will continue to increase as rice production continues to rise to provide food for a still growing global population. The AD of RS is a viable option for the recovery of energy and nutrients. The nutrients (N, P, and K) are mostly conserved in the digestate during the AD, and their significant portion is available in LD. The effective use of nutrients available in the digestate is a way to partially substitute fertilizers for the amendment of agricultural land.

The AD of RS at thermophilic conditions results in higher biogas/methane yield and higher CH_4 content in biogas as compared to mesophilic conditions. In contrast, the mesophilic process is more robust and less sensitive to environmental changes. Using a suitable inoculum and co-substrate in an appropriate amount enhances biogas/methane yield in the AD of RS by providing nutrients balance, dilute toxic compounds, adjust moisture content, improve buffer capacity, and stability of the system. Optimal selection of mixing (speed, time, and intervals), OLR, recycling LD, and TEs (with appropriate amount) also enhances the biogas/methane yield.

Pilot-scale plants and experiments are necessary to design and define operational parameters for farm-scale biogas plant treating RS. Technically, the implementation of farm-scale biogas plants utilizing RS together with other suitable substrates such as PWW or other livestock manure is possible. It will provide energy and nutrients recovery to the local farm. However, the economic feasibility of such a plant should be evaluated.

Author Contributions: F.M. reviewed the literature and wrote the first draft of the paper. A.F., V.S.R. reviewed and edited to enhance the scientific quality of the paper. All authors have read and agreed to the published version of the manuscript.

Funding: Furqan Muhayodin is a doctoral student and funded by the Ministry of Higher Education Commission (HEC), Pakistan (SAP 50020935). This work was supported by the German Federal Ministry of Education and Research under the project "BIORIST"- A joint research project for an innovative process technology for biogas production from rice straw (01LY1508A).

Acknowledgments: The authors extend their thanks and gratitude to Thi Quynh Trang Hoang for providing critical constructive comments to improve the quality of this paper. We acknowledge support by the German Research Foundation and the Open Access Publication Fund of TU Berlin. Our sincere thanks for the two anonymous reviewers who provided critical constructive comments on the manuscript, which resulted in a significantly improved article.

Conflicts of Interest: The authors declare no conflict of interest.

Abbreviations

AD	Anaerobic digestion
AS	Alfalfa silage
AS	Anaerobic sludge
C/N	Carbon-to-nitrogen ratio
CFD	Computational fluid dynamics
CM	Chicken manure
CoM	Cow manure

CS	Corn stover/straw
CSTR	Continuously stirred tank reactor
DM	Dairy manure
FW	Food waste
GM	Goat manure
GS	Grass silage
IRRI	International Rice Research Institute
KW	Kitchen waste
LD	Liquid digestate
LG	Lawn grass
MN	Macronutrients (N and P)
MS	Maize stover/straw
MSa	Maize silage
MSS	Municipal sewage sludge
OFMSW	Organic fraction of municipal solid waste
OLR	Organic loading rate
PASBR	Psychrophilic anaerobic sequencing batch reactor
PM	Pig manure
PMS	Paper mill sludge
PS	Pig slurry
PSa	Phragmites straw
PWW	Piggery wastewater
RM	Ruminal microbiota
rpm	Revolutions per minute
RR	Recovery rate
RS	Rice straw
RSG	Roadside grass
S/I	Substrate-to-inoculum ratio
SD	Solid digestate
SEM-EDS	Scanning electron microscopy
SHW	Slaughterhouse waste
SM	Swine manure
SS	Sewage sludge
TC	Total carbon
TEs	Trace elements
tHR	Hydraulic retention time
TKN	Total Kjeldahl nitrogen
TN	Total nitrogen
TS	Total solids
VFAs	Volatile fatty acids
VS	Volatile solids
WS	Wheat straw

References

1. Arvanitoyannis, I.S.; Tserkezou, P. Corn and rice waste: A comparative and critical presentation of methods and current and potential uses of treated waste. *Int. J. Food Sci. Technol.* **2008**, *43*, 958–988. [CrossRef]
2. Sattar, A.; Arslan, C.; Ji, C.; Sattar, S.; Umair, M.; Zia Bakht, M. Quantification of temperature effect on batch production of bio-hydrogen from rice crop wastes in an anaerobic bio reactor. *Int. J. Hydrog. Energy* **2016**, *41*, 11050–11061. [CrossRef]
3. Koopmans, A.; Koppejan, J. Agricultural and forest residues- generation, utilization and availability. *Reg. Consult. Mod. Appl. Biomass Energy* **1997**.
4. Kadam, K.L.; Forrest, L.H.; Alan Jacobson, W. Rice straw as a lignocellulosic resource: Collection, processing, transportation, and environmental aspects. *Biomass Bioenergy* **2000**, *18*, 369–389. [CrossRef]

5. Purohit, P. Economic potential of biomass gasification projects under clean development mechanism in India. *J. Clean. Prod.* **2009**, *17*, 181–193. [CrossRef]
6. IRRI. Rice Knowledge Bank. Available online: http://www.knowledgebank.irri.org/step-by-step-production/postharvest/rice-by-products (accessed on 11 March 2020).
7. FAOSTAT. Global Rice Production by Year. Food and Agriculture of the United Nations. Available online: http://faostat.fao.org (accessed on 19 March 2019).
8. Gadde, B.; Bonnet, S.; Menke, C.; Garivait, S. Air pollutant emissions from rice straw open field burning in India, Thailand and the Philippines. *Environ. Pollut.* **2009**, *157*, 1554–1558. [CrossRef]
9. Glissmann, K.; Conrad, R. Fermentation pattern of methanogenic degradation of rice straw in anoxic paddy soil. *FEMS Microbiol. Ecol.* **2000**, *31*, 117–126. [CrossRef]
10. Zhang, R.; Zhang, Z. Biogasification of rice straw with an anaerobic-phased solids digester system. *Bioresour. Technol.* **1999**, *68*, 235–245. [CrossRef]
11. Mussoline, W.; Esposito, G.; Giordano, A.; Lens, P. The Anaerobic Digestion of Rice Straw: A Review. *Crit. Rev. Environ. Sci. Technol.* **2013**, *43*, 895–915. [CrossRef]
12. Bakker, R.R.; Jenkins, B.M. Feasibility of collecting naturally leached rice straw for thermal conversion. *Biomass Bioenergy* **2003**, *25*, 597–614. [CrossRef]
13. Huang, Y.F.; Kuan, W.H.; Lo, S.L.; Lin, C.F. Hydrogen-rich fuel gas from rice straw via microwave-induced pyrolysis. *Bioresour. Technol.* **2010**, *101*, 1968–1973. [CrossRef] [PubMed]
14. Iranzo, M.; Cañizares, J.V.; Roca-Perez, L.; Sainz-Pardo, I.; Mormeneo, S.; Boluda, R. Characteristics of rice straw and sewage sludge as composting materials in Valencia (Spain). *Bioresour. Technol.* **2004**, *95*, 107–112. [CrossRef] [PubMed]
15. Abdelhamid, M.T.; Horiuchi, T.; Oba, S. Composting of rice straw with oilseed rape cake and poultry manure and its effects on faba bean (Vicia faba L.) growth and soil properties. *Bioresour. Technol.* **2004**, *93*, 183–189. [CrossRef] [PubMed]
16. Chynoweth, D.P.; Owens, J.M.; Legrand, R. Renewable methane from anaerobic digstion of biomass. *Renew. Energy* **2001**, *22*, 1–8. [CrossRef]
17. Kargbo, F.R.; Xing, J.; Zhang, Y. Pretreatment for energy use of rice straw. A review. *Afr. J. Agric. Res.* **2009**, *4*.
18. Appels, L.; Baeyens, J.; Degrève, J.; Dewil, R. Principles and potential of the anaerobic digestion of waste-activated sludge. *Prog. Energy Combust. Sci.* **2008**, *34*, 755–781. [CrossRef]
19. Raheman, H.; Mondal, S. Biogas production potential of jatropha seed cake. *Biomass Bioenergy* **2012**, *37*, 25–30. [CrossRef]
20. Grimsby, L.K.; Fjørtoft, K.; Aune, J.B. Nitrogen mineralization and energy from anaerobic digestion of jatropha press cake. *Energy Sustain. Dev.* **2013**, *17*, 35–39. [CrossRef]
21. Dinuccio, E.; Balsari, P.; Gioelli, F.; Menardo, S. Evaluation of the biogas productivity potential of some Italian agro-industrial biomasses. *Bioresour. Technol.* **2010**, *101*, 3780–3783. [CrossRef]
22. Mussoline, W.; Esposito, G.; Lens, P.; Spagni, A.; Giordano, A. Enhanced methane production from rice straw co-digested with anaerobic sludge from pulp and paper mill treatment process. *Bioresour. Technol.* **2013**, *148*, 135–143. [CrossRef]
23. Baserga, U. Landwirtschaftliche CO-Vergarungs-Biogasanlagen. Biogas aus organischen Reststoffen und Energiegras. In *FAT-Berichte*; Forschungsanstalt für Agrarwirtschaft und Landtechnik: Tänikon, Schweiz, 1998.
24. Larsen, O.; Rotter, V.S.; Fechter, L.; Spahr, M. BioRist -Entwicklung und Integration eines innovativen Verfahrens zur Biogasherstellung aus Reisstroh in regionale Wertschöpfungsketten im ländlichen Raum in Südostasien unter Berücksichtigung nachhaltiger Entwicklung und Klimaschutz – Beispiel Vietnam 2019.
25. Contreras, L.M.; Schelle, H.; Sebrango, C.R.; Pereda, I. Methane potential and biodegradability of rice straw, rice husk and rice residues from the drying process. *Water Sci. Technol.* **2012**, *65*, 1142–1149. [CrossRef] [PubMed]
26. Wang, X.; Lu, X.; Li, F.; Yang, G. Effects of temperature and carbon-nitrogen (C/N) ratio on the performance of anaerobic co-digestion of dairy manure, chicken manure and rice straw: Focusing on ammonia inhibition. *PLoS ONE* **2014**, *9*, e97265. [CrossRef] [PubMed]
27. Cann, I.; Kobayashi, Y.; Wakita, M.; Hoshino, H. Effects of 3 chemical treatments on in vitro fermentation of rice straw by mixed rumen microbes in the presence or absence of anaerobic rumen fungi. *Reprod. Nutr. Dev.* **1994**, *34*, 47–56. [CrossRef] [PubMed]

28. Li, D.; Liu, S.; Mi, L.; Li, Z.; Yuan, Y.; Yan, Z.; Liu, X. Effects of feedstock ratio and organic loading rate on the anaerobic mesophilic co-digestion of rice straw and cow manure. *Bioresour. Technol.* **2015**, *189*, 319–326. [CrossRef] [PubMed]
29. Mancini, G.; Papirio, S.; Riccardelli, G.; Lens, P.N.L.; Esposito, G. Trace elements dosing and alkaline pretreatment in the anaerobic digestion of rice straw. *Bioresour. Technol.* **2018**, *247*, 897–903. [CrossRef] [PubMed]
30. Himmel, M.E.; Ding, S.H.; Johnson, D.K.; Adney, W.S.; Nimlos, M.R.; Brady, J.W.; Foust, T.D. Biomass Recalcitrance: Engineering Plants and Enzymes for Biofuels Production. *Science* **2007**, *315*, 801–804. [CrossRef]
31. Hernández-Beltrán, J.U.; Hernández-De Lira, I.O.; Cruz-Santos, M.M.; Saucedo-Luevanos, A.; Hernández-Terán, F.; Balagurusamy, N. Insight into Pretreatment Methods of Lignocellulosic Biomass to Increase Biogas Yield: Current State, Challenges, and Opportunities. *Appl. Sci.* **2019**, *9*, 3721. [CrossRef]
32. Ahmed, B.; Aboudi, K.; Tyagi, V.K.; Álvarez-Gallego, C.J.; Fernández-Güelfo, L.A.; Romero-García, L.I.; Kazmi, A.A. Improvement of Anaerobic Digestion of Lignocellulosic Biomass by Hydrothermal Pretreatment. *Appl. Sci.* **2019**, *9*, 3853. [CrossRef]
33. CERT—Technical University of Berlin, Chair of Circular Economy and Recycling Technology. BioRist Project: Joint Research Project for An Innovative Process Technology for Biogas Production from Rice Straw. Available online: http://www.biorist.tu-berlin.de (accessed on 10 March 2020).
34. Sun, Q.; Li, H.; Yan, J.; Liu, L.; Yu, Z.; Yu, X. Selection of appropriate biogas upgrading technology-a review of biogas cleaning, upgrading and utilisation. *Renew. Sustain. Energy Rev.* **2015**, *51*, 521–532. [CrossRef]
35. Angelidaki, I.; Treu, L.; Tsapekos, P.; Luo, G.; Campanaro, S.; Wenzel, H.; Kougias, P.G. Biogas upgrading and utilization: Current status and perspectives. *Biotechnol. Adv.* **2018**, *36*, 452–466. [CrossRef]
36. Mussoline, W.; Esposito, G.; Lens, P.; Garuti, G.; Giordano, A. Electrical energy production and operational strategies from a farm-scale anaerobic batch reactor loaded with rice straw and piggery wastewater. *Renew. Energy* **2014**, *62*, 399–406. [CrossRef]
37. Li, H.; Tan, F.; Ke, L.; Xia, D.; Wang, Y.; He, N.; Zheng, Y.; Li, Q. Mass balances and distributions of C, N, and P in the anaerobic digestion of different substrates and relationships between products and substrates. *Chem. Eng. J.* **2016**, *287*, 329–336. [CrossRef]
38. Gu, Y.; Chen, X.; Liu, Z.; Zhou, X.; Zhang, Y. Effect of inoculum sources on the anaerobic digestion of rice straw. *Bioresour. Technol.* **2014**, *158*, 149–155. [CrossRef] [PubMed]
39. Ye, J.; Li, D.; Sun, Y.; Wang, G.; Yuan, Z.; Zhen, F.; Wang, Y. Improved biogas production from rice straw by co-digestion with kitchen waste and pig manure. *Waste Manag.* **2013**, *33*, 2653–2658. [CrossRef] [PubMed]
40. Li, D.; Mei, Z.; He, W.; Yuan, Y.; Yan, Z.; Li, J.; Liu, X. Biogas production from thermophilic codigestion of air-dried rice straw and animal manure. *Int. J. Energy Res.* **2016**, *40*, 1245–1254. [CrossRef]
41. Li, D.; Liu, S.; Mi, L.; Li, Z.; Yuan, Y.; Yan, Z.; Liu, X. Effects of feedstock ratio and organic loading rate on the anaerobic mesophilic co-digestion of rice straw and pig manure. *Bioresour. Technol.* **2015**, *187*, 120–127. [CrossRef] [PubMed]
42. Yu, Q.; Tian, Z.; Liu, J.; Zhou, J.; Yan, Z.; Yong, X.; Jia, H.; Wu, X.; Wei, P. Biogas Production and Microbial Community Dynamics during the Anaerobic Digestion of Rice Straw at 39–50 °C: A Pilot Study. *Energy Fuels* **2018**, *32*, 5157–5163. [CrossRef]
43. Möller, K.; Müller, T. Effects of anaerobic digestion on digestate nutrient availability and crop growth: A review. *Eng. Life Sci.* **2012**, *12*, 242–257. [CrossRef]
44. Šimon, T.; Kunzová, E.; Friedlová, M. The effect of digestate, cattle slurry and mineral fertilization on the winter wheat yield and soil quality parameters. *Plant Soil Environ.* **2016**, *61*, 522–527. [CrossRef]
45. Alcántara, C.; García-Encina, P.A.; Muñoz, R. Evaluation of mass and energy balances in the integrated microalgae growth-anaerobic digestion process. *Chem. Eng. J.* **2013**, *221*, 238–246. [CrossRef]
46. Massé, D.I.; Croteau, F.; Masse, L. The fate of crop nutrients during digestion of swine manure in psychrophilic anaerobic sequencing batch reactors. *Bioresour. Technol.* **2007**, *98*, 2819–2823. [CrossRef]
47. Marcato, C.E.; Pinelli, E.; Pouech, P.; Winterton, P.; Guiresse, M. Particle size and metal distributions in anaerobically digested pig slurry. *Bioresour. Technol.* **2008**, *99*, 2340–2348. [CrossRef]
48. Schievano, A.; D'Imporzano, G.; Salati, S.; Adani, F. On-field study of anaerobic digestion full-scale plants (part I): An on-field methodology to determine mass, carbon and nutrients balance. *Bioresour. Technol.* **2011**, *102*, 7737–7744. [CrossRef]

49. Ma, G.; Neiberg, J.S.; Harrison, J.H.; Whitefield, E.M. Nutrient contributions and biogas potential of co-digestion of feedstocks and dairy manure. *Waste Manag.* **2017**, *64*, 88–95. [CrossRef]
50. Lukehurst, C.T.; Frost, P.; Al Seadi, T. *Utilisation of Digestate from Biogas Plants as Biofertiliser*; IEA Bioenergy Task 37- Energy from Biogas; IEA Bioenergy: Paris, France, 2010; pp. 1–36.
51. Tambone, F.; Terruzzi, L.; Scaglia, B.; Adani, F. Composting of the solid fraction of digestate derived from pig slurry: Biological processes and compost properties. *Waste Manag.* **2015**, *35*, 55–61. [CrossRef]
52. Tambone, F.; Orzi, V.; D'Imporzano, G.; Adani, F. Solid and liquid fractionation of digestate: Mass balance, chemical characterization, and agronomic and environmental value. *Bioresour. Technol.* **2017**, *243*, 1251–1256.
53. Wild, D.; Kisliakova, A.; Siegrist, H. Prediction of recycle phosphorus loads from anaerobic digestion. *Water Resour.* **1997**, *31*, 2300–2308. [CrossRef]
54. Sathish, S.; Chandrasekaran, M.; Parthiban, A. Effect of co-digestion agricultural-industrial residues: Various slurry temperatures. *Int. J. Ambient Energy* **2018**, *39*, 694–697. [CrossRef]
55. Sanchez, E.; Borja, R.; Weiland, P.; Travieso, L.; Martin, A. Effect of substrate concentration and temperature on the anaerobic digestion of piggery waste in a tropical climate. *Process Biochem.* **2001**, *37*, 483–489. [CrossRef]
56. Zhang, C.; Su, H.; Baeyens, J.; Tan, T. Reviewing the anaerobic digestion of food waste for biogas production. *Renew. Sustain. Energy Rev.* **2014**, *38*, 383–392. [CrossRef]
57. Lianhua, L.; Dong, L.; Yongming, S.; Longlong, M.; Zhenhong, Y.; Xiaoying, K. Effect of temperature and solid concentration on anaerobic digestion of rice straw in South China. *Int. J. Hydrog. Energy* **2010**, *35*, 7261–7266. [CrossRef]
58. Kim, M.; Ahn, Y.H.; Speece, R.E. Comparative process stability and efficiency of anaerobic digestion; mesophilic vs. thermophilic. *Water Res.* **2002**, *36*, 4369–4385. [CrossRef]
59. Li, Y.; Park, S.Y.; Zhu, J. Solid-state anaerobic digestion for methane production from organic waste. *Renew. Sustain. Energy Rev.* **2011**, *15*, 821–826. [CrossRef]
60. Puyuelo, B.; Ponsá, S.; Gea, T.; Sánchez, A. Determining C/N ratios for typical organic wastes using biodegradable fractions. *Chemosphere* **2011**, *85*, 653–659. [CrossRef] [PubMed]
61. Zhu, J.; Zheng, Y.; Xu, F.; Li, Y. Solid-state anaerobic co-digestion of hay and soybean processing waste for biogas production. *Bioresour. Technol.* **2014**, *154*, 240–247. [CrossRef] [PubMed]
62. Xu, R.; Zhang, K.; Liu, P.; Khan, A.; Xiong, J.; Tian, F.; Li, X. A critical review on the interaction of substrate nutrient balance and microbial community structure and function in anaerobic co-digestion. *Bioresour. Technol.* **2018**, *247*, 1119–1127. [CrossRef] [PubMed]
63. Rabii, A.; Aldin, S.; Dahman, Y.; Elbeshbishy, E. A Review on Anaerobic Co-Digestion with a Focus on the Microbial Populations and the Effect of Multi-Stage Digester Configuration. *Energies* **2019**, *12*, 1106. [CrossRef]
64. Shi, X.S.; Yuan, X.Z.; Wang, Y.P.; Zeng, S.J.; Qiu, Y.L.; Guo, R.B.; Wang, L.S. Modeling of the methane production and pH value during the anaerobic co-digestion of dairy manure and spent mushroom substrate. *Chem. Eng. J.* **2014**, *244*, 258–263. [CrossRef]
65. Zhan-jiang, P.; Jie, L.; Feng-mei, S.; Su, W.; Ya-bing, G.; Da-lei, Z. High-solid Anaerobic Co-digestion of Food Waste and Rice Straw for Biogas Production. *J. Northeast Agric. Univ. (Engl. Ed.)* **2014**, *21*, 61–66. [CrossRef]
66. Zhang, T.; Liu, L.; Song, Z.; Ren, G.; Feng, Y.; Han, X.; Yang, G. Biogas production by co-digestion of goat manure with three crop residues. *PLoS ONE* **2013**, *8*, e66845. [CrossRef]
67. Mei, Z.; Liu, X.; Huang, X.; Li, D.; Yan, Z.; Yuan, Y.; Huang, Y. Anaerobic Mesophilic Codigestion of Rice Straw and Chicken Manure: Effects of Organic Loading Rate on Process Stability and Performance. *Appl. Biochem. Biotechnol.* **2016**, *179*, 846–862. [CrossRef]
68. Zhao, M.; Wang, Y.; Zhang, C.; Li, S.; Huang, Z.; Ruan, W. Synergistic and pretreatment effect on anaerobic co-digestion from rice straw and municipal sewage sludge. *Bioresources* **2014**, *9*, 5871–5882. [CrossRef]
69. Haider, M.R.; Zeshan; Yousaf, S.; Malik, R.N.; Visvanathan, C. Effect of mixing ratio of food waste and rice husk co-digestion and substrate to inoculum ratio on biogas production. *Bioresour. Technol.* **2015**, *190*, 451–457. [CrossRef] [PubMed]
70. Deng, Y.; Huang, Z.; Zhao, M.; Ruan, W.; Miao, H.; Ren, H. Effects of co-inoculating rice straw with ruminal microbiota and anaerobic sludge: Digestion performance and spatial distribution of microbial communities. *Appl. Microbiol. Biotechnol.* **2017**, *101*, 5937–5948. [CrossRef] [PubMed]

71. Croce, S.; Wei, Q.; D'Imporzano, G.; Dong, R.; Adani, F. Anaerobic digestion of straw and corn stover: The effect of biological process optimization and pre-treatment on total bio-methane yield and energy performance. *Biotechnol. Adv.* **2016**, *34*, 1289–1304. [CrossRef] [PubMed]
72. Quintero, M.; Castro, L.; Ortiz, C.; Guzmán, C.; Escalante, H. Enhancement of starting up anaerobic digestion of lignocellulosic substrate: Fique's bagasse as an example. *Bioresour. Technol.* **2012**, *108*, 8–13. [CrossRef] [PubMed]
73. Xu, F.; Wang, F.; Long, L.; Yebo, L. Comparison of digestate from solid anaerobic digesters and dewatered effluent from liquid anaerobic digesters as inocula for solid state anaerobic digestion of yard trimmings. *Bioresour. Technol.* **2016**, *200*, 750–760. [CrossRef]
74. Zhang, L.; Lee, Y.W.; Jahng, D. Anaerobic co-digestion of food waste and piggery wastewater: Focusing on the role of trace elements. *Bioresour. Technol.* **2011**, *102*, 5048–5059. [CrossRef]
75. VDI-Gesellschaft Energietechhnik. *Vergärung organischer Stoffe (Fermentation of organic materials) VDI-Richtlinie 4630*; VDI-Handbuch Energietechnik: Düsseldorf, Germany, 2006.
76. Zhou, Y.; Li, C.; Nges, I.A.; Liu, J. The effects of pre-aeration and inoculation on solid-state anaerobic digestion of rice straw. *Bioresour. Technol.* **2017**, *224*, 78–86. [CrossRef]
77. Kim, M.; Yang, Y.; Morikawa-Sakura, M.S.; Wang, Q.; Lee, M.V.; Lee, D.-Y.; Feng, C.; Zhou, Y.; Zhang, Z. Hydrogen production by anaerobic co-digestion of rice straw and sewage sludge. *Int. J. Hydrog. Energy* **2012**, *37*, 3142–3149. [CrossRef]
78. Lv, W.; Schanbacher, F.L.; Yu, Z. Putting microbes to work in sequence: Recent advances in temperature-phased anaerobic digestion processes. *Bioresour. Technol.* **2010**, *101*, 9409–9414. [CrossRef]
79. Liu, X.; Wang, W.; Shi, Y.; Zheng, L.; Gao, X.; Qiao, W.; Zhou, Y. Pilot-scale anaerobic co-digestion of municipal biomass waste and waste activated sludge in China: Effect of organic loading rate. *Waste Manag.* **2012**, *32*, 2056–2060. [CrossRef] [PubMed]
80. Nagao, N.; Tajima, N.; Kawai, M.; Niwa, C.; Kurosawa, N.; Matsuyama, T.; Yusoff, F.M.; Toda, T. Maximum organic loading rate for the single-stage wet anaerobic digestion of food waste. *Bioresour. Technol.* **2012**, *118*, 210–218. [CrossRef] [PubMed]
81. Shen, F.; Li, H.; Wu, X.; Wang, Y.; Zhang, Q. Effect of organic loading rate on anaerobic co-digestion of rice straw and pig manure with or without biological pretreatment. *Bioresour. Technol.* **2018**, *250*, 155–162. [CrossRef] [PubMed]
82. Krakat, N.; Demirel, B.; Anjum, R.; Dietz, D. Methods of ammonia removal in anaerobic digestion: A review. *Water Sci. Technol.* **2017**, *76*, 1925–1938. [CrossRef]
83. Kaparaju, P.; Buendia, I.; Ellegaard, L.; Angelidakia, I. Effects of mixing on methane production during thermophilic anaerobic digestion of manure: Lab-scale and pilot-scale studies. *Bioresour. Technol.* **2008**, *99*, 4919–4928. [CrossRef]
84. Lindmark, J.; Thorin, E.; Bel Fdhila, R.; Dahlquist, E. Effects of mixing on the result of anaerobic digestion: Review. *Renew. Sustain. Energy Rev.* **2014**, *40*, 1030–1047. [CrossRef]
85. Kim, M.; Kim, B.-C.; Choi, Y.; Nam, K. Minimizing mixing intensity to improve the performance of rice straw anaerobic digestion via enhanced development of microbe-substrate aggregates. *Bioresour. Technol.* **2017**, *245*, 590–597. [CrossRef] [PubMed]
86. Wu, B. CFD analysis of mechanical mixing in anaerobic digesters. *Am. Socity Agric. Biol. Eng.* **2009**, *52*, 1371–1382.
87. Shen, F.; Tian, L.; Yuan, H.; Pang, Y.; Chen, S.; Zou, D.; Zhu, B.; Liu, Y.; Li, X. Improving the mixing performances of rice straw anaerobic digestion for higher biogas production by computational fluid dynamics (CFD) simulation. *Appl. Biochem. Biotechnol.* **2013**, *171*, 626–642. [CrossRef]
88. Wu, B. CFD simulation of mixing in egg-shaped anaerobic digesters. *Water Res.* **2010**, *44*, 1507–1519. [CrossRef]
89. Stroot, P.G.; McMahon; Mackie, R.I.; Raskin, L. Anaerobic codigestion of municipal solid waste and biosolids under various mixing conditionds-I. Digester performance. *Water Resour.* **2001**, *35*, 1804–1816.
90. McMahon, K.; Stroot, P.G.; Mackie, R.I.; Raskin, L. Anaerobic codigestion of municipal solid waste and biosolids under various mixing conditions-II: Microbial population dynamics. *Water Resour.* **2001**, *35*, 1817–1827. [CrossRef]
91. Ong, H.K.; Greenfield, P.F.; Pullammanappallil, P.C. Effect of mixing on biomethanation of cattle-manure slurry. *Environ. Technol.* **2002**, *23*, 1081–1090. [CrossRef]

92. Tian, L.; Zou, D.; Yuan, H.; Wang, L.; Zhang, X.; Li, X. Identifying proper agitation interval to prevent floating layers formation of corn stover and improve biogas production in anaerobic digestion. *Bioresour. Technol.* **2015**, *186*, 1–7. [CrossRef] [PubMed]
93. Kozłowski, K.; Mazurkiewicz, J.; Chełkowski, D.; Jeżowska, A.; Cieślik, M.; Brzoski, M.; Smurzyńska, A.; Dongmin, Y.; Wei, Q. The Effect of Mixing During Laboratory Fermentation of Maize Straw with Thermofilic Technology. *J. Ecol. Eng.* **2018**, *19*, 93–98. [CrossRef]
94. Hu, Y.; Shen, F.; Yuan, H.; Zou, D.; Pang, Y.; Liu, Y.; Zhu, B.; Chufo, W.A.; Jaffar, M.; Li, X. Influence of recirculation of liquid fraction of the digestate (LFD) on maize stover anaerobic digestion. *Biosyst. Eng.* **2014**, *127*, 189–196. [CrossRef]
95. Gerin, P.A.; Vliegen, F.; Jossart, J.-M. Energy and CO2 balance of maize and grass as energy crops for anaerobic digestion. *Bioresour. Technol.* **2008**, *99*, 2620–2627. [CrossRef]
96. Zeb, I.; Ma, J.; Frear, C.; Zhao, Q.; Ndegwa, P.; Yao, Y.; Kafle, G.K. Recycling separated liquid-effluent to dilute feedstock in anaerobic digestion of dairy manure. *Energy* **2017**, *119*, 1144–1151. [CrossRef]
97. Shahriari, H.; Warith, M.; Hamoda, M.; Kennedy, K.J. Effect of leachate recirculation on mesophilic anaerobic digestion of food waste. *Waste Manag.* **2012**, *32*, 400–403. [CrossRef] [PubMed]
98. Pezzolla, D.; Di Maria, F.; Zadra, C.; Massaccesi, L.; Sordi, A.; Gigliotti, G. Optimization of solid-state anaerobic digestion through the percolate recirculation. *Biomass Bioenergy* **2017**, *96*, 112–118. [CrossRef]
99. Nges, I.A.; Wang, B.; Cui, C.; Liu, J. Digestate liquor recycle in minimal nutrients-supplemented anaerobic digestion of wheat straw. *Biochem. Eng. J.* **2015**, *94*, 106–114. [CrossRef]
100. Peng, X.; Nges, I.A.; Liu, J. Improving methane production from wheat straw by digestate liquor recirculation in continuous stirred tank processes. *Renew. Energy* **2016**, *85*, 12–18. [CrossRef]
101. Sibiya, N.T.; Tesfagiorgis, H.B.; Muzenda, E. Influence of digestate recirculation and recirculation percentage on biogas production from lawn grass via anaerobic digestion. *Proc. World Congr. Eng. Comput. Sci.* **2015**, *II*.
102. Nordberg, A.; Jarvis, A.; Stenberg, B.; Mathisen, B.; Svensson, B.H. Anaerobic digestion of alfalfa silage with recirculation of process liquid. *Bioresour. Technol.* **2007**, *98*, 104–111. [CrossRef]
103. Qian, M.; Zhang, Y.; Li, R.; Nelles, M.; Stinner, W.; Li, Y. Effects of Percolate Recirculation on Dry Anaerobic Co-digestion of Organic Fraction of Municipal Solid Waste and Corn Straw. *Energy Fuels* **2017**, *31*, 12183–12191. [CrossRef]
104. Demirel, B.; Scherer, P. Trace element requirements of agricultural biogas digesters during biological conversion of renewable biomass to methane. *Biomass Bioenergy* **2011**, *35*, 992–998. [CrossRef]
105. Zandvoort, M.H.; van Hullebusch, E.D.; Gieteling, J.; Lens, P.N.L. Granular sludge in full-scale anaerobic bioreactors: Trace element content and deficiencies. *Enzym. Microb. Technol.* **2006**, *39*, 337–346. [CrossRef]
106. Chen, Y.; Cheng, J.J.; Creamer, K.S. Inhibition of anaerobic digestion process: A review. *Bioresour. Technol.* **2008**, *99*, 4044–4064. [CrossRef]
107. Lin, C.Y.; Chen, C.C. Effect of heavy metals on the methanogenic UASB granule. *Water Res.* **1998**, *33*, 409–416. [CrossRef]
108. Zayed, G.; Winter, J. Inhibition of methane production from whey by heavy metals - protective effect of sulfide. *Appl. Microbiol Biotechnol.* **2000**, *53*, 726–731. [CrossRef]
109. Cai, Y.; Zhao, X.; Zhao, Y.; Wang, H.; Yuan, X.; Zhu, W.; Cui, Z.; Wang, X. Optimization of Fe2+ supplement in anaerobic digestion accounting for the Fe-bioavailability. *Bioresour. Technol.* **2018**, *250*, 163–170. [CrossRef] [PubMed]
110. Mancini, G.; Papirio, S.; Lens, P.; Esposito, G. A Preliminary Study of the Effect of Bioavailable Fe and Co on the Anaerobic Digestion of Rice Straw. *Energies* **2019**, *12*, 577. [CrossRef]
111. Hinken, L.; Urban, I.; Haun, E.; Weichgrebe, D.; Rosenwinkel, K.-H. The valuation of malnutrition in the mono-digestion of maize silage by anaerobic batch tests. *Water Sci. Technol.* **2008**, *58*, 1453–1459. [CrossRef]
112. Evranos, B.; Demirel, B. The impact of Ni, Co and Mo supplementation on methane yield from anaerobic mono-digestion of maize silage. *Environ. Technol.* **2015**, *36*, 1556–1562. [CrossRef]
113. Liu, C.; Yuan, H.; Zou, D.; Liu, Y.; Zhu, B.; Li, X. Improving Biomethane Production and Mass Bioconversion of Corn Stover Anaerobic Digestion by Adding NaOH Pretreatment and Trace Elements. *Biomed Res. Int.* **2015**, *2015*, 125241. [CrossRef]
114. Khatri, S.; Wu, S.; Kizito, S.; Zhang, W.; Li, J.; Dong, R. Synergistic effect of alkaline pretreatment and Fe dosing on batch anaerobic digestion of maize straw. *Appl. Energy* **2015**, *158*, 55–64. [CrossRef]

115. Zhang, H.; Tian, Y.; Wang, L.; Mi, X.; Chai, Y. Effect of ferrous chloride on biogas production and enzymatic activities during anaerobic fermentation of cow dung and Phragmites straw. *Biodegradation* **2016**, *27*, 69–82. [CrossRef] [PubMed]
116. Tian, Y.; Zhang, H.; Chai, Y.; Wang, L.; Mi, X.; Zhang, L.; Ware, M.A. Biogas properties and enzymatic analysis during anaerobic fermentation of Phragmites australis straw and cow dung: Influence of nickel chloride supplement. *Biodegradation* **2017**, *28*, 15–25. [CrossRef] [PubMed]
117. Hao, H.; Tian, Y.; Zhang, H.; Chai, Y. Copper stressed anaerobic fermentation: Biogas properties, process stability, biodegradation and enzyme responses. *Biodegradation* **2017**, *28*, 369–381. [CrossRef] [PubMed]
118. Ortner, M.; Rameder, M.; Rachbauer, L.; Bochmann, G.; Fuchs, W. Bioavailability of essential trace elements and their impact on anaerobic digestion of slaughterhouse waste. *Biochem. Eng. J.* **2015**, *99*, 107–113. [CrossRef]
119. Marcato, C.-E.; Pinelli, E.; Cecchi, M.; Winterton, P.; Guiresse, M. Bioavailability of Cu and Zn in raw and anaerobically digested pig slurry. *Ecotoxicol. Environ. Saf.* **2009**, *72*, 1538–1544. [CrossRef] [PubMed]
120. Ortner, M.; Rachbauer, L.; Somitsch, W.; Fuchs, W. Can bioavailability of trace nutrients be measured in anaerobic digestion? *Appl. Energy* **2014**, *126*, 190–198. [CrossRef]
121. González-Suárez, A.; Pereda-Reyes, I.; Oliva-Merencio, D.; Suárez-Quiñones, T.; José da Silva, A.; Zaiat, M. Bioavailability and dosing strategies of mineral in anaerobic mono-digestion of maize straw. *Eng. Life Sci.* **2018**, *18*, 562–569. [CrossRef]
122. Ahlberg-Eliasson, K.; Nadeau, E.; Levén, L.; Schnürer, A. Production efficiency of Swedish farm-scale biogas plants. *Biomass Bioenergy* **2017**, *97*, 27–37. [CrossRef]
123. Kaparaju, P.; Rintala, J. Mitigation of greenhouse gas emissions by adopting anaerobic digestion technology on dairy, sow and pig farms in Finland. *Renew. Energy* **2011**, *36*, 31–41. [CrossRef]
124. Mussoline, W.; Esposito, G.; Lens, P.; Garuti, G.; Giordano, A. Design considerations for a farm-scale biogas plant based on pilot-scale anaerobic digesters loaded with rice straw and piggery wastewater. *Biomass Bioenergy* **2012**, *46*, 469–478. [CrossRef]
125. Chiumenti, A.; da Borso, F.; Limina, S. Dry anaerobic digestion of cow manure and agricultural products in a full-scale plant: Efficiency and comparison with wet fermentation. *Waste Manag.* **2018**, *71*, 704–710. [CrossRef]
126. André, L.; Zdanevitch, I.; Pineau, C.; Lencauchez, J.; Damiano, A.; Pauss, A.; Ribeiro, T. Dry anaerobic co-digestion of roadside grass and cattle manure at a 60 L batch pilot scale. *Bioresour. Technol.* **2019**, *289*, 121737. [CrossRef]
127. Wandera, S.M.; Qiao, W.; Algapani, D.E.; Bi, S.; Yin, D.; Qi, X.; Liu, Y.; Dach, J.; Dong, R. Searching for possibilities to improve the performance of full scale agricultural biogas plants. *Renew. Energy* **2018**, *116*, 720–727. [CrossRef]

© 2020 by the authors. Licensee MDPI, Basel, Switzerland. This article is an open access article distributed under the terms and conditions of the Creative Commons Attribution (CC BY) license (http://creativecommons.org/licenses/by/4.0/).

applied sciences

MDPI

Review

A Review on Anaerobic Digestion of Lignocellulosic Wastes: Pretreatments and Operational Conditions

Tahseen Sayara [1] and Antoni Sánchez [2,*]

1 Department of Environment and Sustainable Agriculture, Faculty of Agricultural Sciences and Technology, Palestine Technical University-Kadoorie, 7 Tulkarm, Palestine; tsayara@yahoo.com
2 Departament d'Enginyeria Química, Biològica i Ambiental, Escola d'Enginyeria, Universitat Autònoma de Barcelona, 08193 Bellaterra, Barcelona, Spain
* Correspondence: antoni.sanchez@uab.cat

Received: 30 August 2019; Accepted: 15 October 2019; Published: 1 November 2019

Abstract: Anaerobic digestion (AD) has become extremely popular in the last years to treat and valorize organic wastes both at laboratory and industrial scales, for a wide range of highly produced organic wastes: municipal wastes, wastewater sludge, manure, agrowastes, food industry residuals, etc. Although the principles of AD are well known, it is very important to highlight that knowing the biochemical composition of waste is crucial in order to know its anaerobic biodegradability, which makes an AD process economically feasible. In this paper, we review the main principles of AD, moving to the specific features of lignocellulosic wastes, especially regarding the pretreatments that can enhance the biogas production of such wastes. The main point to consider is that lignocellulosic wastes are present in any organic wastes, and sometimes are the major fraction. Therefore, improving their AD could cause a boost in the development in this technology. The conclusions are that there is no unique strategy to improve the anaerobic biodegradability of lignocellulosic wastes, but pretreatments and codigestion both have an important role on this issue.

Keywords: feedstock and degradation pathway; AD systems; pretreatment technologies; process stability; codigestion

---

1. Introduction

Organic wastes are produced in large amounts worldwide. Mismanagement of such wastes will lead to various problems for both human beings and the environment. Among the different technologies used to treat this type of waste is the anaerobic digestion process (AD) [1]. This process is a biological treatment that allows the decomposition and stabilization of a wide spectrum of organic wastes, from complex lignocellulosic materials to easily degradable food waste, while simultaneously producing renewable energy, recovering fibers and nutrients for soil amendments, and offsetting greenhouse gas emissions [2,3].

The AD process is a well-established technology that transforms the organic fraction into renewable fuels, such as methane. However, the process is recognized as a complex one, and many research studies have indicated that hydrolysis and methanogenesis can be the rate-limiting steps in AD because of the accumulation of undesirable volatile fatty acids (VFA) [4–6]. Nevertheless, controlling and optimizing the major operational parameters can improve the process efficiency and obtain stable methane generation. Recently, numerous studies have been performed to enhance the biogas production and to mitigate the inhibition effect caused by various compounds and situations. In this regard, pretreatment methods of feedstock, including physical, chemical, biological, thermal, and combined approaches, were investigated to overcome kinetic disadvantages and increase process potential, mainly for lignocellulosic biomass, as this kind of waste represents a promising source of alternative energy since it has high potential for biogas production [7,8].

Monodigestion of substrates has been applied in AD, and this approach proves to work effectively for a narrow range of generated wastes, especially for easily degradable fraction waste. However, introducing a mixture of different substrates together, which is known as codigestion, is able to improve the process performance and make it the more attractive alternative in waste management hierarchy. In this context, codigestion provides a great chance for recycling a wide spectrum of wastes, supporting the synergic effects of microbial activities through improving nutrient balance (mainly carbon/nitrogen (C/N) ratio), and consequently, the enhancing the process stability in general [9]. Various mixtures of agricultural wastes, municipal wastes, industrial wastes, sewage sludge, etc., have been investigated by researchers using the codigestion approach and satisfactory results have been obtained [10]. Importantly, selecting an adequate co-substrate and mixing ratio, organic loading rate, co-substrate characteristics, and co-substrate-prompted inhibitions are essential factors to be considered, as they may greatly influence the process performance [11].

This current review provides a comprehensive overview of the process of AD, which is considered to be one of the most viable options for recycling the organic fraction of solid waste. Feedstock and degradation pathways, AD systems, pretreatment technologies, process stability, and codigestion for lignocellulosic materials are introduced and discussed thoroughly in this review.

2. Overview of AD Process

2.1. Feedstock for AD

A wide variety of feedstock can be processed through AD. However, lignocellulosic biomass, as it has a high potential for biogas production, represents a promising source for alternative energy [8]. This type of feedstock is abundantly available and can be found in different wastes, such as agricultural residues (crop residues), wood, and grass. About 181.5 billion tons of lignocellulosic biomass are produced annually worldwide [12]. Nevertheless, process inhibition is commonly noticed during the digestion of such materials caused by the complexity of their structure, which contains about 10–25% lignin [13]. Accordingly, pretreatment is often needed before use in AD for biogas production, or alternatively it can be codigested with other feedstock [14]. Solid-state AD (ss-AD) digesters are ideally suited for such feedstock [15]. Consequently, more attention is being directed to this process, as it provides promising solutions either for waste management or for introducing a renewable energy source.

2.2. Feedstock Degradation Pathway

AD plays a key role in the recovery of renewable energy in the form of biogas and nutrients from waste materials. The degradation process of feedstock passes through sequential and distinct phases or steps, as shown in Figure 1. During these phases, the rate and characteristics of the generated gas reflect the microbial processes taking place in the reactor. The four main phases successive phases are set out below.

2.2.1. (a) Hydrolysis

Macromolecules, which include proteins, carbohydrates, and lipids, are the main components of organic substrates, and under anaerobic conditions they are initially broken down to monosaccharides, namely amino acids, long chain fatty acids, and glycerol, respectively. The efficiency of this step is based on the presence and active action of hydrolytic and fermentative microbes (e.g., *Clostridium, Proteus Vulgaris, Bacillus, Bacteriodes, Micrococcus, Staphylococcus*) to excrete extracellular enzymes. The presence of a variety of extracellular enzymes, including amylase, cellulase, protease, and lipase, which are exerted by bacteria, are mandatory as catalyzers for the decomposition of each macromolecule. For instance, hydrolytic enzymes (e.g., cellulase, β-glucosidase, xylanase) or complex enzyme systems (e.g., cellulosome) attack polysaccharides, protease degrades proteins, and lipase is suitable for breaking down lipids [16,17]. Hydrolysis is known to be a rate-limiting step, especially for lignocellulosic substrates [18], because of the presence of lignin, which along with cellulose and hemicellulose units forms a rigid three-dimensional complex compound. This physical barrier protects the biomass from the enzymatic attack. Thus, the existence of microbes with augmented enzymatic activity is mandatory for an efficient decomposition, especially in the case of lignocellulosic materials [19].

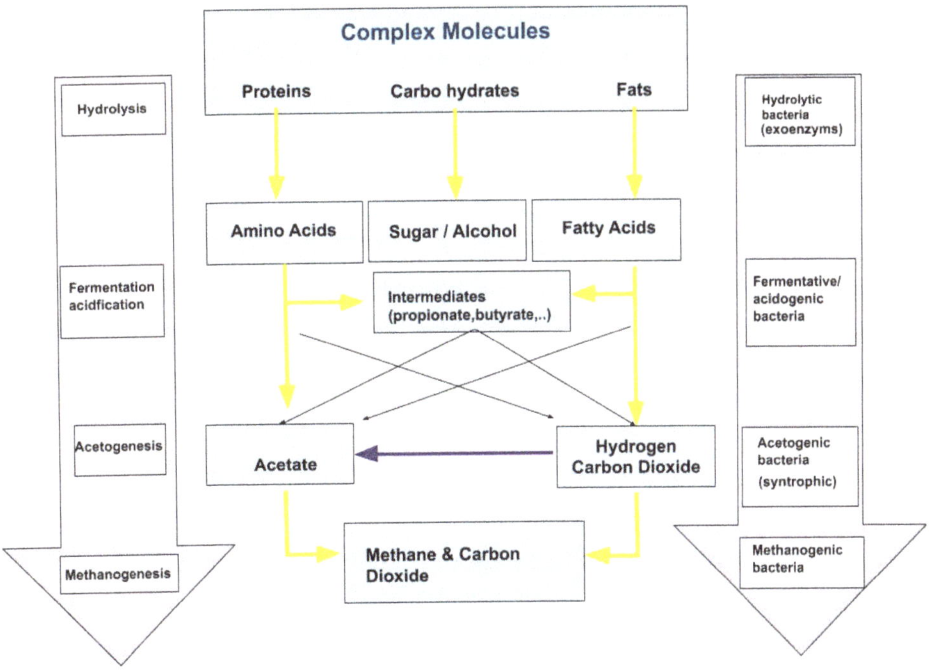

Figure 1. Schematic diagram of substrate decomposition through anaerobic digestion.

2.2.2. (b) Acidogenesis

Acidogenesis is the second step of AD, where the hydrolyzed products of macromolecules are subjected to the fermentation step following different metabolic pathways. Different facultative and obligate anaerobic bacteria (*Lactobacillus, Escherichia, Staphylococcus, Bacillus, Sarcina, Veillonella, Desulfobacter, Desulforomonas, Eubacterium limosum, etc.*) are involved in this step to produce volatile fatty acids (VFA), hydrogen, carbon dioxide, and alcohols. Sugars and amino acids are the main substrates of this step. Results of glycerol fermentation are propionate production and biomass generation [19,20]. Acidogenic microorganisms produce high concentrations of hydrogen during this phase, which may cause inhibition of the production of acetate.

2.2.3. (c) Acetogenesis

During acetogenesis, different syntrophic oxidation reactions convert the longer fatty acids into acetate (by homoacetogenic microorganisms) and hydrogen and carbon dioxide, and products of acidogenesis are utilized by hydrogen producing acetogens, using carbon dioxide and hydrogen ions as electron acceptors (e.g., *Syntrophomonas wolfei*). This bioconversion process is not exergonic, and thus, a syntrophic relationship with methanogens (as they remove hydrogen during the production of methane) is mandatory to keep the hydrogen partial pressure low in order for acetogenic reactions to be energetically favorable [19,21].

2.2.4. (d) Methanogenesis

In this final step, acetotrophic and hydrogenotrophic methanogens (*Methanobacterium, Methanobrevibacerium, Methanoplanus, Methanospirillum, Methanosaeta, Methanosarcina*) work on converting acetate (by acetoclastic methanogens) and hydrogen and carbon dioxide (carbon dioxide-reducing and hydrogen-oxidizing methanogens) into methane and carbon dioxide. The larger portion of methane is derived from the conversion of acetate and the rest is primarily produced from H_2 and CO_2 [20]. Alternatively, hydrogen and carbon dioxide may be produced through syntrophic acetate oxidation coupled with hydrogenotrophic methanogenesis. Extended methane production can be conducted via the hydrogenotrophic pathway based on process characteristics [22,23].

Generally, maintaining a synchronized state between different degradation steps is crucial throughout the process and any fluctuation or decrease in the activity of one or several microbial groups could severely influence the process performance and efficiency, and could even lead to process failure.

2.3. *Anaerobic Digestion of Substrates under Thermophilic and Mesophilic Conditions*

In the AD process, temperature is recognized as one of the key important factors that affects the entire process performance [20]. As general rule, the performance of the process is increased with the increase in temperature [20,24]. For instance, the rate of anaerobic degradation processes, and particularly enzymatic and coenzymatic activities responsible for the rates of hydrolysis and methanogenesis, are highly affected by the dominant temperature. Additionally, some other factors, such as gas transfer rates and settling characteristics of biosolids, are significantly affected by the temperature [25,26].

For the AD process, two main temperature regimes are usually applied—mesophilic (37 °C) and thermophilic (55 °C). Mesophilic bacteria are supposedly more robust and can tolerate higher changes in environmental parameters, including temperature. Because mesophilic digestion takes place at a lower temperature, digestion in this temperature regime requires longer retention time and yields less biogas; however, mesophilic digesters remain attractive because of their stability and lower heating energy costs compared to thermophilic digesters [14,24]. On the other hand, thermophilic digestion operates at a higher temperature, and consequently reaction rates and kinetics are enhanced, leading to the possibility of higher loading rates, thus increasing biogas production [14,27,28]. Moreover, thermophilic digestion is known to have higher levels of pathogen removal [29]. Nevertheless, this method, is more sensitive to toxic substances and changes of operation parameters [24].

2.4. *Solid-State (ss) and Wet-State Anaerobic Digestion*

Wet AD and solid-state AD (ss-AD) are considered the two main types of anaerobic technologies that have been employed to digest solid organic matter for biogas production. The classification between the two technologies is based principally on the particular content of total solids (TS) concentration at below or higher than 15%, respectively [30,31].

2.4.1. Wet-State AD

In wet digestion processes, the feedstock is slurried with the addition of water to the mixture in order to be within the desired level (TS less than 15%), and normally this can be achieved by adding water either by recirculation of the liquid effluent fraction or by codigestion with wetter wastes. The latter is more preferable, as it provides a chance to mix various waste streams, such as sewage sludge or manure and OFMSW [32–35]. The less complicated mechanical equipment needed and the opportunity to reduce the inhibitory effect of some substances as they become diluted by water are considered advantages when applying this process. On the other hand, complicated pre-treatment, high consumption of water and energy for heating, and the reduction of working volume caused by the sedimentation of inert materials remain as obstacles in this process [36–38]. Continuous stirred tank reactors (CSTR) are generally used in wet digestion processes, in which mechanical mixers or a combination of mechanical mixing and biogas injection are employed [37].

2.4.2. Solid-State AD (ss-AD)

Solid-state anaerobic digestion (ss-AD) provides the ability to process high-solid-content substrates; typically, with a solid content above 15%. The process is recognized as robust and advantageous compared to wet AD because it requires less energy for heating, less water, smaller digester volume, minimal material handling, and because of its low moisture content, the digestate of ss-AD is more coherent and can be used as fertilizer or pelletized fuel, which is much easier to handle than the effluent produced by wet AD [14,34,35]. These advantages contributed to the large industrial development of ss-AD, with the rapid emergence of full-scale plants [31]. Nonetheless, it should be noted that regardless of the various advantages of ss-AD technology and advancements in system design, there are still a number of aspects that need to be deeply investigated and improved for further commercialization of the technology.

Water content is noted as a key factor in deriving microbial activity, and low content of this parameter in ss-AD would lead to accumulation of the microbial end products, and thus inhibition of methanogenesis [38]. In this regard, and because of slow mass transportation, longer retention time (up to three times) has been documented for ss-AD compared to wet AD [38–40]. According to [41], water availability presents two distinct forms: (i) "free water", which can act as a solvent for soluble compounds and salts; and (ii) "bound water", which is unable to act as a solvent, as it has more structural bonding (chemical and physical interactions) than liquid or free water. The variation in the composition and structure of the organic matter has a major role in determining the dominant form of water and consequently the rate of degradation. In this context, it should be considered that when the TS content increases, the quantity of free water decreases; thus, rheological behavior of the substrate is modified and the diffusion of metabolic compounds within the substrate at a macroscopic level is lowered [42,43], meaning the transport of soluble content within the substrate can become a limiting factor [43].

As illustrated in the literature, substrate degradation and consequently biogas production are reduced by increasing the TS content [38,39,42]. One study [44] demonstrated that an increase from 20% to 30% in TS content was associated with a decrease in organic waste degradation by 17%. Also, another study [45] reported reduced methane yields and cellulose hydrolysis rates of lignocellulosic biomass in ss-AD compared to those of wet AD. The limitation of ss-AD might be explained by the slow diffusion of dissolved inhibitory products inside the organic matrix causing local accumulation at the microbial scale [46–48]. In a recent study [38], methanogenesis inhibition was observed at high solid content levels, which was attributed to gas transfer limitation, and more particularly by a local accumulation of gases (H_2 and CO_2) leading to VFA accumulation. In this context, and because of thermodynamic reasons, anaerobic oxidation (acetate formation) can only take place at low partial hydrogen pressure, meaning the balance between the hydrogen-forming and hydrogen-consuming microorganisms is very important. In order to avoid stress in the acidogenic bacteria, the partial pressure of the hydrogen gas should always be less than 1–4 atm [49].

Solid-state AD under mesophilic and thermophilic conditions has been investigated, and mesophilic AD is considered more stable and better established than thermophilic AD [14]. However, for lignocellulosic wastes, thermophilic conditions usually show better initial performance than mesophilic AD, since high temperatures improve the hydrolysis of substrates, being "the major rate-limiting step" [14]. It was found that VFA generation and feedstock degradation were faster in a thermophilic digester than in a mesophilic digester, and that 95% of the methane yield potential was achieved in 11 and 27 days under thermophilic and mesophilic conditions, respectively [50]. The fermentative microbial activities are enhanced under thermophilic ss-AD; therefore, more care must be taken, as this behavior may cause accumulation of VFAs and result in inhibitory effects on the AD process [51].

High energy input for heating under thermophilic ss-AD should be considered, as it may offset the higher methane production yields and rates [7,52]. Accordingly, net energy production and the performance of ss-AD under different temperatures should be evaluated and optimized for feasible process operation. In this regard, even though thermophilic ss-AD of switch grass showed greater methane yield than mesophilic ss-AD, the latter had higher net energy production than thermophilic conditions because of reduced initial heating requirements after inoculation [52].

Throughout the literature, the feasibility of two-stage reactors has been confirmed regarding energy balance. For instance, in a two-stage AD system treating OFMSW, better energy balance was obtained, where higher gross energetic potential was related to the higher performance in the methanogenic reactor rather than the hydrogen production from the first stage [1]. Furthermore, in a similar reactor treating thin stillage, optimizing the process was able to increase the energy balance by 18.5% [53,54], and better energy balance with a surplus of 2.17 kJ/day was also documented in a two-stage reactor treating sewage sludge as compared to a single-stage reactor [27].

3. Pretreatment of Feedstock to Enhance Biogas Production

Some wastes are found in large quantities in the waste stream, but they are difficult to degrade via microorganisms because of the complexity of their structure [13]. These wastes include lignocellulosic substrates, keratins, or high ammonia content materials. However, these wastes, when properly pretreated, can be a valuable source for biogas production. Appropriate pretreatment facilitates and increases the microbial accessibility to nutrients found within the mixture, thus speeding up biomass utilization during the anaerobic digestion process [55]. Different technologies or approaches have been applied in an attempt to overcome the problem and to increase the biogas production. According to [49], the selection criteria for any pretreatment technology should consider the following important points: (1) cost-effectiveness, (2) the microbial accessibility to nutrients, (3) any inhibition in biogas production, (4) energy requirements, and (5) eco-friendliness. In the following subsections, the main pretreatment technologies are discussed, and Table 1 provides a summary of research conducted in this field.

3.1. Physical or Mechanical Pretreatment

For lignocellulosic materials, physical or mechanical pretreatment methods are usually employed to promote microbial accessibility to hydrolysable polymers and to reduce the cellulose crystallinity and its degree of polymerization. This includes a change in the structure or appearance of the materials using different techniques [56,57]. The process includes crushing or breakdown of the feedstock particles, resulting in increased surface area that provides more contact between the substrate and the microorganisms, and presumably increases the process efficiency [57]. Comminution, such as ball milling, is a physical pretreatment method that increases the feedstock surface area and decrystallizes cellulose. Indeed, ball milling was found to be an effective pretreatment, with similar glucose yield and superior carbohydrate yield to steam explosion pretreatment. In this context, the effect of grinding on wheat straw was investigated, and the results showed that ball milling samples yield 46% total carbohydrates and 72% glucose as a result of the reduction in the cellulose crystallinity from 22% to 13% [58]. Ball milling was also applied to non-degraded digestate in order to feed it back into the digestion process [59]. Enhanced methane production of 9% was reported in the case of two-stage maize silage digestate, and an increase of 17% was detected when using two-stage hay/straw digestate. Irradiation is also an effective pretreatment method that disorganizes the structure of the biomass cell wall and decreases the crystallinity of the cellulose [60]. Solubilization of microalgae as a substrate for biogas production was evaluated using microwave pretreatment. The biogas production rate was increased by 27–75% at specific energy rates of 21.8, 43.6, and 65.4 MJ/kg TS.

The advantages of this method are that there is no risk of forming inhibitory compounds, and there is improvement in the methane yield in some cases caused by size reduction. However, the main disadvantage is the high energy requirements [55]. In some cases, this method is not viable as a stand-alone pretreatment for industrial applications, but it can be combined with other pretreatments that are more cost-effective.

3.2. Chemical Pretreatment

This method depends mainly on breakdown or destruction of organic compounds using acids, alkalis, and oxidants, and its efficiency is highly affected by the substrate characteristics [61]. Importantly, this type of pretreatment is less recommended for easily biodegradable substrates, as it can be associated with accumulation of VFAs, which ultimately leads to inhibition of the methanogenesis step [57].

During acid pretreatment, hemicellulose is hydrolyzed into monosaccharides, while the lignin condensates and precipitates [55,61,62]. The process is effective for substrates with high lignocellulosic content as it breaks down the lignin, and also because the hydrolytic microbes are capable of acclimating to acidic conditions [61]. Strong acidic pretreatment may result in the production of inhibitory by-products, such as furfural and hydroxymethylfurfural (HMF) [62–64]. The use of diluted acids combined with thermal methods is normally recommended to avoid the formation of such inhibitory products [57]. In this regard, and with mesophilic anaerobic digestion of wheat plant, diluted sulfuric acid pretreatment at 121 °C confirmed that methane yield could be increased by 15.5% higher than that of the untreated wheat plant after pretreatment for 120 min [65]. Also, in semi-continuous digestion experiments conducted for 12 days at 35 °C using HCl at pH 2 for subsequent digestion of waste-activated sludge (WAS), an increase of 14.3% in methane yield was achieved compared to untreated substrates [66].

The aim of alkali pretreatment is to enhance the enlargement of the solid particles, thus increasing the surface area, meaning the substrate becomes more accessible for microbial activities. Solvation and saponification are the main reactions during this process [55,57,62,63,67]. This process is considered more effective for materials with low lignin contents, such as agricultural residues, herbaceous crops, and hardwoods, than on softwood, which has high lignin contents [68]. Basic compounds, such as calcium oxide, ammonia, and sodium hydroxide, are used to solubilize lignin in this process [62]. The effect of alkaline (NaOH) pretreatment on ensiled sorghum forage was investigated in semi-continuous digesters. The researchers observed that pretreatment with 10 g NaOH/100 g TS increased the methane yield by 25% compared to untreated sorghum, without experiencing any inhibition of the process [69]. Additionally, the effect of trace element (TE) addition and NaOH pretreatment on the AD of rice straw was investigated in batch tests. Co, Ni, and Se were added to the raw rice straw at different dosages. The NaOH pretreatment was applied to the rice straw, both alone and in combination with the addition of TEs, in order to evaluate potential synergistic effects of the pretreatment and the TE supplementation on the biogas production yields. The obtained results showed that the alkaline pretreatment was more effective than the TE addition in increasing the cumulative biogas production, causing a 21.4% enhancement of the final biomethane yield, whereas the increase due to TE dosing was not statistically significant. The analysis of volatile fatty acids (VFAs) confirmed that the NaOH pretreatment resulted in a higher production of VFAs, indicating increased hydrolysis, while TE addition did not cause significant changes in the VFA concentrations [70].

3.3. Biological Pretreatment

Biological pretreatments mainly depend on using microorganisms, such as bacteria and fungi, to degrade the lignin fraction [3,62,71]. Such pretreatment is gaining popularity because of its cost-effectiveness as a natural process that does not require chemicals or energy, in addition to being an environmentally friendly method. However, careful selection of the suitable microbial consortium for efficient pretreatment of biomass is a critical step [72]. Additionally, it is important to highlight that the process is slow and requires longer residence times, which is often not practical for large-scale applications [3,62,71,73].

The effect of bacterial-based biological pretreatment on liquefaction of microalgae *Chlorella vulgaris* with cellulose-secreting bacteria prior to anaerobic digestion was studied [74]. The results showed that bacterial pretreatment enhanced the bioavailability of biomass, and hence methane generation, with methane yield being almost double that of control. Furthermore, the BMP test to evaluate the potential effect of enzymatic pretreatment on pulp mill biosludge with protease from *B. licheniformis* found that an increase of 26% in biogas yield could be achieved under the studied conditions [75].

3.4. Thermal Treatments

This approach enhances the anaerobic process through hydrolyzing complex organic constituents, where the structure of the materials (mainly the cell membrane) is disintegrated under a certain temperature and pressure. Nowadays, different thermal pretreatment techniques using various temperatures (50–250 °C) are being studied and employed to improve the hydrolysis rate of AD feedstock. Among them is liquid hot water treatment, where the water is maintained as a liquid at high temperatures (160 to 230 °C) and under high pressures (>5 MPa) [55,62]. Another option is the use of thermal pretreatment as "an isochoric or constant volume process", where the material is placed in a sealed container and heated without applying extra external pressure [72]. On the other hand, steam treatment involves exposure of the biomass to high temperatures of up to 240 °C and pressure for a few minutes, thereby causing disruption in the structure of the material [55,73]. Another method is autohydrolysis, which is fundamentally based on hydrolyzing hemicelluloses using highly pressurized liquid hot water at 200 °C.

Besides high energy demand and operation at high pressure, the main disadvantage of these processes is the potential formation of inhibitors, such as furfural and soluble phenolic compounds, which inhibit the production of methane [55]. Concerning the lignocellulosic substrates, temperatures exceeding 160 °C cause not only the solubilization of hemicellulose but also the solubilization of lignin. The released compounds are mostly phenolic compounds that are usually inhibitory to anaerobic microbial populations [55]. The formation of inhibitors during liquid hot water pretreatment is relatively low because of the high water input [76]. In order to minimize the formation of these inhibitors, the addition of external compounds is needed to keep the pH in the range of 4 to 7 [55]. Thermal pretreatment at high temperatures (>170 °C) might lead to the creation of chemical bonds and result in the agglomeration of the particles [77]. One of the most known phenomena is the Maillard reaction, which occurs between carbohydrates and amino acids, resulting in the formation of complex substrates that are difficult to biodegrade. This reaction can occur in extreme thermal treatment at temperatures exceeding 150 °C, or with longer treatment times at lower temperatures (<100 °C) [1,53,54,78].

Table 1. AD pretreatment methods.

Pretreatment Process	Substrate	Conditions	Results	Ref.
Physical	Straw	Milling straw particles to different sizes of 0.25 mm, 1 mm, and 10 mm over 62 days.	Highest methane production for straw with 10 mm particle size (192 ± 25 Nm L/g VS), which was associated with straw biodegradability of 43%.	[40]
	Fruit and vegetable waste	Three sonication times of 9, 18, and 27 min, operating at 20 kHz and amplitude of 80 μm on the substrate.	Highest methane yield at 18 min sonication with specific energy if 2380 kJ/kg TS for 12-day batch period, while longer exposure to sonication led to lower methane yield.	[79]
	Olive mill solid residue	Microwave irradiation at a power of 800 W and temperature if 50 °C.	Maximum methane yield of 395 ml CH_4/g VS for an applied specific energy of 7660 kJ/kg TS.	[80]
	OFMSW	Extrusion pretreatment.	Biogas yield of 800 L/kg VS, containing about 60% methane content.	[81]

Table 1. *Cont.*

Pretreatment Process	Substrate	Conditions	Results	Ref.
Chemical	Cotton stalk residues	Alkali hydrogen peroxide (AHP), treatment at 37 °C for 61 days.	Highest methane yield of 192.4 mL/g VS after 3% AHP pretreatment, yield improved by 254.3% over the untreated waste.	[82]
	Agriculture straw	3% H_2O_2 and 8% $Ca(OH)_2$, treatment at 37 ± 1 °C for 35 days.	Highest methane yields of 216.7 and 206.6 mL CH_4/g VS using acid and alkaline, which are 115.4% and 105.3% greater than untreated waste, respectively.	[83]
	Sunflower oil cake	170 °C, 1% (weight) sulfuric acid concentration. Mesophilic method.	Biogas yield of 302 ± 10 mL CH_4/g VS at 170 °C), 50% greater than untreated waste.	[84]
Biological	Food waste and brown water	Composting or microaeration.	Microaeration pretreatment improved methane yield by 10–21%, caused by enhancement of hydrolysis.	[85]
	Chicken feathers	Aerobically pretreated for 2–8 days with *Bacillus* sp. *C4*, a bacterium that produces both α- and β-keratinases.	445 ml CH_4/g VS methane, 124% more than untreated feather.	[86]
	Napier grass	Microbial consortium: WSD-5.	CH_4 yield was increased by 49%.	[87]
	Organic waste	Soft rot fungi *Trichoderma viride*, 4 days treatment at 25 °C.	Up to 400% increase in methane production compared with controls.	[88]
	Paddy straw	White rot fungi *Fusarium* sp., 10 days treatment at 30 °C (70 % moisture).	53.8% increase in biogas production.	[89]
	Organic waste	Soft rot fungi *Trichoderma viride*, 10 days treatment at 22 °C (70% moisture).	More than two-fold increase in methane production.	[90]
Thermal	Wheat straw	Different thermal pretreatment temperatures of 120, 140, 160, and 180 °C.	Highest biogas yield of 615 ml/g VS (53% increase) and volatile solids reduction of 69% at 180 °C, compared to untreated material.	[91]
	Activated sludge	Steam explosion at 170 °C, autoclaved at 70, 100, and 125 °C.	A linear relationship was found between the thermal treatment temperature in the autoclave and biogas production of aerobic granules, but steam explosion was more effective.	[92]
	Hay	Steam explosion at 160–220 °C.	CH_4 yield increased by 16%.	[93]

The aforementioned technologies have been applied as pretreatment methods in anaerobic digestion, and Table 1 provides some examples where these technologies have been successfully employed. However, it is worth mentioning that a combination of different technologies has been employed in many cases.

4. AD Stability or Inhibition

The anaerobic process is a sensitive process, where any variation or fluctuation in the optimum operational conditions could affect or even inhibit the process. In this regard, and in order to ensure stable conditions, it is important to achieve or maintain a balance between the acidogenic and methanogenic microorganisms. Usually, the inhibition of the anaerobic process is evidenced by the decrease in steady-state methane yield rates.

4.1. *Effect of Ammonia*

Ammonia, which is the end-product of anaerobic digestion of proteins, urea, and nucleic acids [94], plays a major role in the performance and stability of the anaerobic digestion process [6]. Optimal ammonia concentration ensures sufficient buffer capacity for the methanogenic medium, especially for nitrogen-rich organic feedstock. However, excess ammonia concentration is usually reported as the fundamental cause of digester failure when it exceeds the inhibition threshold levels [95–99]. Inhibition is caused by total ammonia nitrogen (TAN) concentration, which is a combination of free (unionized) ammonia nitrogen (FAN) and ionized ammonium nitrogen (NH_4^+) [6,100–102].

TAN concentrations between 1500–7000 mg N/L were found to be inhibitory for anaerobic digestion [98,103]. This broad range of inhibiting concentrations might be explained by different causes, including differences in substrates composition, inoculum origin, environmental and operational conditions (temperature, pH), and acclimation periods [97]. However, it should be noted that any probable inhibition by ammonia in the anaerobic digestion process should not only be directly attributed to the TAN concentration, but to the FAN levels, as this is believed to be the prime cause of inhibition of methanogenic microflora [104,105].

4.1.1. Mechanism of Inhibition by Ammonia

Microorganism growth and maintenance of an appropriate level of alkalinity within the AD reactor are both fundamentally based on ammonia concentration, and as such it is noted as a decisive element in the process stability [6]. The reason behind considering FAN as the main cause of inhibition is related to its high permeability to bacterial cell membranes [106]. The diffusion of FAN into microbial cells upsets the balance of the intracellular pH of methanogens, leading to lower levels or inhibition of enzymatic reactions and abnormal material transportation [107]. Another possibility is caused by the hydrophobic ammonia molecule, which may diffuse passively into bacterial cells, causing proton imbalance or potassium deficiency [108]. In general, and to avoid toxicity from ammonia, it is recommended that the concentration should never reach the range of 1500–3000 mg/L [109].

4.1.2. Mitigating Ammonia Inhibitory Effects

Even though the ammonia inhibition mechanism has been determined by different researchers, unfortunately until now there has been no direct approach to avoid ammonia toxicity, especially when it exceeds the threshold inhibition level [6]. However, some measures could be employed to mitigate ammonia inhibition, which are described below.

(a) Inoculum Selection and Adaptation of Microorganisms

The inoculum source and adaptation of methanogenic consortia to high ammonia levels have been recognized as effective and feasible tactics for enhancing the anaerobic digestion process. In this context, acclimation of acetoclastic methanogens to high concentrations of TAN reduce their susceptibility to any increase in TAN and improve their tolerance to a wider range of pH levels [96]. Methane was successfully produced in dry fermentation of chicken manure at about 8000 to 14,000 mg TAN/kg after adaptation time of 254 days at 37 °C and pH of 7.3–8.8. It was concluded that spontaneous acclimation of the methanogenic biomass to high concentrations of ammonia could occur and result in production of methane, even under a high TS content (25%) and high concentration of ammonia [6,35,96,97,110].

The research conducted in [111] investigated the effect of six different inoculum sources on the digestion of rice straw using BMP tests under the same conditions, and significant differences in the biogas yields were observed, caused by the different inocula sources. Methane yields were 180 mL/g VS, 160 mL/g VS, 75 mL/g VS, 125 mL/g VS, 15 mL/g VS, and 5 mL/g VS for dairy manure, swine manure, chicken manure, granular sludge, municipal sludge, and paper mill sludge inocula, respectively. Reactors inoculated with digested manure had higher methane yields compared to others, and this was attributed to the availability of macro- and micronutrients in manure, which enhanced the digestion of rice straw. One study [112] demonstrated that using diary manure digester effluent as the inoculum improved methane yields from corn stover up to 30% and 100% compared to inoculum obtained from food waste and sewage sludge digesters, respectively. This variation was attributed to the higher populations of hydrolytic microbes in dairy manure effluent. Also, best inoculum selection was able to reduce the lag phase from 20–30 days to 2–5 days [113].

(b) Substrate/Inoculum Ratio (S/I)

Providing an optimal substrate to inoculum (S/I) ratio is considered to be important in AD process. However, this ratio may vary with types of substrates under consideration, because while the digestion of food waste may be susceptible to VFA accumulation because of its composition and high biodegradability, other substrates may have buffering capacities that reduce the potential VFA accumulation [114,115]. In ss-AD of lignocellulosic biomass, a wide range of S/I ratios have been investigated under both mesophilic and thermophilic conditions. Ratios between 2 to 3 (VS basis) were found to be relatively robust for ss-AD mesophilic conditions, whereas a relatively high S/I ratio was preferred under thermophilic conditions. For instance, the highest biogas yield for corn stover in mesophilic ss-AD was obtained with an S/I ratio of 2.43 (VS basis), while the highest yield under thermophilic conditions was achieved at an S/I ratio of 4.58 [7,116–118]. According to [115], and in accordance with the proposed guidelines by Holliger et al. [119], a minimum S/I of 0.5 is recommended for highly degradable substrates, such as food waste, while an S/I of 1 may be used for less degradable substrates, such as lignocellulosic waste. As will be illustrated in the following sections, other parameters, such as pH and C/N ratio, also directly affect ammonia toxicity [6,120,121]. Importantly, other different approaches have been studied to control ammonia inhibition during the AD process, for example anammox, which is fundamentally based on a reaction driven by a specialized group of planctomycete-like bacteria that convert (oxidize) ammonium to nitrogen gas, where nitrite is used as the electron acceptor under anoxic conditions [122]. Other approaches are struvite precipitation [123] and the use of zeolite as support media for the immobilization of microorganisms in different high-rate reactor configurations. Furthermore, zeolite has also been used as an ion exchanger for the removal of ammonium in anaerobic digestion due to the presence of Na^{+}, Ca^{2+}, and Mg^{2+} cations in its crystalline structure. Also, carbon fiber textiles (CFT) were used as supporting materials for better retention of microorganisms. Nevertheless, these approaches were reported to be expensive at large scale [124,125].

4.2. *Effect of Organic Loading Rate (OLR)*

Providing an adequate OLR helps to ensure a stable digestion process and an optimum biogas yield during the process [126]. Feedstock composition will determine the level of biochemical activities that will take place in the digester. Consequently, overload or undersupply will affect the process performance [6].

Anaerobic stability depends on the harmonic relations between acid formers and methane formers [127], and these two types of microorganisms differ widely in terms of physiology and nutritional needs [128]. In the case of easily degradable substrates with low buffering capacity, high loading rate will lead to formation of VFAs, pH will be dropped, and methanogenic bacteria will be inhibited [19,22,23,49,129]. For commercial biogas production, protein-rich feedstocks are preferred over other materials as they have high bio-methane potential (BMP) [130,131]. Unfortunately, high loads of such materials are often associated with process instability because of the release of ammonia nitrogen (NH_3-N) [6,97,132], causing inhibition of the process.

4.3. Carbon to Nitrogen (C/N) Ratio

In order to optimize biogas production, introducing substrates with an optimal C/N ratio is essential for the entire process and basically for providing sufficient and adequate nutrients levels for microorganisms [133]. Higher concentrations of ammonia are formed with low C/N values, and eventually this will result in hindrance of microbial growth. On the other hand, during the fermentative stage, a C/N ratio above the optimal value will lead to production of large amounts of VFAs. Furthermore, the C/N ratio can reduce the TAN level only if the TAN concentration is slightly more than the threshold inhibitory level; action should be taken promptly, before the inhibition of the process [6].

Table 2 provides the C/N ratios of different substrates. The optimal C/N ratios of various substrates attained from different AD processes will likely be different, and the AD process is more stable when the C/N ratio ranges from 20 to 30 [14,134]. During the anaerobic codigestion process, substrates are added to maintain the C/N proportion within digesters [135]. To unify a specific range, including the existing carbon of the easily degradable part and excluding the carbon that is not specifically affected by microorganisms, an available C/N ratio is proposed [126]. By maintaining the C/N ratio at 17:1 [136], enhanced methane generation by around 3.8- and 1.5-fold was achieved compared with perennial ryegrass alone and waste activated sludge (WAS) alone, respectively.

Table 2. C/N ratio of some substrates [137].

Substrate	C/N Ratio	Substrate	C/N
Kitchen waste	26:30	Grass cutting waste	11:15
Food waste	2:18	Rice straw	50:68
Cattle manure	15:26	Wheat straw	51:151
Poultry manure	4:16	Corn straw	51:57
Sheep manure	20:34	Sawdust	199:501
Vegetable waste	8:36	Algae	74:101
Slaughterhouse waste	21:36	Sugar cane waste	139:151

Additionally, it is worth mentioning that organic wastes used for biogas production are generally rich in lignocellulose-type resistant materials [14]. Thus, special pretreatments (Section 3) are required to utilize such wastes under short retention times with anaerobic organisms [138]. Therefore, significant amounts of "invalid carbon" in these materials affect the calculation of C/N. Thus, C/N calculation shows only the general characteristics of organic waste materials, not the actual substances utilized by anaerobic microbes.

4.4. pH Value

Growth and enzymatic activities of the microorganisms and solubilization of organic matter are affected by pH value [127,139–143]. Providing an optimal pH value during the different phases was the key reason some reactors were divided into two phases, with an acidogenic and methanogenic phase [144,145]. Moreover, pH affects the distribution of total ammonia nitrogen (TAN) between toxic NH_3 and innocuous NH_4^+ [146]. Optimum performance of the anaerobic microorganisms will be reached with neutral pH (6.8–7.2) [147]. Importantly, pH affects the distribution of total ammonia. It was noticed that at a higher pH of 8.5, unionized ammonia was highly (2473 mg/L) associated with poor biogas production, which confirms the ammonia toxicity, in particular for leather fleshing waste [148]. By contrast, it was reported that at pH of 4.5 the ammonia nitrogen was minimal (510 mg/L), but high VFA (26,803 mg/L) inhibited the methanogens [6].

4.5. Concentration of Ions

Some metal ions, including sodium (Na^+), potassium (K^+), calcium (Ca^{2+}), ferric (Fe^{3+}), and magnesium (Mg^{2+}), are formed because of the breakdown of organic matter or introduced with the pH adjustment (alkali and acid reagents used in pretreatment). Excessive concentrations of such ions would lead to decreased microbial growth or would even cause toxicity, as they cause dehydration of bacterial cells due to osmotic pressure [149,150].

Mg^{2+} was found to be inhibitory to methanogenesis when it reaches 720 mg/L. Also, high concentrations (> 100 mM) of the magnesium ion can cause disaggregation of methanogens; thus, the conversion of acetate is inhibited [151]. Nevertheless, Ca^{2+} was moderately inhibitory to microorganisms in concentrations above 300 mg/L. However, excessive amounts of Ca^{2+} can cause precipitation of carbonates and phosphates, which results in scaling of the reactors, pipes, and biomass; thus, it reduces the specific methanogenic activity and results in a loss of buffer capacity [152–154]. K^+ concentrations above 3 mg/L are toxic to microorganisms [97], and Fe^{3+} was reported to inhibit 52–82% of methanogenesis activities with concentrations of 21 mg/L or above, because Fe^{3+} could deactivate enzymes of microorganisms by reacting with their functional groups [155].

4.6. Sulfide

In the anaerobic process, sulfate is reduced to sulfide by sulfate-reducing bacteria [156]. The produced sulfide could inhibit the process, either by toxicity from the competition for a common organic matter substrate that cause stress for methanogenesis, or through the toxicity of sulfide to various bacteria population [127,157,158].

5. Codigestion of AD

As discussed in the aforementioned sections, providing optimal operating conditions is a crucial point for an effective AD. It is well-known that the deficiency of some nutrients could lead to low performance or even the inhibition of the process [53]. In this regard, codigestion arises as an efficient and cost-effective way to overcome some of the difficulties during this process [14,134,135]. The codigestion process simply involves mixing two or more substrates together, which would improve the utilization of nutrients through balancing them (mainly the C/N ratio), enhancing the microbial diversity, improve buffer capacity, and dilute inhibitors; consequently, codigestion may achieve positive synergistic effects and lead to higher methane yield than monodigestion of substrates [53,159,160]. In this context, lignocellulosic biomass is categorized as a great prospect for anaerobic digestion due to its high content of cellulose and hemicellulose. However, its elevated C/N ratio (Table 1) and the complexity of its biomass structure (contains 10–25% lignin) inhibits the anaerobic digestion process when it is used individually [13,161,162]. On the other hand, animal waste, such as manure, contains high amounts of organic nitrogen [163,164], which would lead to accumulation of ammonia, and potentially inhibition of the microbial activities because of a low C/N ratio [134,148,163,165]. Therefore, monodigestion of lignocellulosic biomass or animal wastes would result in less stable performance and low methane production as a result of the inappropriate C/N ratio. Creating an optimum C/N ratio through codigestion of lignocellulosic biomass with appropriate animal waste can maximize methane production and make the AD process more stable [162,164].

Cellulose, hemicellulose, and C/N ratio potential values for rice straw, wheat straw, corn stover, and switch grass are 32%, 24%, and 47; 38%, 21%, and 60; 37%, 22%, and 63; and 38%, 26%, and 93, respectively. These values indicate the high potential of these substrates to produce biogas, however, they are still considered difficult to digest, as they contain a significant fraction of lignin, whereas the codigestion of these substrates with a highly nitrogen-rich substrate, such as manure, would result in a successful way to utilize these materials in an anaerobic digestion process, since the materials complement each other when used together and the potential for inhibition is reduced [166]. Additionally, for the codigestion of hay with soybean processing waste at 25:75 ratio under solid-state

AD process, the methane yield was 258 L/kg VS, which is 50% and 148% higher than the amount obtained by each substrate individually [117]. Co-digestion of barley straw with pig manure produced 233.4 L/kgVS methane, but it produced only 192 L/kg VS methane when it was codigested with cow manure [167]. Corn stover and dog food at an equal ratio (1:1) produced a methane yield of 304.4 L/kg VS, which led to an increase of 129% compared to corn stover alone and 9% compared to dog food digestion alone [116]. Rice straw with swine manure, wheat straw with chicken manure, corn stover with chicken manure, and crop silage with cow manure produced 350, 235, 223–298, and 249 L/kg VS methane, respectively, in an AD codigestion process with optimum C/N ratio. In this context, codigestion resulted in a 31% increase of the expected yield, which was calculated from the methane potential of the individual fractions [168]. Food waste codigestion with sewage sludge can produce a high yield of biogas of 708 L/kg VS [169]. In Table 3, other examples of codigestion experiments are resumed.

Table 3. Codigestion of lignocellulosic biomass with other feedstock for methane production under ss-AD.

Lignocellulosic Biomass	Co-Substrate	Mixing Ratio	Temp (°C)	CH_4 (L/kg VS)	Ref
Yard waste	Food waste	4:1	36	120	[46]
Swine manure	Corn stover	NA	35	350	[170]
Hay	Soybean processing waste	1:1.3	37	258	[117]
Agricultural wastes	Chicken manure	7:3	55	502	[163]
Cattle manure	Palm pressed	1:3	37	346.2	[171]
Cattle manure	Silage of cardoon	85:15	37	308	[172]
Cow manure	Rice straw	1:1	37	383.5	[173]
Cattle manure	Corn stover	NA	37	228	[167]
Cow manure	Kitchen waste	1:1	35	179.8	[174]
Cattle manure	Wheat straw	95:5	55	351	[175]

6. Conclusions

In summary, the results compiled in this review clearly demonstrated that lignocellulosic wastes can be important sources of renewable methane through anaerobic digestion. Among all the proposals used to overcome the problems of these wastes in this process, which are mainly related to the hydrolysis step, pretreatments and codigestion seems to be the more suitable strategies. Regarding the operational conditions, inhibition must be considered, but solid-state anaerobic digesters have proved to be a great alternative to classical AD stirred reactors. It is evident that the state-of-the-art of this approach is still not completely developed, but it seems to be the correct approach for lignocellulosic wastes. Finally, it is clear that these wastes are the most abundant sources of organic matter on earth, and therefore, all approaches to convert them into renewable energy are a step toward sustainability and climate change mitigation.

Author Contributions: Writing—original draft preparation, T.S.; writing—review and editing, A.S.

Funding: Authors wish to thank the financial support of Spanish Ministerio de Economía y Competitividad (Project CTM2015-69513-R).

Acknowledgments: Sayara would like to thank the Palestine Technical University for the administrative support.

Conflicts of Interest: The authors declare no conflict of interest.

Abbreviation

	Full Name
AD	Anaerobic Digestion
BMP	Biochemical Methane Potential
FAN	Free Ammonia Nitrogen
OFMSW	Organic Fraction of Municipal Solid Waste
ss-AD	Solid State Anaerobic Digestion
TAN	Total Ammonia Nitrogen
TS	Total Solids
VFA	Volatile Fatty Acids
VS	Volatile Solids

References

1. Escamilla-Alvarado, C.; Rios-Leal, E.; Ponce-Noyola, M.T.; Poggi-Varaldo, H.M. Gas biofuels from solid substrate hydrogenogenic-methanogenic fermentation of the organic fraction of municipal solid waste. *Process Biochem.* **2012**, *47*, 1572–1587. [CrossRef]
2. Pantaleo, A.; Gennaro, B.D.; Shah, N. Assessment of optimal size of anaerobic co-digestion plants: An application to cattle farms in the province of Bari (Italy). *Renew. Sustain. Energy Rev.* **2013**, *20*, 57–70. [CrossRef]
3. Wagner, A.O.; Lackner, N.; Mutschlechner, M.; Prem, E.M.; Markt, R.; Illmer, P. Biological pretreatment strategies for second-generation lignocellulosic resources to enhance biogas production. *Energies* **2018**, *11*, 1797. [CrossRef] [PubMed]
4. Siegert, I.; Banks, C. The effect of volatile fatty acid addition on the anaerobic digestion of cellulose and glucose in batch reactors. *Process Biochem.* **2005**, *40*, 3412–3418. [CrossRef]
5. Metcalf, E.; Tchobanoglous, G.; Stensel, H.D.; Tsuchihashi, R.; Burton, F. *Wastewater Engineering: Treatment and Resource Recovery*, 5th ed.; McGraw-Hill: New York, NY, USA, 2014.
6. Rajagopal, R.; Massé, D.I.; Singh, G. A critical review on inhibition of anaerobic digestion process by excess ammonia. *Bioresour. Technol.* **2013**, *143*, 632–641. [CrossRef]
7. Li, Y.; Park, S.Y.; Zhu, J. Solid-state anaerobic digestion for methane production from organic waste. *Renew. Sustain. Energy Rev.* **2011**, *15*, 821–826. [CrossRef]
8. Hu, F.; Ragauskas, A. Pretreatment and lignocellulosic chemistry. *Bioenergy Res.* **2012**, *5*, 1043–1066. [CrossRef]
9. Henard, C.A.; Smith, H.K.; Guarnieri, M.T. Phosphoketolase overexpression increases biomass and lipid yield from methane in an obligate methanotrophic biocatalyst. *Metab. Eng.* **2017**, *41*, 152–158. [CrossRef]
10. Tasnim, F.; Iqbal, S.A.; Chowdhury, A.R. Biogas production from anaerobic codigestion of cow manure with kitchen waste and Water Hyacinth. *Renew. Energy* **2017**, *109*, 434–439. [CrossRef]
11. Xie, S.; Hai, F.I.; Zhan, X.; Guo, W.; Ngo, H.H.; Price, W.E.; Nghiem, L.D. Anaerobic co-digestion: A critical review of mathematical modeling for performance optimization. *Bioresour. Technol.* **2016**, *222*, 498–512. [CrossRef]
12. Kumar, R.; Singh, S.; Singh, O.V. Bioconversion of lignocellulosic biomass:biochemical and molecular perspectives. *J. Ind. Microbiol. Biotechnol.* **2008**, *35*, 377–391. [CrossRef] [PubMed]
13. Sawatdeenarunat, C.; Surendra, K.C.; Takara, D.; Oechsner, H.; Khanal, S.K. Anaerobic digestion of lignocellulosic biomass: Challenges and opportunities. *Bioresour. Technol.* **2015**, *178*, 178–186. [CrossRef] [PubMed]
14. Yang, L.; Xu, F.; Ge, X.; Li, Y. Challenges and strategies for solid-state anaerobic digestion of lignocellulosic biomass. Renew. *Sustain. Energy Rev.* **2015**, *44*, 824–834. [CrossRef]
15. Zhou, H.; Wen, Z. Solid-state anaerobic digestion for waste management and biogas production. *Adv. Biochem. Eng. Biotechnol.* **2019**, *169*, 147–168.
16. Mshandete, A.; Bjornsson, L.; Kivaisi, A.K.; Rubindamayugi, S.T.; Mattiasson, B. Enhancement of anaerobic batch digestion of sisal pulp waste by mesophilic aerobic pre-treatment. *Water Res.* **2005**, *39*, 1569–1575. [CrossRef]

17. Azman, S.; Khadem, A.F.; Van'er, J.B.; Zeeman, G.; Plugge, C.M. Presence and role of anaerobic hydrolytic microbes in conversion of lignocellulosic biomass for biogas production. *Crit. Rev. Environ. Sci. Technol.* **2015**, *3389*, 2523–2564. [CrossRef]
18. Sträuber, H.; Schröder, M.; Kleinsteuber, S. Metabolic and microbial community dynamics during the hydrolytic and acidogenic fermentation in a leach-bed process. *Energy Sustain. Soc.* **2012**, *2*, 13. [CrossRef]
19. Tsapekos, P. Enhancing Biogas Production from Recalcitrant Lignocellulosic Residue. Ph.D. Thesis, Technical University of Denmark, Lyngby, Denmark, 2017.
20. Angelidaki, I.; Karakashev, D.; Batstone, D.J.; Plugge, C.M.; Stams, A.J.M. Biomethanation and its potential. In *Methods in Enzymology: Methods in Methane Metabolism*, 1st ed.; Rosenzweig, A.C., Ragsdale, S.W., Eds.; Elsevier Academic Press: San Diego, CA, USA, 2011; Volume 494, pp. 327–351.
21. Treu, L.; Campanaro, S.; Kougias, P.G.; Zhu, X.; Angelidaki, I. Untangling the effect of fatty acid addition at species level revealed different transcriptional responses of the biogas microbial community members. *Environ. Sci. Technol.* **2016**, *50*, 6079–6090. [CrossRef]
22. Wirth, R.; Kovács, E.; Maróti, G.; Bagi, Z.; Rákhely, G.; Kovács, K.L. Characterization of a biogas-producing microbial community by short-read next generation DNA sequencing. *Biotechnol. Biofuels* **2012**, *5*, 41. [CrossRef]
23. Campanaro, S.; Treu, L.; Kougias, P.G.; De Francisci, D.; Valle, G.; Angelidaki, I. Metagenomic analysis and functional characterization of the biogas microbiome using high throughput shotgun sequencing and a novel binning strategy. *Biotechnol. Biofuels* **2016**, *9*, 26. [CrossRef]
24. Mata-Alvarez, J. *Fundamentals of the Anaerobic Digestion Process Biomethanization of the Organic Fraction of Municipal Solid Wastes*; IWA Publishing Company: London, UK, 2002.
25. Stronach, S.M.; Rudd, T.; Lester, J.N. *Anaerobic Digestion Processes in Wastewater Treatment*; Springer: Berlin/Heidelberg, Germany, 1986.
26. Eddy, M.; Tchobanoglous, G. *Wastewater Engineering: Treatment and Reuse*, 4th ed.; McGraw-Hill: New York, NY, USA, 2003.
27. Penaud, V.; Delgenes, J.P.; Moletta, R. Thermo-chemical pretreatment of a microbial biomass: Influence of sodium hydroxide addition on solubilization and anaerobic biodegradability. *Enzym. Microb. Technol.* **1999**, *25*, 258–263. [CrossRef]
28. Lu, J.; Gavala, H.N.; Skiadas, I.V.; Mladenovska, Z.; Ahrin, B.K. Improving anaerobic sewage sludge digestion by implementation of a hyper-thermophilic prehydrolysis step. *J. Environ. Manag.* **2008**, *88*, 881–889. [CrossRef]
29. Pinnekamp, J. Effects of thermal pretreatment of sewage sludge on anaerobic digestion. *Water Sci. Technol.* **1989**, *21*, 97–108. [CrossRef]
30. Rapport, J.; Zhang, R.; Jenkins, B.M.; Williams, R.B. *Current Anaerobic Digestion Technologies Used for Treatment of Municipal Organic Solid Waste*; California Environmental Protection Agency: Sacramento, CA, USA, 2008.
31. De Baere, L.; Mattheeuws, B.; Velghe, F. State of the art of anaerobic digestion in Europe. In Proceedings of the 12th World Congress on Anaerobic Digestion (AD12), Guadalajara, Mexico, 31 October–4 November 2010.
32. Luning, L.; Van Zundert, E.H.M.; Brinkmann, A.J.F. Comparison of dry and wet digestion for solid waste. *Water Sci. Technol.* **2003**, *48*, 15–20. [CrossRef] [PubMed]
33. Hartmann, H.; Ahring, B.K. Strategies for the anaerobic digestion of the organic fraction of municipal solid waste: an overview. *Water Sci. Technol.* **2006**, *53*, 7–22. [CrossRef] [PubMed]
34. Forster-Carneiro, T.; Perez, M.; Romero, L.I. Influence of total solid and inoculum contents on performance of anaerobic reactors treating food waste. *Bioresour. Technol.* **2008**, *99*, 6994–7002. [CrossRef]
35. Abouelenien, F.; Nakashimada, Y.; Nishio, N. Dry mesophilic fermentation of chicken manure for production of Methane by repeated batch culture. *J. Biosci. Bioeng.* **2009**, *107*, 293–295. [CrossRef]
36. Vandevivere, P.; De Baere, L.; Verstraete, W. Types of anaerobic digesters for solid wastes. In *Biomethanization of the Organic Fraction of Municipal Wastes*; Mata Alvarez, J., Ed.; IWA Press: London, UK, 2002; pp. 111–137.
37. Banks, C.J.; Stentiford, E.I. Biodegradable municipal solid waste: Biotreatment options. *Water Resour. Manag.* **2007**, *160*, 11–18. [CrossRef]
38. Abbassi-Guendouz, A.; Brockmann, D.; Trably, E.; Dumas, C.; Delgene's, J.; Steyer, J.; Escudie, R. Total solids content drives high solid anaerobic digestion via mass transfer limitation. *Bioresour. Technol.* **2012**, *111*, 55–61. [CrossRef]

39. Motte, J.C.; Trably, E.; Escudié, R.; Hamelin, J.; Steyer, J.P.; Bernet, N.; Delgenes, J.P.; Dumas, C. Total solids content: A key parameter of metabolic pathways in dry anaerobic digestion. *Biotechnol. Biofuels* **2013**, *6*, 164. [CrossRef]
40. Motte, J.C.; Escudié, R.; Hamelin, J.; Steyer, J.P.; Bernet, N.; Delgenes, J.P. Substrate milling pretreatment as a key parameter for solid-state anaerobic digestion optimization. *Bioresour. Technol.* **2014**, *173*, 185–192. [CrossRef] [PubMed]
41. Pommier, S.; Chenu, D. A logistic model for the prediction of the influence of water on the solid waste methanization in landfills. *Biotechnol. Bioeng.* **2007**, *97*, 473–482. [CrossRef] [PubMed]
42. García-Bernet, D.; Buffière, P.; Latrille, E.; Steyer, J.-P.; Escudié, R. Water distribution in biowastes and digestates of dry anaerobic digestion technology. *Chem. Eng. J.* **2011**, *172*, 924–928. [CrossRef]
43. Bollon, J.; Benbelkacem, H.; Gourdon, R.; Buffiere, P. Measurement of diffusion coefficients in dry anaerobic digestion media. *Chem. Eng. Sci.* **2013**, *89*, 115–119. [CrossRef]
44. Fernández, J.; Pérez, M.; Romero, L.I. Kinetics of mesophilic anaerobic digestion of the organic fraction of municipal solid waste: influence of initial total solid concentration. *Bioresour. Technol.* **2010**, *101*, 6322–6328. [CrossRef]
45. Brown, D.; Shi, J.; Li, Y. Comparison of solid-state to liquid anaerobic digestion of lignocellulosic feedstocks for biogas production. *Bioresour. Technol.* **2012**, *124*, 379–386. [CrossRef]
46. Brown, D.; Li, Y. Solid state anaerobic co-digestion of yard waste and food waste for biogas production. *Bioresour. Technol.* **2013**, *127*, 275–280. [CrossRef]
47. Martin, D.J. The site of reaction in solid-state digestion: A new hypothesis. *Process Saf. Environ. Prot.* **2001**, *79*, 29–37. [CrossRef]
48. Staley, B.F.; de Los Reyes, F.L.; Barlaz, M.A. Effect of spatial differences in microbial activity, pH, and substrate levels on methanogenesis initiation in refuse. *Appl. Environ. Microbiol.* **2011**, *77*, 2381–2391. [CrossRef]
49. Patinvoh, R.J.; Osadolor, O.A.; Chandolias, K.; Sárvári Horváth, I.; Taherzadeh, M.J. Innovative pretreatment strategies for biogas production. *Bioresour. Technol.* **2017**, *224*, 13–24. [CrossRef]
50. Hegde, G.; Pullammanappallil, P. Comparison of thermophilic and mesophilic one-stage, batch, high-solids anaerobic digestion. *Environ. Technol.* **2007**, *284*, 361–369. [CrossRef] [PubMed]
51. Lin, L.; Wan, C.; Liu, X.; Lee, D.J.; Lei, Z.; Zhang, Y. Effect of initial pH on mesophilic hydrolysis and acidification of swine manure. *Bioresour. Technol.* **2013**, *136*, 302–308. [CrossRef] [PubMed]
52. Sheets, J.P.; Yang, L.; Ge, X.; Wang, Z.; Li, Y. Beyond land application: Emerging technologies for the treatment and reuse of anaerobically digested agricultural and food waste. *Waste Manag.* **2015**, *44*, 94–115. [CrossRef] [PubMed]
53. Nasr, N.; Elbeshbishy, E.; Hafez, H.; Nakhla, G.; Hesham, M. Comparative assessment of single-stage and two-stage anaerobic digestion for the treatment of thin stillage. *Bioresour. Technol.* **2012**, *111*, 122–126. [CrossRef] [PubMed]
54. Carrere, H.; Dumas, C.; Battimelli, A.; Batstone, D.J.; Delgenes, J.P.; Steyer, J.P.; Ferrer, I. Pretreatment methods to improve sludge anaerobic degradability: A review. *J. Hazard. Mater.* **2010**, *183*, 1–15. [CrossRef] [PubMed]
55. Hendriks, A.T.; Zeeman, G. Pretreatments to enhance the digestibility of lignocellulosic biomass: A review. *Bioresour. Technol.* **2009**, *100*, 10–18. [CrossRef]
56. Yu, Q.; Liu, R.; Li, K.; Ma, R. A review of crop straw pretreatment methods for biogas production by anaerobic digestion in China. *Renew. Sustain. Energy Rev.* **2019**, *107*, 51–58. [CrossRef]
57. Jain, S.; Jain, S.; Wolf, I.T.; Lee, J.; Tong, Y.W. A comprehensive review on operating parameters and different pretreatment methodologies for anaerobic digestion of municipal solid waste. *Renew. Sustain. Energy Rev.* **2015**, *52*, 142–154. [CrossRef]
58. Silva, G.G.; Couturier, M.; Berrin, J.G.; Buléon, A.; Rouau, X. Effects of grinding processes on enzymatic degradation of wheat straw. *Bioresour. Technol.* **2012**, *103*, 192–200. [CrossRef]
59. Lindner, J.; Zielonka, S.; Oechsner, H.; Lemmer, A. Effects of mechanical treatment of digestate after anaerobic digestion on the degree of degradation. *Bioresour. Technol.* **2015**, *178*, 194–200. [CrossRef]
60. Johnson, D.K.; Elander, R.T. Pretreatments for enhanced digestibility of feedstocks. In *Biomass Recalcitrance*; Blackwell Publishing Ltd.: Hoboken, NJ, USA, 2009; pp. 436–453.
61. Salihu, A.; Alam, M.Z. Pretreatment methods of organic wastes for biogas production. *J. Appl. Sci.* **2016**, *16*, 124–137. [CrossRef]

62. Sousa, L.D.; Chundawat, S.P.; Balan, V.; Dale, B.E. 'Cradle-to-grave' assessment of existing lignocellulose pretreatment technologies. *Curr. Opin. Biotechnol.* **2009**, *203*, 339–347. [CrossRef] [PubMed]
63. Modenbach, A.A.; Nokes, S.E. The use of high-solids loadings in biomass pretreatment–A review. *Biotechnol. Bioeng.* **2012**, *109*, 1430–1442. [CrossRef] [PubMed]
64. Mussoline, W.; Esposito, G.; Giordano, A.; Lens, P.N.L. The anaerobic digestion of rice straw: A review. *Crit. Rev. Environ. Sci. Technol.* **2013**, *43*, 895–915. [CrossRef]
65. Taherdanak, M.; Zilouei, H.; Karimi, K. The influence of dilute sulfuric acid pretreatment on biogas production from wheat plant. *Int. J. Green Energy* **2016**, *13*, 1129–1134. [CrossRef]
66. Devlin, D.C.; Esteves, S.R.R.; Dinsdale, R.M.; Guwy, A.J. The effect of acid pretreatment on the anaerobic digestion and dewatering of waste activated sludge. *Bioresour. Technol.* **2011**, *102*, 4076–4082. [CrossRef]
67. Torres, M.L.; Lloréns, M.C.E. Effect of alkaline pretreatment on anaerobic digestion of solid wastes. *Waste Manag.* **2008**, *28*, 2229–2234. [CrossRef]
68. Yi, Z.; Zhongli, P.; Ruihong, Z. Overview of biomass pretreatment for cellulosic ethanol production. *Int. J. Agric. Biol. Eng.* **2009**, *2*, 51–68.
69. Sambusiti, C.; Ficara, E.; Malpei, F.; Steyer, J.P.; Carrère, H. Benefit of sodium hydroxide pretreatment of ensiled sorghum forage on the anaerobic reactor stability and methane production. *Bioresour. Technol.* **2013**, *144*, 149–155. [CrossRef]
70. Mancini, G.; Papirio, S.; Riccardelli, G.; Lens, P.N.L.; Esposito, G. Trace elements dosing and alkaline pretreatment in the anaerobic digestion of rice straw. *Bioresour. Technol.* **2018**, *247*, 897–903. [CrossRef]
71. Sindhu, R.; Binod, P.; Pandey, A. Biological pretreatment of lignocellulosic biomass—An overview. *Bioresour. Technol.* **2016**, *199*, 76–82. [CrossRef] [PubMed]
72. Sharma, H.K.; Xu, C.C.; Qin, W. Biological pretreatment of lignocellulosic biomass for biofuels and bioproducts: An overview. *Waste Biomass Valor.* **2019**, *10*, 235–251. [CrossRef]
73. Agbor, V.B.; Cicek, N.; Sparling, R.; Berlin, A.; Levin, D.B. Biomass pretreatment: Fundamentals toward application. *Biotechnol. Adv.* **2011**, *29*, 675–685. [CrossRef] [PubMed]
74. Kavitha, S.; Subbulakshmi, P.; Banu, J.R.; Gopi, M.; Yeom, I.T. Enhancement of biogas production from microalgal biomass through cellulolytic bacterial pretreatment. *Bioresour. Technol.* **2017**, *233*, 34–43. [CrossRef]
75. Bonilla, S.; Choolaei, Z.; Meyer, T.; Edwards, E.A.; Yakunin, A.F.; Allen, D.G. Evaluating the effect of enzymatic pretreatment on the anaerobic digestibility of pulp and paper biosludge. *Biotechnol. Rep.* **2018**, *17*, 77–85. [CrossRef]
76. Sims, R.E.H. *Bioenergy Options for a Cleaner Environment: In Developed and Developing Countries*; Elsevier: Amsterdam, The Netherlands, 2004; p. 198.
77. Bougrier, C.; Albasi, C.; Delgenes, J.P.; Carrere, H. Effect of ultrasonic, thermal and ozone pretreatments on waste activated sludge solubilization and anaerobic biodegradability. *Chem. Eng. Process* **2006**, *45*, 711–718. [CrossRef]
78. Elliot, A.; Mahmood, T. Comparison of mechanical pretreatment methods for the enhancement of anaerobic digestion of pulp and paper waste. *Water Sci. Technol.* **2012**, *84*, 497–505. [CrossRef]
79. Zeynali, R.; Khojastehpour, M.; Ebrahimi-Nik, M. Effect of ultrasonic pre-treatment on biogas yield and specific energy in anaerobic digestion of fruit and vegetable wholesale market wastes. *Sustain. Environ. Res.* **2017**, *27*, 259–264. [CrossRef]
80. Rincón, B.; Bujalance, L.; Fermoso, F.G.; Martín, A.; Borja, R. Biochemical methane potential of two-phase olive mill solid waste: Influence of thermal pretreatment on the process kinetics. *Bioresour. Technol.* **2013**, *140*, 249–255. [CrossRef]
81. Novarino, D.; Zanetti, M.C. Anaerobic digestion of extruded OFMSW. *Bioresour. Technol.* **2012**, *104*, 44–50. [CrossRef]
82. Zhang, H.; Ning, Z.; Khalid, H.; Zhang, R.; Liu, G.; Chen, C. Enhancement of methane production from Cotton Stalk using different pretreatment techniques. *Sci. Rep.* **2018**, *8*, 3463. [CrossRef] [PubMed]
83. Song, Z.; Gaihe, Y.; Liu, X.; Yan, Z.; Yuan, Y.; Liao, Y. Comparison of seven chemical pretreatments of Corn straw for improving Methane yield by anaerobic digestion. *PLoS ONE* **2014**, *9*, e93801. [CrossRef] [PubMed]
84. Monlau, F.; Latrille, E.; Da Costa, A.C.; Steyer, J.P.; Carrere, H.A. Enhancement of methane production from sunflower oil cakes by dilute acid pretreatment. *Appl. Energy* **2013**, *102*, 1105–1113. [CrossRef]

85. Lim, J.W.; Wang, J.Y. Enhanced hydrolysis and methane yield by applying microaeration pretreatment to the anaerobic co-digestion of brown water and food waste. *Waste Manag.* **2013**, *33*, 813–819. [CrossRef] [PubMed]
86. Patinvoh, R.J.; Feuk-Lagerstedt, E.; Lundin, M.; Sárvári Horváth, I.; Taherzadeh, M.J. Biological pretreatment of chicken feather and biogas production from total broth. *Appl. Biochem. Biotechnol.* **2016**, *180*, 1401–1415. [CrossRef]
87. Wen, B.; Yuan, X.; Li, Q.X.; Liu, J.; Ren, J.; Wang, X.; Cui, Z. Comparison and evaluation of concurrent saccharification and anaerobic digestion of Napier grass after pretreatment by three microbial consortia. *Bioresour. Technol.* **2015**, *175*, 102–111. [CrossRef]
88. Wagner, A.O.; Schwarzenauer, T.; Illmer, P. Improvement of methane generation capacity by aerobic pre-treatment of organic waste with a cellulolytic Trichoderma viride culture. *J. Environ. Manag.* **2013**, *129*, 357–360. [CrossRef]
89. Phutela, U.G.; Sahni, N. Effect of Fusarium sp. on paddy straw digestibility and biogas production. *J. Adv. Lab. Res. Biol.* **2012**, *3*, 9–12.
90. Mutschlechner, M.; Illmer, P.; Wagner, A.O. Biological pre-treatment: Enhancing biogas production using the highly cellulolytic fungus Trichoderma viride. *Waste Manag.* **2015**, *43*, 98–107. [CrossRef]
91. Rajput, A.A.; Zeshan; Visvanathan, C. Effect of thermal pretreatment on chemical composition, physical structure and biogas production kinetics of wheat straw. *J. Environ. Manag.* **2018**, *221*, 45–52. [CrossRef]
92. Liu, Y.; Nilsen, P.J.; Maulidiany, N.D. Thermal pretreatment to enhance biogas production of waste aerobic granular sludge with and without calcium phosphate precipitates. *Chemosphere* **2019**, *234*, 725–732. [CrossRef] [PubMed]
93. Bauer, A.; Lizasoain, J.; Theuretzbacher, F.; Agger, J.W.; Rincón, M.; Menardo, S.; Saylor, M.K.; Enguídanos, R.; Nielsen, P.J.; Potthast, A.; et al. Steam explosion pretreatment for enhancing biogas production of late harvested hay. *Bioresour. Technol.* **2014**, *166*, 403–410. [CrossRef] [PubMed]
94. González-Fernández, C.; García-Encina, P.A. Impact of substrate to inoculum ratio anaerobic digestion of swine slurry. *Biomass Bioenergy.* **2009**, *33*, 1065–1069. [CrossRef]
95. Liu, T.; Sung, S. Ammonia inhibition on thermophilic aceticlastic methanogens. *Water Sci. Technol.* **2002**, *45*, 113–120. [CrossRef]
96. Sung, S.; Liu, T. Ammonia inhibition on thermophilic anaerobic digestion. *Chemosphere* **2003**, *53*, 43–52. [CrossRef]
97. Chen, Y.; Cheng, J.J.; Creamer, K.S. Inhibition of anaerobic digestion process: A review. *Bioresour. Technol.* **2008**, *99*, 4044–4064. [CrossRef]
98. Hejnfelt, A.; Angelidaki, I. Anaerobic digestion of slaughterhouse by-products. *Biomass Bioenergy* **2009**, *33*, 1046–1054. [CrossRef]
99. Zhang, C.; Su, H.; Baeyens, J.; Tan, T. Reviewing the anaerobic digestion of food waste for biogas production. *Renew. Sustain. Energy Rev.* **2014**, *38*, 383–392. [CrossRef]
100. Hansen, K.H.; Angelidaki, I.; Ahring, B.K. Anaerobic digestion of swine manure: inhibition by ammonia. *Water Res.* **1998**, *32*, 5–12. [CrossRef]
101. Nakakubo, R.; Moller, H.B.; Nielsen, A.M.; Matsuda, J. Ammonia inhibition of methanogenesis and identification of process indicators during anaerobic digestion. *Environ. Eng. Sci.* **2008**, *25*, 1487–1496. [CrossRef]
102. Procházka, J.; Dolejs, P.; Maca, J.; Dohanyos, M. Stability and inhibition of anaerobic processes caused by insufficiency or excess of ammonia nitrogen. *Appl. Microbiol. Biotechnol.* **2012**, *93*, 439–447. [CrossRef] [PubMed]
103. McCarty, P.L. Anaerobic waste treatment fundamentals III. *Public Works* **1964**, *95*, 91–94.
104. Fernandes, T.V.; Keesman, K.J.; Zeeman, G.; Van Lier, J.B. Effect of ammonia on the anaerobic hydrolysis of cellulose and tributyrin. *Biomass Bioenergy* **2012**, *47*, 316–323. [CrossRef]
105. Ho, L.; Ho, G. Mitigating ammonia inhibition of thermophilic anaerobic treatment of digested piggery wastewater: Use of pH reduction, zeolite, biomass and humic acid. *Water Res.* **2012**, *46*, 4339–4350. [CrossRef] [PubMed]
106. Müller, T.; Walter, B.; Wirtz, A.; Burkovski, A. Ammonium toxicity in bacteria. *Curr. Microbiol.* **2006**, *52*, 400–406. [CrossRef] [PubMed]

107. Wittmann, C.; Zeng, A.P.; Deckwer, W.D. Growth inhibition by ammonia and use of a pH-controlled feeding strategy for the effective cultivation of Mycobacterium chloropheolicum. *Appl. Microbiol. Biotechnol.* **1995**, *44*, 519–525. [CrossRef]
108. Gallert, C.; Bauer, S.; Winter, J. Effect of ammonia on the anaerobic degradation of protein by a mesophilic and thermophilic biowaste population. *Appl. Microbiol. Biotechnol.* **1998**, *50*, 495–501. [CrossRef]
109. Bitton, G. *Wastewater Microbiology*, 3rd ed.; John Wiley & Sons Inc.: Hoboken, NJ, USA, 2005.
110. Angenent, L.T.; Sung, S.; Raskin, L. Methanogenic population dynamics during startup of a full-scale anaerobic sequencing batch reactor treating swine waste. *Water Res.* **2002**, *36*, 4648–4654. [CrossRef]
111. Gu, Y.; Chen, X.; Liu, Z.; Zhou, X.; Zhang, Y. Effect of inoculum sources on the anaerobic digestion of rice straw. *Bioresour. Technol.* **2014**, *158*, 149–155. [CrossRef]
112. Xu, F.; Shi, J.; Lv, W.; Yu, Z.; Li, Y. Comparison of different liquid anaerobic digestion effluents as inocula and nitrogen sources for solid-state batch anaerobic digestion of corn stover. *Waste Manag.* **2013**, *33*, 26–32. [CrossRef]
113. Forster-Carneiro, T.; Pérez, M.; Romero, L.I.; Sales, D. Dry-thermophilic anaerobic digestion of organic fraction of the municipal solid waste: Focusing on the inoculum sources. *Bioresour. Technol.* **2007**, *98*, 3195–3203. [CrossRef] [PubMed]
114. Kafle, G.K.; Bhattarai, S.; Kim, S.H.; Chen, L. Effect of feed to microbe ratios on anaerobic digestion of Chinese cabbage waste under mesophilic and thermophilic conditions: biogas potential and kinetic study. *J. Environ. Manag.* **2014**, *133*, 293–301. [CrossRef] [PubMed]
115. Ohemeng-Ntiamoah, J.; Datta, T. Perspectives on variabilities in biomethane potential test parameters and outcomes: A review of studies published between 2007 and 2018. *Sci. Total Environ.* **2019**, *664*, 1052–1062. [CrossRef] [PubMed]
116. Xu, F.; Li, Y. Solid-state co-digestion of expired dog food and corn stover for methane production. *Bioresour. Technol.* **2012**, *118*, 219–226. [CrossRef] [PubMed]
117. Zhu, J.; Zheng, Y.; Xu, F.; Li, Y. Solid-state anaerobic co-digestion of hay and soybean processing waste for biogas production. *Bioresour. Technol.* **2014**, *154*, 240–247. [CrossRef] [PubMed]
118. Ge, X.; Xu, F.; Li, Y. Solid-state anaerobic digestion of lignocellulosic biomass: Recent progress and perspectives. *Bioresour. Technol.* **2016**, *205*, 239–249. [CrossRef]
119. Holliger, C.; Alves, M.; Andrade, D.; Angelidaki, I.; Astals, S.; Baier, U.; Bougrier, C.; Buffière, P.; Carballa, M.; de Wilde, V.; et al. Towards a standardization of biomethane potential tests. *Water Sci. Technol.* **2016**, *74*, 2515–2522. [CrossRef]
120. Kayhanian, M. Ammonia inhibition in high-solids biogasification: An overview and practical solutions. *Environ. Technol.* **1999**, *20*, 355–365. [CrossRef]
121. Massé, D.I.; Masse, L.; Croteau, F. The effect of temperature fluctuations on psychrophilic anaerobic sequencing batch reactors treating swine manure. *Bioresour. Technol.* **2003**, *89*, 57–62. [CrossRef]
122. Egli, K.R. On the Use of Anammox in Treating Ammonium-Rich Wastewater. Ph.D. Thesis, Universität Zürich, Zürich, Switzerland, 2003.
123. Nelson, N.O.; Mikkelsen, R.L.; Hesterberg, D.L. Struvite precipitation in anaerobic swine lagoon liquid: Effect of pH and Mg:P ratio and determination of rate constant. *Bioresour. Technol.* **2003**, *89*, 229–236. [CrossRef]
124. Sasaki, K.; Morita, M.; Hirano, S.I.; Ohmura, N.; Igarashi, Y. Decreasing ammonia inhibition in thermophilic methanogenic bioreactors using carbon fiber textiles. *Appl. Microbiol. Biotech.* **2011**, *90*, 1555–1561. [CrossRef] [PubMed]
125. Sawayama, S.; Tada, C.; Tsukahara, K.; Yagishita, T. Effect of ammonium addition on methanogenic community in a fluidized bed anaerobic digestion. *J. Biosci. Bioeng.* **2004**, *97*, 65–70. [CrossRef]
126. Wang, M.; Li, W.; Li, P.; Yan, S.; Zhang, Y. An alternative parameter to characterize biogas materials: Available carbon-nitrogen ratio. *Waste Manag.* **2017**, *62*, 76–83. [CrossRef] [PubMed]
127. Bitton, G. *Wastewater Microbiology*, 4th ed.; Wiley-Blackwell: Hoboken, NJ, USA, 2011.
128. Tchobanoglous, G.; Burton, F.L.; Stensel, H.D. *Wastewater Engineering: Treatment and Reuse, Metcalf & Eddy*, 4th ed.; McGraw-Hill: New York, NY, USA, 2004.
129. Angelidaki, I.; Ellegaard, L.; Ahring, B.K. A comprehensive model of anaerobic bioconversion of complex substrates to biogas. *Biotechnol. Bioeng.* **1999**, *63*, 363–372. [CrossRef]

130. Ek, A.; Hallin, S.; Vallin, L.; Schnürer, A.; Karlsson, M. Slaughterhouse waste co-digestion—Experiences from 15 years of full-scale operation. In Proceedings of the World Renewable Energy Congress—Sweden, Linköping, Sweden, 8–13 May 2011.
131. Nordell, E.; Moestedt, J.; Karlsson, M. Biogas Producing Laboratory Reactor. SE Patent 1150954-4, 4 October 2011.
132. Yenigün, O.; Demirel, B. Ammonia inhibition in anaerobic digestion: A review. *Process Biochem.* **2013**, *48*, 901–911. [CrossRef]
133. Reilly, M.; Dinsdale, R.; Guwy, A. Enhanced biomethane potential from wheat straw by low temperature alkaline calcium hydroxide pre-treatment. *Bioresour. Technol.* **2015**, *189*, 258–265. [CrossRef]
134. Resch, C.; Wörl, A.; Waltenberger, R.; Braun, R.; Kirchmayr, R. Enhancement options for the utilisation of nitrogen rich animal by-products in anaerobic digestion. *Bioresour. Technol.* **2011**, *102*, 2503–2510. [CrossRef]
135. Zhang, J.; Li, W.; Lee, J.; Loh, K.; Dai, Y.; Tong, Y.W. Enhancement of biogas production in anaerobic co-digestion of food waste and waste activated sludge by biological co-pretreatment. *Energy* **2017**, *137*, 479–486. [CrossRef]
136. Dai, B.; Xu, J.; He, Y.; Xiong, P.; Wang, X.; Deng, Y.; Wang, Y.; Yin, Z. Acid inhibition during anaerobic digestion of biodegradable kitchen waste. *J. Renew. Sustain. Energy* **2015**, *7*, 023118. [CrossRef]
137. Guillaume, S.; Lendormi, T. Anaerobic co-digestion of dairy cattle slurry and agro-industrial fats: Effect of fat ratio on the digester efficiency. *Can. J. Chem. Eng.* **2015**, *93*, 304–308. [CrossRef]
138. Kabir, M.M.; Forgacs, G.; Horvath, I.S. *Biogas from Lignocellulosic Materials, Lignocellulose-Based Bioproducts*; Springer: Berlin/Heidelberg, Germany, 2015; pp. 207–251.
139. Feng, L.; Luo, J.; Chen, Y. *Dilemma of Sewage Sludge Treatment and Disposal in China*; ACS Publications: Washington, DC, USA, 2015.
140. Dai, X.; Hu, C.; Zhang, D.; Chen, Y. A new method for the simultaneous enhancement of methane yield and reduction of hydrogen sulfide production in the anaerobic digestion of waste activated sludge. *Bioresour. Technol.* **2017**, *243*, 914–921. [CrossRef] [PubMed]
141. Neshat, S.A.; Mohammadi, M.; Najafpour, G.D.; Lahijani, P. Anaerobic codigestion of animal manures and lignocellulosic residues as a potent approach for sustainable biogas production. *Renew. Sustain. Energy Rev.* **2017**, *79*, 308–322. [CrossRef]
142. Kusch, S.; Schumacher, B.; Oechsner, H.; Schäfer, W. Methane yield of oat husks. *Biomass Bioenergy* **2011**, *35*, 2627–2633. [CrossRef]
143. Yao, Y.; Chen, S.; Kafle, G.K. Importance of "weak-base" poplar wastes to process performance and methane yield in solid-state anaerobic digestion. *J. Environ. Manag.* **2017**, *193*, 423–429. [CrossRef]
144. Anggarini, S.; Hidayat, N.; Sunyoto, N.M.S.; Wulandari, P.S. Optimization of hydraulic retention time (HRT) and inoculums addition in wastewater treatment using anaerobic digestion system. *Agric. Agric. Sci. Procedia* **2015**, *3*, 95–101. [CrossRef]
145. Garoma, T.; Pappaterra, D. An investigation of ultrasound effect on digestate solubilization and methane yield. *Waste Manag.* **2018**, *71*, 728–733. [CrossRef]
146. Borja, R.; Sánchez, E.; Weiland, P. Influence of ammonia concentration on thermophilic anaerobic digestion of cattle manure in upflow anaerobic sludge blanket (UASB) reactors. *Process Biochem.* **1996**, *31*, 477–483. [CrossRef]
147. Lemmer, A.; Merkle, W.; Baer, K.; Graf, F. Effects of high-pressure anaerobic digestion up to 30 bar on pH-value, production kinetics and specific methane yield. *Energy* **2017**, *138*, 659–667. [CrossRef]
148. Shanmugam, P.; Horan, N.J. 2009 Optimising the biogas production from leather fleshing waste by co-digestion with MSW. *Bioresour. Technol.* **2009**, *100*, 4117–4120. [CrossRef]
149. Soto, M.; Mendez, R.; Lema, J.M. Methanogenic and nonmethanogenic activity tests. Theoretical basis and experimental set-up. *Water Res.* **1993**, *27*, 1361–1376. [CrossRef]
150. Yerkes, D.W.; Boonyakitsombut, S.; Speece, R.E. Antagonism of sodium toxicity by the compatible solute betaine in anaerobic methanogenic systems. *Water Sci. Technol.* **1997**, *36*, 15–24. [CrossRef]
151. Schmidt, J.E.; Ahring, B.K. Effects of magnesium on thermophilic acetate-degrading granules in upflow anaerobic sludge blanket (UASB) reactors. *Enzym. Microb. Technol.* **1993**, *15*, 304–310. [CrossRef]
152. Schmidt, J.E.; Ahring, B.K. Effects of hydrogen and formate on the degradation of propionate and butyrate in thermophilic granules from an Upflow Anaerobic Sludge Blanket reactor. *Appl. Environ. Microbiol.* **1993**, *59*, 2546–2551. [PubMed]

153. Yu, H.Q.; Fang, H.H.P.; Tay, J.H. Enhanced sludge granulation in upflow anaerobic sludge blanket (UASB) reactors by aluminum chloride. *Chemosphere* **2001**, *44*, 31–36. [CrossRef]
154. Ariunbaatar, J.; Panico, A.; Frunzo, L.; Esposito, G.; Lens, P.N.; Pirozzi, F. Enhanced anaerobic digestion of food waste by thermal and ozonation pretreatment methods. *Environ. Manag.* **2014**, *146*, 142–149. [CrossRef]
155. Zhang, X.Y.; Cheng, S.A.; Wang, X.; Huang, X.; Logan, B.E. Separator characteristics for increasing performance of microbial fuel cells. *Environ. Sci. Technol.* **2009**, *43*, 8456–8461. [CrossRef]
156. Iman, W.; Koster, A.C. Inhibition of methanogenesis from acetate in granular sludge by long-chain Fatty acids. *Appl. Environ. Microb.* **1987**, *53*, 403–409.
157. Harada, H.; Uemura, S.; Monomoi, K. Interactions between sulphate-reducing bacteria and methane-producing bacteria in UASB reactors fed with low strength wastes containing different levels of sulphate. *Water Res.* **1994**, *28*, 355–367. [CrossRef]
158. Colleran, E.; Pender, S. Mesophilic and thermophilic anaerobic digestion of sulphatecontaining wastewaters. *Water Sci. Technol.* **2002**, *45*, 231–235. [CrossRef]
159. Astals, S.; Nolla-Ardèvol, V.; Mata-Alvarez, J. Anaerobic co-digestion of pig manure and crude glycerol at mesophilic conditions: Biogas and digestate. *Bioresour. Technol.* **2012**, *110*, 63–70. [CrossRef]
160. Girault, R.; Bridoux, G.; Nauleau, F.; Poullain, C.; Buffet, J.; Steyer, J.P.; Sadowski, A.; Beline, F. A waste characterisation procedure for ADM1 implementation based on degradation kinetics. *Water Res.* **2012**, *46*, 4099–4110. [CrossRef] [PubMed]
161. Giuliano, A.; Bolzonella, D.; Pavan, P.; Cavinato, C.; Cecchi, F. Co-digestion of 328 livestock effluents, energy crops and agro-waste: Feeding and process optimization in 329 mesophilic and thermophilic conditions. *Bioresour. Technol.* **2013**, *128*, 612–618. [CrossRef] [PubMed]
162. Ye, J.Q.; Li, D.; Sun, Y.M.; Wang, G.H.; Yuan, Z.H.; Zhen, F.; Wang, Y. Improved biogas production from rice straw by co-digestion with kitchen waste and pig manure. *Waste Manag.* **2013**, *33*, 2653–2658. [CrossRef] [PubMed]
163. Abouelenien, F.; Namba, Y.; Kosseva, M.R.; Nishio, N.; Nakashimada, Y. Enhancement of methane production from co-digestion of chicken manure with agricultural wastes. *Bioresour. Technol.* **2014**, *159*, 80–87. [CrossRef] [PubMed]
164. Li, Y.; Zhang, R.; He, Y.; Zhang, C.; Liu, X.; Chen, C.; Lui, G. Anaerobic co-digestion of chicken manure and corn stover in batch and continuously stirred tank reactor (CSTR). *Bioresour. Technol.* **2014**, *156*, 342–347. [CrossRef] [PubMed]
165. Ogunwande, G.A.; Osunade, J.A.; Adekalu, K.O.; Ogunjimi, L.A.O. Nitrogen loss in chicken litter compost as affected by carbon to nitrogen ratio and turning frequency. *Bioresour. Technol.* **2008**, *99*, 7495–7503. [CrossRef]
166. Karthikeyan, O.P.; Visvanathan, C. Bio-energy recovery from high-solid organic substrates by dry anaerobic bio-conversion processes: A review. *Rev. Environ. Sci. Biotechnol.* **2012**, *12*, 257–284. [CrossRef]
167. Wei, Y.; Li, X.; Yu, L.; Zou, D.; Yuan, H. Mesophilic anaerobic co-digestion of cattle manure and corn stover with biological and chemical pretreatment. *Bioresour. Technol.* **2015**, *198*, 431–436. [CrossRef]
168. Pagés-Díaz, J.; Pereda-Reyes, I.; Taherzadeh, M.; Sárvári-Horváth, I.; Lundin, M. Anaerobic co-digestion of solid slaughter-house wastes with agro-residues: Synergistic and antagonistic interactions determined in batch digestion assays. *Chem. Eng. J.* **2014**, *245*, 89–98. [CrossRef]
169. Liu, J.; Luo, J.; Zhou, J.; Liu, Q.; Qian, G.; Xu, Z.P. Inhibitory effect of high-strength ammonia nitrogen on bio-treatment of landfill leachate using EGSB reactor under mesophilic and atmospheric conditions. *Bioresour. Technol.* **2012**, *113*, 239–243. [CrossRef]
170. You, Z.; Wei, T.; Cheng, J.J. Improving anaerobic codigestion of corn stover using sodium hydroxide pretreatment. *Energy Fuels* **2013**, *28*, 549–554. [CrossRef]
171. Bah, H.; Zhang, W.; Wu, S.; Qi, D.; Kizito, S.; Dong, R. Evaluation of batch anaerobic co-digestion of palm pressed fiber and cattle manure under mesophilic conditions. *Waste Manag.* **2014**, *34*, 1984–1991. [CrossRef] [PubMed]
172. Kalamaras, S.; Kotsopoulos, T. Anaerobic co-digestion of cattle manure and alternative crops for the substitution of maize in South Europe. *Bioresour. Technol.* **2014**, *172*, 68–75. [CrossRef] [PubMed]
173. Li, D.; Liu, S.; Mi, L.; Li, Z.; Yuan, Y.; Yan, Z.; Liu, X. Effects of feedstock ratio and organic loading rate on the anaerobic mesophilic co-digestion of rice straw and cow manure. *Bioresour. Technol.* **2015**, *189*, 319–326. [CrossRef] [PubMed]

174. Zhai, N.; Zhang, T.; Yin, D.; Yang, G.; Wang, X.; Ren, G.; Feng, Y. Effect of initial pH on anaerobic co-digestion of kitchen waste and cow manure. *Waste Manag.* **2015**, *38*, 126–131. [CrossRef]
175. Xavier, C.A.N.; Moset, V.; Wahid, R.; Moller, H.B. The efficiency of shredded and briquetted wheat straw in anaerobic co-digestion with dairy cattle manure. *Biosyst. Eng.* **2015**, *139*, 16–24. [CrossRef]

© 2019 by the authors. Licensee MDPI, Basel, Switzerland. This article is an open access article distributed under the terms and conditions of the Creative Commons Attribution (CC BY) license (http://creativecommons.org/licenses/by/4.0/).

applied sciences

MDPI

Review

Insight into Pretreatment Methods of Lignocellulosic Biomass to Increase Biogas Yield: Current State, Challenges, and Opportunities

Javier Ulises Hernández-Beltrán, Inty Omar Hernández-De Lira, Mónica María Cruz-Santos, Alexia Saucedo-Luevanos, Fernando Hernández-Terán and Nagamani Balagurusamy *

Laboratorio de Biorremediación, Facultad de Ciencias Biológicas, Universidad Autónoma de Coahuila, Torreón 27000, Coahuila, Mexico
* Correspondence: bnagamani@uadec.edu.mx; Tel.: +52-871-7571-785

Received: 29 July 2019; Accepted: 26 August 2019; Published: 6 September 2019

Abstract: Lignocellulosic biomass is recalcitrant due to its heterogeneous structure, which is one of the major limitations for its use as a feedstock for methane production. Although different pretreatment methods are being used, intermediaries formed are known to show adverse effect on microorganisms involved in methane formation. This review, apart from highlighting the efficiency and limitations of the different pretreatment methods from engineering, chemical, and biochemical point of views, will discuss the strategies to increase the carbon recovery in the form of methane by way of amending pretreatments to lower inhibitory effects on microbial groups and by optimizing process conditions.

Keywords: lignocellulosic biomass; pretreatment methods; limitations; anaerobic digestion; biogas

1. Introduction

Rapid growth in human population, biodiversity, and natural resources losses, environmental pollution, and climate change [1] have led researchers and policy makers to pay attention to sustainable energy policies for reduction the effect of greenhouse gas emissions [2]. In addition, depletion of fossil reserves is occurring at faster rate as there is an accelerated increase in the consumption of fossil fuels due to industrialization and motorization of the world [3].

Today, fossil fuels still serve as the primary global energy resource and account for more than 88% of the primary energy usage [4] and their combustion result in the emission of greenhouse gases, especially CO_2 [5]. Thus, replacing fossil fuels with renewable sources of energy provides the opportunity to tackle these phenomena by limiting the increase of global temperature [6]. Renewable energy is defined as the contribution of renewable energies to the total supply of primary energy, including the equivalent of primary energy from hydroelectric sources (excluding storage by pumping), geothermal, solar, wind, tidal, and wave energy. It also includes the energy derived from solid biofuels, biodiesel, other liquid biofuels, biogas and the renewable fraction of municipal waste. Likewise, biofuels are defined as fuels derived directly or indirectly from biomass [7]. Biomass as referred as the fourth largest energy source (after coal, oil, and natural gas), and the largest and most important renewable energy option today and can be used to produce different forms of energy [8]. Biomass has the advantage to be considered as carbon neutral because the quantity of CO_2 released during combustion is the same as that absorbed by the plant during photosynthesis [9]. The largest potential feedstock for bioenergy production is lignocellulosic biomass, which represents the most economical and highly renewable natural source in the world.

1.1. Classification of Types of Lignocellulose

Lignocellulosic biomass is a biological material obtained from living or recently living organisms [7]. A brief summary of different types of lignocellulosic residues generated in Mexico are presented in Table 1.

Table 1. A brief summary of lignocellulosic residues generated in Mexico.

Lignocellulosic Residues	Amounts of Residues Produced per Year in México
Straw from rice, wheat, soy, oats & corn	78,534,119.08 tons
Stump, sawdust, plant & tree branches	8.2 million m^3 as rolls
Sugar beet & Sugarcane residues	16,935,954.03 tons
Urban solid residues	102,895 tons
Municipal sewage sludge	7200 million m^3
Residues from nuts, pistachios, peanuts & pine nuts	74,249.575 tons
Manure from cows, pigs, poultry, turkey & sheep	142,139,944.5 tons

The amount of residues produced per year in Mexico were obtained from the data on crop production, which was multiplied with the coefficient of waste generated for each residue. Data on production was consulted from the web portal of Agri-Food and Fisheries Information Service (SIAP), Government of Mexico [10].

1.2. Chemical Composition of Lignocellulosic Biomass

During the growth, plant biomass produce primary and secondary walls; primary wall is involved in structural functions like protection, signal transduction, and interactions with neighboring cells; besides, it contains low proportion of cellulose but a greater presence of pectin, meanwhile, secondary wall is composed principally of lignocellulose [11] (see Figure 1).

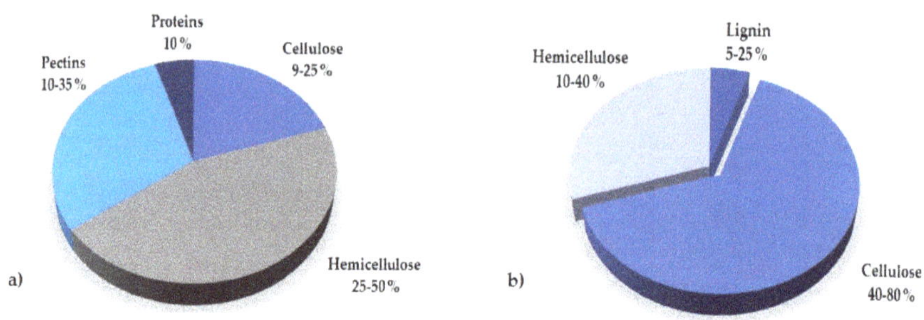

Figure 1. Composition of the cell wall in plants. (**a**) primary wall, (**b**) secondary wall.

Lignocellulosic biomass is composed of three main biopolymers: cellulose, hemicellulose and lignin [12]. The individual composition of different lignocellulosic substrates on dry basis is presented in Table 2 and the structure is presented in Figure 2.

Table 2. Composition of different lignocellulosic substrates; cellulose, hemicellulose, and lignin content (%) on dry basis.

Lignocellulosic Substrates	Composition (% Dry Basis)			
	Cellulose	Hemicellulose	Lignin	References
Bamboo stem	43.04	22.13	27.14	[13]
Birch	40.1 ± 0.6	17.5 ± 0.2	24.2 ± 0.1	[14]
Corn cob	42.0 ± 0.1	45.9 ± 0.9	2.8 ± 0.2	[15]
Corn stalk	36.4 ± 0.1	30.3 ± 0.1	6.9 ± 1.4	[16]
Corn stover	42.21	22.28	19.54	[17]
Corn straw	49.3 ± 1.8	28.8 ± 1.4	7.5 ± 0.4	[18]
Cotton Stalk	41.6 ± 0.5	23.6 ± 0.4	23.3 ± 0.7	[19]
Eucalyptus	52.07 ± 2.6	24.51 ± 1.1	25.2 ± 1.1	[20]
Empty fruit bunch	34.9	26.64	31.1	[21]
Giant reed	41.5 ± 2.6	20.5 ± 0.6	18.4 ± 1.4	[22]
Grass	47.12 ± 3.2	36.01 ± 3.17	11.55 ± 0.3	[23]
Maize straw	38.33 ± 0.8	29.76 ± 1.35	3..82 ± 0.5	[24]
Meadow grass	41.28 ± 5.3	28.14 ± 3.2	30.14 ± 7.9	[25]
Miscanthus	36.3 ± 2.1	22.16 ± 1.9	22.55 ± 2.5	[26]
Oat straw	35.0	28.2	4.1	[27]
Oil palm empty fruit bunch	38.5 ± 1.9	26.1 ± 1.1	11.6 ± 1.6	[28]
Pinewood	38.2 ± 0.3	24.1 ± 0.7	34.4 ± 0.3	[29]
Poplar	46.0 ± 0.1	16.7 ± 0.1	26..6 ± 0.3	[14]
Rice hulls	36.0	12.0	26.0	[30]
Rice straw	37.8 ± 0.2	29.6 ± 0.7	14.8 ± 0.4	[31]
Rye Straw	36.5 ± 0.1	NR	21.3 ± 0.1	[32]
Sawdust waste	31.5 ± 1.3	26.1 ± 2.1	24.9 ± 1.7	[33]
Sorghum straw	26.93 ± 1.2	32.57 ± 1.9	10.16 ± 1.8	[34]
Spruce	24.7 ± 0.2	10.2 ± 0.1	35.0 ± 0.3	[35]
Sugarcane bagasse	46.1 ± 0.7	20.1 ± 0.9	20.3 ± 0.6	[36]
Sunflower stalk	34 ± 0.6	20.8 ± 0.8	29.7 ± 0.6	[37]
Water hyacinth	36.84 ± 0.8	27.7 ± 0.2	10.7 ± 0.4	[38]
Wheat straw	43.4	26.9	22.2	[39]
Willow sawdust	35.6 ± 0.9	21.5 ± 0.9	28.7 ± 0.2	[40]

NR: Not Reported.

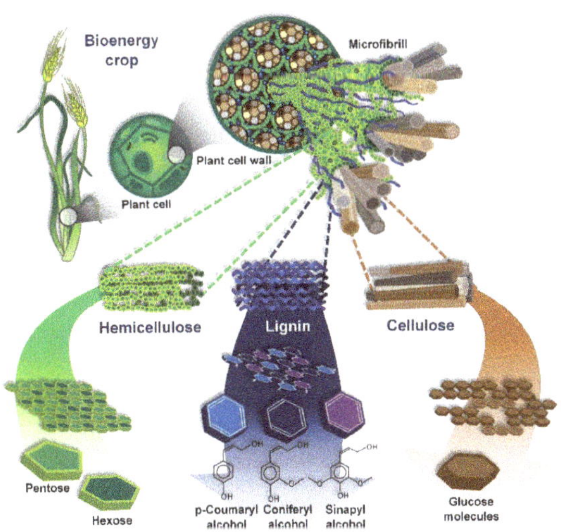

Figure 2. Structure of lignocellulosic biomass and its biopolymers; cellulose, hemicellulose, and lignin.

Cellulose is the major component of plant cell structure and is a homo-polymer made of β-1,4-glycosidic bonds, which give stiffness and stability, characteristic to l structure of the cell wall [41]. The principal unit of cellulose, D-pyran glucose, makes the crystalline and amorphous regions in the cellular arrangement. The orientation and direction that have the chains of cellulose is of vital importance and determine their function. In the primary wall are the first type of alpha helical, (lower molecular weight than the secondary wall) allowing the cell storing sugars and fluids [42]. On the other hand, secondary wall cellulose chains are rigid and straight to give the molecular structure to the cell [43].

Lignin is formed from the oxidation of phydroxycinnamyl alcohols: p-coumaryl, coniferyl, and sinapyl. The formation of lignin involves three routes of biosynthesis: the shikimate pathway, the phenylpropanoid pathway and the synthesis of monolignols [46]. The units of lignin formation are: guaiacyl (G), syringyl (S), and p-hydroxyphenyl (H) [47]. The structure formed by three-dimensional and amorphous heteropolymers of lignin helps the cell in situations of stress, whether due to structural or metabolic damage, also serves as protection of the cell wall against pathogens [48].

2. Biogas as a Sustainable Energy

The implementation of biomass as sources of renewable energy technology is considered as sustainable technology in meeting energy needs and as well as to minimize the emission of greenhouse gases. In addition, biomass use presents the advantage of cost–benefit viability and reduce the waste flow into the environment [49].

The production of biogas from lignocellulosic substrates through the transformation of volatile organic solids (VS) by anaerobic digestion is proven to be an alternative source of energy. The potential of biogas from waste can substitute current use of natural gas in many regions [50]. Biogas is produced through a bioprocess involving four steps—i.e., hydrolysis, acidogenesis, acetogenesis, and methanogenesis—using a microbial consortium containing many different types of bacteria and archaea [51] as shown in Figure 3. Composition of biogas slightly varies depending on the type of feedstock used in anaerobic digestion. It is mainly composed by CH_4 (40–75%) and CO_2 (25–60%), with minor impurities such as H_2S, NH_3, among others [52]. Furthermore, CO_2, the second major component can be sequestered and used to produce chemicals [53].

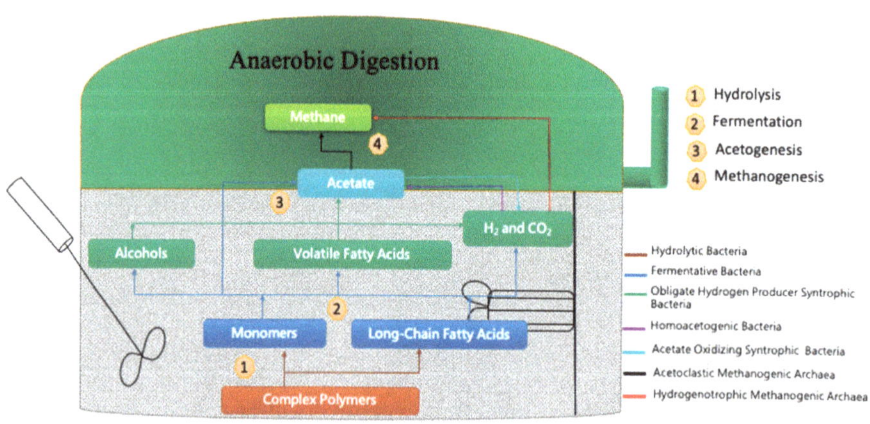

Figure 3. Steps and microorganisms involved in biogas production.

Hemicellulose can also be found in the primary cell wall in contrast to the cellulose which is present mainly within the secondary wall [44]. It is in turn the second most abundant polysaccharide compound in nature. It has as sub units D-xylose, mannose, L-arabinose, galactose, and glucuronic acid which have links "β-1,4-glycosidic bonds in main chains; "β-1.2 -,"β - 1.3-, "β-1.6-glycosidic bonds;

with a number, less than 200 units of polymerization (polyxylose, galactoglucomannan, glucomannan). Xylan is the main polysaccharide present in hemicellulose in plants [45].

However, the heterogeneity in chemical composition of lignocellulosic biomass implies an important technological and operational challenge [54] and use of either mechanical, chemical, physiochemical, or biological pretreatments has been shown to improve their biodegradability [55].

3. Pretreatment Methods in the Chain of Biogas Production

Pretreatment or fractionation is an important tool to alter the structure of lignocellulosic biomass to make the holocellulose (cellulose + hemicellulose) bioavailable for bioconversion [56].

The main objective of pretreatment is the disruption of the physical barriers of the cell wall to depolymerize and to reduce the cellulose crystallinity [57]. Regardless of the type of biomass, pretreatment has been identified as the crucial step, both technically and economically, in the bioconversion of lignocellulosic biomass for its use in biorefinery [58]. The pretreatment method must have to be economical as both for operating and capital costs could be more than 40% of the total processing cost [59].

A large number of pretreatment methods for biomass have been studied, which can be broadly classified into (1) mechanical processes and this refers to reduce particle size; (2) chemical processes through the use of diluted acids, alkalis, or organic solvents; (3) physicochemical processes such as steam explosion and hot water; and (4) biological processes through the use of microbial consortia or by enzymatic means [60] as shown in Figure 4.

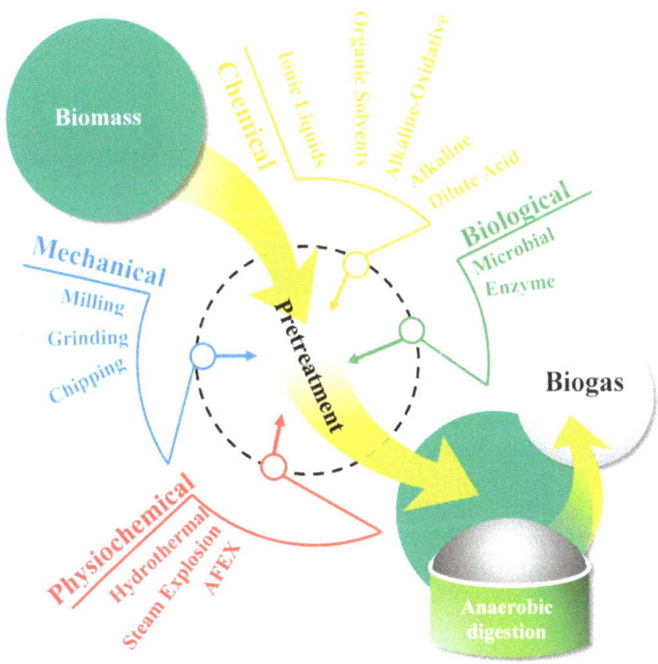

Figure 4. Pretreatment methods to increase the bioavailability of lignocellulosic biomass.

3.1. Mechanical

The mechanical pretreatment is the reduction of the particle size by methods such as milling, chipping, or grinding [61]. Mechanical pretreatment is well-known method to improve biogas production, however it is yet considered to be an expensive method, due its high energy requirements [62]. The size reduction of lignocellulosic biomass is an essential step to increase the accessible surface area and the porosity of the particles, besides reducing the crystallinity of the cellulose and improves the efficiency of the next processing step and global production chain [63]. One advantage of mechanical pretreatment is that it does not produce any secondary inhibitory substances, which suggest that could be suitable for methane production or any other bioprocess [64]. Dahunsi [64] reported that mechanical pretreatment applied to six different lignocelluloses caused breakdown of structural material and increased the methane yield up to 22%. However, it is observed that excessive size reduction of lignocellulosic biomass can lead to a lower efficiency of methane production [61]. Process efficiency in terms of cost-energy, operational aspects such as the right particle size, processing time of mechanical pretreatment, and the mechanical speed for size reduction must be standardized. It can be observed from literature that there is no universal particle size suitable for biomethanation process and it varies according to the type of the substrate and the process used for biomethanation. Dumas et al. [65] found negligible differences in biogas yield of wheat straw having the particle size in the range from 0.7 to 0.2 mm, and that the methane yield did not show any increase at particle size of 0.048 mm. Sharma et al. [66] compared the effect of different particle size, viz., 0.088, 0.40, 1.0, and 6.0 mm of various agricultural residues—such as wheat straw, rice straw, *Mirabilis* leaves, cauliflower leaves, *Ipomoea fistulosa* leaves, dhub grass, and banana peel—and observed that the highest biogas yield was observed with residues having a particle size 0.88 mm. They further reported that there was no statistical difference in biogas yield between 0.4 and 0.088 mm. However, De la Rubia et al. [67] observed the highest methane yield in sunflower oil cake at particle size 1.4–2.0 mm, which was 17% higher than the particle size in the range of 0.355–0.55 mm. Chandra et al. [68] also observed the particle size of 0.30 mm of wheat and rice straw did not show any significant increase in methane yield in comparison to the particle size 0.75 mm of both straws. However, both particle sizes showed an increase than the untreated particle size of 1.5 mm.

Tsapekos et al. [69] tested six different mechanical pretreatments, which generated particle size in the range of 0 to >20 cm. They reported that the mechanical pretreatments which generated particle size larger than >20 cm, equivalent to 33% and 41% of the total of biomass did not show any significant increase in yield than untreated biomass. The methane yield was significant higher and improved by 25% where the amount of biomass with particle size greater than >20 cm was the about only 22% of the total of biomass. Similarly, Herrmann et al. [70] found a close correlation between particle length and methane yield while comparing different particle length of the same crop material. Krause et al. [71] studied two different particle sizes, <2 mm and >20–100 mm, in different substrates like softwood, hardwood, and cotton and found that increase in methane yield was positively correlated with reduced particle size.

On the other side, the processing time on particle reduction is not only important in energy terms, but also influence the efficiency of the bioconversion of lignocellulosic biomass [59]. Rodríguez et al. [63] compared 30 and 60 min of mechanical pretreatment, and reported a 21% increase in methane yield for 60 min and that 30 min pretreatment did not improve the methane yield. For the cost reduction of the energy involved in mechanical pretreatments, the speed of the milling equipment must be considered [72]. Tsapekos et al. [25] evaluated different speeds of milling, in the range of 200 to 1200 rpm on methane yield from meadow grass and observed that there was no difference due to the milling speeds tested. They observed 27% increase in methane yield in treatments than the untreated biomass.

The methane yield from different mechanically pretreated lignocellulosic substrates is presented in Table 3. It can be observed that different particle sizes were found to be optimum for different types of lignocellulosic biomass. In the case of wheat straw, particle size of 2 mm increased the methane yield to an order of 83% [73]. Likewise, barley straw presented an increase of 54% and 41% of methane yield at particle size of 5 mm and 20 mm, respectively [73]. It can also be observed that particle sizes larger and smaller than the range between 2 and 5 mm, presented a decrease in methane yield [68,73,74]. The observed variations on the effect of particle size, time, and speed of milling pointed out that apparently the best strategy for the mechanical pretreatment depends on the type the lignocellulosic material and a universal strategy cannot be recommended.

Table 3. Effect of mechanical pretreatments on methane yield from lignocellulosic biomass.

Substrates	Particle Size	Methane Yield (Untreated)	Methane Yield (After Treatment)	Reaction System	References
Barley Straw	5 mm	240 mL/g VS	370 mL/g VS	Glass reactor 2 L	[73]
	20 mm	240 mL/g VS	339 mL/g VS	Glass reactor 2 L	[73]
	50 mm	240 mL/g VS	286 mL/g VS	Glass reactor 2 L	[73]
Crop feedstocks (winter rye, sorghum, forage rye, maize, triticale)	6–33 mm	278 mL/g ODM	403 mL/g ODM	Stirred tank reactor 3 L	[70]
Maize stalks	2 mm	246 mL/g VS	272 mL/g VS	Glass reactor 2 L	[73]
	20 mm	246 mL/g VS	254 mL/g VS	Glass reactor 2 L	[73]
Meadow grass	2 mm	297 mL/g VS	376 mL/g VS	Bottle 0.5 L	[25]
	<200 mm	-	347 mL/g VS (increase 20%)	Bottle 0.5 L	[72]
	0–200 mm	303 mL/g VS	372 mL/g VS	Bottle 0.5 L	[69]
Rice straw	0.3 mm	58.1 mL/g VS	62.7 mL/g VS	-	[68]
	0.75 mm	58.1 mL/g VS	65.7 mL/g VS	-	[68]
	50 mm	197 mL/g VS	203 mL/g VS	Glass reactor 2 L	[73]
Switchgrass	2–10 mm	127.4 mL/g VS	170.7 mL/g VS	Bottle 0.5 L	[75]
Water hyacinth	0.001 mm	-	Increase 20% (from 50 to 70%)	Digester 0.45 L	[76]
	0.05 mm	-	Increase 16% (from 50 to 66%)	Digester 0.45 L	[76]
	1.0 mm	-	Increase 10% (from 50 to 60%)	Digester 0.45 L	[76]
	2.5 mm	-	Increase 5% (from 50 to 55%)	Digester 0.45 L	[76]
Wheat straw	0.3 mm	167.8 mL/g VS	245.6 mL/g VS	Reactor 2 L	[74]
	1.2 mm	167.8 mL/g VS	264.7 mL/g VS	Reactor 2 L	[74]
	0.3 mm	67.1 mL/g VS	70.3 mL/g VS	-	[68]
	0.75 mm	67.1 mL/g VS	93.1 mL/g VS	-	[68]
	0.088–0.759 mm	183.4 mL/g VS	252.8 mL/g VS	Lab Flask	[65]
	2 mm	182 mL/g VS	334 mL/g VS	Glass reactor 2 L	[73]
	50 mm	182 mL/g VS	285 mL/g VS	Glass reactor 2 L	[73]
Waste paper	60 min of beating time	132 mL/g VS	215 mL/g VS	Flask 0.5 L	[63]

3.2. Chemical Pretreatments

Chemical pretreatment of lignocellulosic biomass with acids, alkalis, and organic solvents [77] are considered as one of the most promising, since they can be quite effective in degrading more complex-structured substrates [78]; further improves the bioavailability of carbohydrates by removing lignin and/or by decreasing the degree of polymerization and cellulose crystallinity [79]. Chemical pretreatment has gained larger attention because they usually less expensive and result in faster rates and better efficiencies in enhancing the degradation complex organic molecules.

3.2.1. Acid

The main objective of dilute acid treatment is to achieve greater access to carbohydrate fractions of cellulose by hydrolyzing the hemicellulose [80]. Low acid concentration at 0.2% to 2.5% *v*/*v* and temperatures between 130 °C and 210 °C are employed [81], a common practice is the use of autoclave at 121 °C for 1 h [82,83]. On the contrary, low temperature (room temperature) is used while using high reagent acid concentration (above 30% *v*/*v*). High acid concentrations could cause corrosion problems and as a result high maintenance costs [84] and further, generate inhibitor compounds at high rate and therefore increase the purification costs. Many studies confirmed that the dilute acid treatment is the most promising technology for lignocellulose pretreatment [85]. According to Syaichurrozi et al. [86], pretreatment of *Salvinia molesta* pretreatment by 2%, 4%, and 6% *v*/*v* of sulfuric acid (H_2SO_4) increased methane yield around 100% in comparison to untreated, however no yield differences were reported between the tested concentrations. Similarly, Song et al. [18] observed that 2% sulfuric acid pretreatment of corn straw resulted in the highest methane concentration of 175 mL/g VS. Further, they reported that use of 2% hydrochloric acid and 4% acetic acid yielded lower methane production than sulfuric acid, of 163 and 145 mL/g VS respectively. Germec et al. [87] found that the optimum acid concentration for pretreatment of spent tea leaves was 1.58% *v/v*. In relation to the temperature and reaction time, Germec et al. [87] observed that 131°C for 20 min was efficient than 121 °C for 1 h, which was also suggested by Saha and Cotta [83]. On the other hand, Syaichurrozi et al. [86] recommended the dilute acid pretreatment at room temperature of 30 °C for 2 days.

3.2.2. Alkaline

Alkaline pretreatment shows high efficiency [88], especially in the delignification process [89]. Alkaline pretreatment produces a swelling reaction in the cell wall allowing an increase in the internal surface area, and simultaneously decrease of polymerization degree and crystallinity of cellulose [84]. To take full advantage of these alkaline effects, critical process parameter such as alkaline loading, reaction temperature, and quantity must be optimized [90].

High concentrations of alkaline reagent lead to degradation and decomposition of polysaccharides. Low concentrations at low temperature and at atmospheric pressure are recommended [30]. Furthermore, it does not generate toxic compounds such as furfurals and hydroxymethylfurfurals (HMF), and hence higher efficiency is observed in the biomethanation process.

Although, there is still discrepancy on the choice of alkaline reagent, in terms of the greatest advantages, considering the cost of the reagent, performance, and manipulation [91], sodium hydroxide is preferred because it catalyzes under mild conditions, effectively attacks the linkage between lignin and hemicellulose in lignin-carbohydrate; in particular it cleaves the ether and ester bonds in the lignocellulose structure and also effective in the cleavage of the ester and carbon-to-carbon bonds in lignin molecules [92–94]. Jiang et al. [22] and Xihui et al. [95] optimized the NaOH loading rate to increase the methane yield. Jiang et al. [22] varied the NaOH concentration between 0.5% and 2% w/v for pretreatment of giant reed, and observed an increase in methane yield from 217 mL/g VS to 355 mL/g VS after pretreating at 2% w/v. Likewise, Xihui et al. [95] achieved the highest methane yield of 201 mL/g VS from *Pennisetum* hybrid at 2% w/w of NaOH, and they tested the alkali pretreatment at concentrations of 2%, 4%, 6%, and 8% w/w.

However, sodium discharge might be environmentally harmful as they can lead to negative impacts such as soil salinization. KOH, although is three times more expensive than NaOH, it represents is an alternative solution. Romero-Güiza et al. [96] reported an increase of 128% methane yield from 0.7% w/w KOH pretreated wheat straw than the untreated wheat straw. Moset et al. [97] compared NaOH and KOH pretreatment (0.5 to 8%) of wheat straw on biogas yield and reported the highest methane yield of 374 mL/g VS and 370 mL/g VS with 2% NaOH and 0.5% KOH respectively, and untreated straw recorded a methane yield of 303 mL/g VS. These results indicated that requirement of lower concentration of KOH could offset its high cost. In addition, KOH is easily recovered, and its discharge might have a potential soil amendment value since potassium is a nutrient required for plant growth [96]. Thomas et al. [26] suggested the alternative option of using CaO. However, they observed that NaOH was more efficient than CaO and increased the methane production in the range of 24 and 55% from the pretreated *Miscanthus x giganteus*, *M. sacchariflorus*, and *M. sinensis*, while the increase was between 19 and 30% for Cao treated samples. However, CaO is an interesting alternative alkali reagent as it is already used to mitigate soil acidity and is also less expensive than NaOH.

Calcium hydroxide is another common alkali reagent used, Gu et al. [98] tested five different concentrations of $Ca(OH)_2$: 6%, 8%,10%, 12%, and 15%, in which the highest biogas production from obtained at 8% and 10% pretreated rice straw and recorded a biogas production of 564.7 mL/g VS and 574.5 mL/g VS, respectively, which is 34.3% and 36.7% higher than untreated biomass.

3.2.3. Organosolv

Organosolv pretreatment emerges as the only technology capable of isolating each component of the lignocellulosic biomass, attaining relatively pure lignin that can be sold as a by-product or converted into higher-value products in a biorefinery concept [99]. Organic solvents—such as methanol, ethanol, acetone, acetic acid, peracetic acid, and so on—are used with or without the addition of a catalyst reagent [100]. The catalysts used include mineral acids (hydrochloric, sulfuric, and phosphoric acids) and organic acids (oxalic, acetylsalicylic, and salicylic acids) [101]. Although catalyst addition aids in higher pretreatment efficiency, use of a catalyst could have a negative impact on the environment [102], i.e., chemical catalyst as acid causes acid-catalyzed degradation of the monosaccharides into furfural and 5-hydroxymethyl furfural followed by condensation reactions between lignin and the reactive aldehydes [103]. Acid catalysts are added in order to increase the rate of lignin removal and decrease pretreatment temperature, because acid catalysts cleave acid-labile bonds (α–aryl ether and arylglycerol–β-aryl ether bonds), which help stabilization of lignin fragments [101]. However, the use of chemical catalysts involves issues, such as equipment corrosion and the need of processing downstream effluents, resulting in high water consumption [103]. Formic acid is a promising catalyst as it is believed to have fewer corrosive effects than stronger mineral acids and further to avoid production of inhibitors that are normally formed when acetic acid is employed [104].

The organosolv process causes hydrolysis of the internal bonds in lignin and between lignin and hemicellulose. The organic solvents also cause hydrolysis of the glyosidic bonds in hemicelluloses and to a smaller extent in cellulose [103]. The preferred conditions of organosolv process is generally in the following ranges: a cooking temperature of 180–195 °C, a cooking time of 30–90 min and liquor to solid ratio ranging from 4:1 to 10:1 [105].

As mentioned previously, organic acids have been suggested as an alternative to inorganic ones, to avoid corrosion and decrease the energy demand during acid recovery. Amnuaycheewa et al. [106] pretreated the rice straw with organic acids—such as acetic acid, citric acid, and oxalic acid—and make a comparison with hydrochloric acid as inorganic acid. It was found that pretreatment with oxalic acid reagent yielded 2.68 and 1.25 times higher reducing sugars than untreated and HCl-treated rice straw, respectively. Further, the accumulated methane yields obtained from pretreatments of organic acids were 2.47–2.53 times higher than the inorganic acid one.

A critical issue with organosolv pretreatment is the risk of operating at high pressure and the solvents volatility due to their low boiling point of the organic solvents. A system of solvent recovery is necessary to reduce the cost of operation and to avoid their residual inhibitory effect on microorganisms. Teghammar et al. [107] recovered more than 98% of organic solvent N-methylmorpholine-N-oxide (NMMO or NMO), which was used for pretreatment of softwood spruce, rice straw, and triticale straw at 130 °C for 1–15 h. Further, they reported an increase in methane production between 400% and 1200%. Kabir et al. [108] studied the use of recycled NMMO for pretreatment of barley straw and forest residues and observed similar performance as the fresh NMMO on barley straw. However, pretreatment of forest residues with recycled NMMO resulted in 55% reduction of methane yield.

Among solvents, low boiling point alcohols, especially ethanol, have attracted more attention because of their high effectiveness on lignocellulosic biomass, as well as the simplicity of their recovery by distillation [109]. Other advantages are lower cost and lower toxicity to humans compared to solvents as methanol [104,110].

3.2.4. Ionic Liquids

Ionic liquids are environmental friendly and represent a new class of solvents which have high polarities, a low melting point, non-volatility, and design-ability [111]. Ionic liquid-based pretreatment gained attention due to their effectiveness on a range of biomass types, even in the scale-up from lab scale to small pilot scale, and their ability to disrupt lignin and decrystallize cellulose [112].

Ionic liquids have unique properties compared to traditional solvents, including a low vapor pressure, which avoids atmospheric emissions by eliminating solvent losses through evaporation.

They also have a relatively high thermal stability, low flammability, and possibility of recycling, which contributes to the reduction of waste [111].

Ionic liquids can be easily prepared using different cations and anions, resulting in hydrophobic or hydrophilic types, according to the desired application [113]. Cations of methylimidazolium and methylpyridinium with allyl-, ethyl-, or butyl-chains are commonly used. Meanwhile, the most effective anions are halogens, formates, acetates, amides, imides, thiocynates, phosphates, sulfites, sulfonates, and dichloroaluminates [114]. Kim et al. [115] studied the transition of the crystalline structure when lignocelluloses such as cotton stalks, hemp stalks, and acacia pruning were pretreated with 1-ethyl-3-methylimidazolium acetate (EmimOAC) or 1-ethyl-3-methylmidazolium chloride (EmimCl), and the results indicated that type of anion in the ionic liquid significantly influenced the efficiency of lignin and hemicellulose removal. Acetate anion showed greater cellulose extraction from lignocellulose substrates.

There are various parameters affecting the process of biomass pretreatment using ionic liquids: for instance, physiochemical properties of ionic liquid, reaction time and temperature, ratio of biomass to ionic liquid, biomass type, and water content of sample. Physiochemical property like viscosity is probably the most negative parameter of the process because extremely viscous solution is formed during pretreatment and significantly affect the industrial application of ionic liquids [116].

A disadvantage of ionic liquids is their high cost, therefore, a strategy to follow is the recycling to compensate their high production costs [117]. To make ionic liquid pretreatment economically feasible, [118] a recovery of >97% of ionic liquid used and >90% waste heat recovery is recommended. Xu et al. [20] evaluated the efficiency of reuse of ionic liquids 1-allyl-3-methylimidazolium chloride (amimCl) and 1-butyl-3-methylimidazolium acetate (bmimOAc) on eucalyptus and reported that four reusing cycles were effective, where hydrolysis efficiencies of 54.3% for (amimCl) and 72.8% for (bmimOAc) were observed. After their fourth reuse, ionic liquids showed deteriorations by relatively lower sugar conversion and lignin removal. Gao et al. [35] evaluated the water hyacinth pretreated with 1-N-butyl-3-methyimidazolium chloride [BmimCl]/dimethyl sulfoxide (DMSO) at 120 °C for 120 min and reported an increase in methane yield by 97.6% compared with non-pretreated water hyacinth. The ionic liquids and co-solvents were successfully recovered.

On the other hand, Xie et al. [119] simulated a process for biogas upgrading using three imidazolium-based ionic liquids 1-hexyl-3-methylimidazolium bis(trifluoromethylsulfonyl)imide [hmim][Tf2N], 1-Butyl-3-methylimidazolium bis(trifluoromethanesulfonyl)imide [bmim][Tf2N] and 1-Butyl-3-methylimidazolium hexafluorophosphate [bmim][PF6] and reported a 11% reduction in energy consumption in [bmim][Tf2N], which suggest the promising potential of ionic liquid technology for biogas upgrading.

3.2.5. Alkaline Hydrogen Peroxide

The alkaline reagent can be combined with hydrogen peroxide. The use of H_2O_2 improves the enzymatic digestibility in a wide range of lignocellulosic biomass with the main intention of improving the enzymatic hydrolysis yields [120]. H_2O_2 is used as a bleaching agent of lignocellulosic biomass. During the pretreatment, many free radicals are released but has the advantage of not leaving residues in the biomass because it degrades into oxygen and water and hardly forms secondary products. Thus, an oxidative fragmentation occurs, as well as lignin removal from the lignocellulosic matrix by attacking lignin side chains [121]. Alkaline hydrogen peroxide improves the performance in terms of delignification and enzymatic digestibility than single alkali reagent [122]. However, this method is limited due to the need to maintain constant pH during the process to avoid hemicellulose elimination, and is one of the most important parameter to achieve process efficiency [117].

Alkaline hydrogen peroxide is a delignification process, but not only improves the depolymerization of lignin, but also has the advantage of requiring mild conditions of temperature and pressure [123], which reduces the formation of inhibitors [124]. Ayeni et al. [125] studied the operating conditions of alkaline hydrogen peroxide pretreatment of sugarcane bagasse and reported optimum pretreatment conditions were 0.3% hydrogen peroxide concentration, 100 °C for 4.6 h. Siciliano et al. [126] confirmed the use of low concentration of oxidant at room temperature significantly improved the anaerobic treatability of olive mill residues and reported methane yield up to 0.328 L/g of COD (chemical oxygen demand) removed. Similarly, Katukuri et al. [127] reported 49% increase in methane yield over untreated substrate of *Miscanthus floridulus* by using 0.8% of H_2O_2.

On the other hand, Alencar et al. [128] evaluated the H_2O_2 recycling strategy and tested the efficiency of recycled H_2O_2 in five successive reuse cycles and reported a decrease in efficiency at each cycle of reuse than the previous.

Methane yield from different residues after pretreatment by acid, alkali, organosolv, and H_2O_2 treatment is presented in Table 4.

Table 4. Effect of different chemical pretreatments on methane yield from different lignocellulosic biomass.

Subastrates	Pretreatment Conditions	Methane Yield (Untreated)	Methane Yield (After Pretreament)	Reaction System	References
Corn Straw	Sulfuric acid 2% *v/v*	100.6 mL/g VS	175.6 mL/g VS	Flask 1 L	[18]
	Hydrochloric acid 2% *v/v*	100.6 mL/g VS	163.4 mL/g VS	Flask 1 L	[18]
	Acetic acid 4% *v/v*	100.6 mL/g VS	145.1 mL/g VS	Flask 1 L	[18]
	Hydrogen peroxide 3% *v/v*	100.6 mL/g VS	216.7 mL/g VS	Flask 1 L	[18]
	Sodium hydroxide 8% *v/v*	100.6 mL/g VS	163.5 mL/g VS	Flask 1 L	[18]
	Calcium hydroxyde 8% *v/v*	100.6 mL/g VS	206.6 mL/g VS	Flask 1 L	[18]
	Ammonia 10% *v/v*	100.6 mL/g VS	168.3 mL/g VS	Flask 1 L	[18]
Cotton gin waste	Citric acid 0.5 mmol/g VS substrate	95.4 mL/g VS	147.1 mL/g VS	Flask 0.25 L	[78]
	Hydrogen peroxide 0.5 mmol/g VS substrate	178.0 mL/g VS	247.5 mL/g VS	Flask 0.25 L	[78]
	Ethanol 0.5 mmol/g VS substrate	172.5 mL/g VS	241.5 mL/g VS	Flask 0.25 L	[78]
Juice industry waste	Citric acid 0.5 mmol/g VS substrate	180.4 mL/g VS	353.5 mL/g VS	Flask 0.25 L	[78]
	Hydrogen peroxide 0.5 mmol/g VS substrate	275.0 mL/g VS	385.6 mL/g VS	Flask 0.25 L	[78]
	Ethanol 0.5 mmol/g VS substrate	214.3 mL/g VS	332.8 mL/g VS	Flask 0.25 L	[78]
Olive pomace	Citric acid 0.5 mmol/g VS substrate	200.1 mL/g VS	183.2 mL/g VS	Flask 0.25 L	[78]
	Hydrogen peroxide 0.5 mmol/g VS substrate	178.7 mL/g VS	172.3 mL/g VS	Flask 0.25 L	[78]
	Ethanol 0.5 mmol/g VS substrate	193.0 mL/g VS	157.5 mL/g VS	Flask 0.25 L	[78]
Salvinia molesta	Silfuric acid 2% *v/v*	11.2 mL/g VS	16.6 mL/g VS	Bottle 0.6 L	[86]
	Silfuric acid 4% *v/v*	11.2 mL/g VS	17.4 mL/g VS	Bottle 0.6 L	[86]
	Silfuric acid 6% *v/v*	11.2 mL/g VS	17.8 mL/g VS	Bottle 0.6 L	[86]
Switchgrass	Sodium hydroxide 7 g/L	112.4 mL/g VS	132.5 112.4 mL/g VS	Bottle 0.5 L	[75]
Water hyacinth	Silfuric acid 5% *v/v*; 60 min residence time	58 mL/g VS	64 mL/g VS	Flask 1 L	[129]
Wheat straw	NMMO 85% 120 °C; 3 h	274 mL/g VS	304 mL/g VS	Serum bottles 0.125 L	[130]
	Ethanol 50% 180 °C; 1 h	274 mL/g VS	316 mL/g VS	Serum bottles 0.125 L	[130]
	Sodium hydroxide 1.6% w/w 30C; 24 h	274 mL/g VS	315 mL/g VS	Serum bottles 0.125 L	[130]
Winert waste	Citric Acid 0.5 mmol/g VS substrate	159.1 mL/g VS	163.7 mL/g VS	Flask 0.25 L	[78]
	Hydrogen 0.5 mmol/g VS substrate peroxide	171.3 mL/g VS	190.2 mL/g VS	Flask 0.25 L	[78]
	Ethanol 0.5 mmol/g VS substrate	190.7 mL/g VS	239.5 mL/g VS	Flask 0.25 L	[78]

3.3. Physicochemical Pretreatments

Physicochemical methods are used to solubilize lignocellulosic components of the structure based on temperature and moisture content and to make the lignocellulosic material easily exposed for hydrolysis step, and avoiding the formation of inhibitors. Although these methods are more complicated to implement, their significant effect on the pretreatment of lignocellulose feedstock promises high yield in the subsequent bioprocesses [131]. In general, physical pre-treatment requires greater energy expenditure, making it an expensive process and consequently not very profitable on an industrial scale. Therefore, this review focusses on process optimization and the conditions to reduce energy-time-costs of this process.

3.3.1. Steam Explosion

Steam explosion consists of exposure to hot steam for few minutes, followed by an explosive decompression of the biomass, which is effective in breakage of the fibrous rigid structure of straw and woody biomass [132,133]. To facilitate auto hydrolysis reactions, the biomass is treated with saturated steam at a temperature of 160–260 °C and at a corresponding pressure of 0.69–4.83 MPa for several seconds up to few min [59,133]. The difference between pretreatment with steam and the explosion with steam, is the rapid depressurization and cooling of biomass at the end of the steam explosion, which causes hydrolysis of hemicellulose into water soluble oligomers or to individual sugars [134]. Nges et al. [135] used steam explosion pretreatment on *Miscanthus lutarioriparius* under conditions of 0.3 M NaOH and at 0.5 mm particle size, and reported 57% increase in methane yield in comparison with the untreated substrate. Li et al. [136] evaluated steam explosion pretreatment on *Miscanthus lutarioriparius* under five different conditions: 0.5 MPa 153 °C 5 min, 1.0 MPa 180 °C 5 min, 1.5 MPa 198 °C 3 min, 1.5 MPa 198 °C 5 min, and 1.5 MPa 198 °C 10 min to improve the anaerobic biodegradability and reported 5.9%, 19.9%, 51.3%, 49.7%, and 49.8% increase in biochemical methane potential respectively.

3.3.2. Hydrothermal

Of late, hydrothermal pretreatment has gained importance, as it is efficient in penetration of the biomass, cellulose hydration, and removal of hemicellulose and part of lignin. The major advantages are that there is no requirement of chemicals and corrosion-resistant material for the reactor [137]. Typically, it can remove most of hemicellulose and part of lignin in biomass by degrading them into soluble fractions and loosen the recalcitrant structure as well [138]. The temperature is the most important factor influencing the pretreatment effect [139]. Typically, hydrothermal pretreatment is conducted in range of 90–260 °C [130]. Phuttaro et al. [140] suggested hydrothermal pretreatment at 175 °C for 15 min of reaction time on Napier grass increased the methane yield by 25% than the untreated. They also reported the formation of inhibitors like 5-hydroxymethylfural and furfural at 200 °C, which significantly inhibited methanogenesis. Similarly, Rajput et al. [141] tested 120, 140, 160, and 180 °C for pretreatment of wheat straw and established that the optimal temperature for hydrothermal pretreatment of wheat straw was 180 °C, which showed a 53% increase in the methane yield in comparison with untreated. Hashemi et al. [142] pretreated hydrothermally Safflower straw at 120, 150, and 180 °C for 1, 2, and 5 h and the highest biomethane yield was obtained at the least severe pretreatment conditions (120 °C for 1 h), which showed 98.3% improvement in comparison to the untreated straw. On the other hand, Luo et al. [143] examined hydrothermal pretreatment of rice straw at different temperatures from 90 to 130 °C. Both 100 °C and 130 °C pretreatments presented similar methane yield of 127.6 and 124.6 mL/g VS, respectively, which were 22.90% and 19.83% higher than untreated rice straw. Antwi et al. [144] evaluated hydrothermal pretreatment on cocoa pods residues at temperatures in the range of 155–220°C and at reaction time between 5 and 15 min and reported an optimum methane yield of 526.38 mL/g VS at 150 °C for 15 min and opined that higher severity conditions resulted in lower biogas yield.

3.3.3. Ammonia Fiber Explosion (AFEX)

AFEX is a physicochemical pretreatment under varying water loading, ammonia loading, reaction temperature, and residence time. In this pretreatment, liquid ammonia and the steam explosion process are applied. The advantage of this process is that it does not require small particle size for efficiency and further, inhibitors are not formed during the process. However, the limitation is that less efficiency has been reported in the pretreatment of high lignin containing biomass [61].

To reduce costs and protect the environment, ammonia is recycled after pretreatment. Wang et al. [145] achieved 99.3% of efficiency in ammonia removal after pretreatment of wheat straw at 0.70% ammonia and at 105 °C. They observed the methane yield of 538.1 mL/g VS after pretreatment, which was 31.9% higher than untreated wheat straw. Hashemi et al. [146] carried out a detailed study using 10% *v/v* of aqueous ammonia pretreatment on sugarcane bagasse and at varying temperatures in the range of 50 to 70 °C, reaction time of 12 and 24 h and ethanol addition at different concentrations of 5, 25, and 50% *v/v*. They reported that the highest methane yield of 299.3 mL/g VS compared with 105.6 mL/g VS of untreated material was obtained under process conditions of 70 °C for 12 h with 10% ammonia and 50% ethanol.

The effect of physiochemical treatment on biogas yield is presented in Table 5. It can be observed that steam explosion pretreatment of *Miscanthus lutarioriparius* at 198 °C for 3 min recorded an increase in methane yield of 51%, but no differences were found by increasing the pretreatment time to 5 and 10 min [135] Luo et al. [143] employed hydrothermal pretreatment of the rice straw at 100 °C for 10 min and observed 222% increase in methane yield.

Table 5. Effect of different physicochemical pretreatments on methane yield from different lignocellulosic biomass.

Substrates	Pretreatment Conditions	Methane Yield (Untreated)	Methane Yield (After Treatment)	Reaction System	References
Cocoa pods residues	Hydrothermal 150 °C	196.3 mL/g VS	289.3 mL/g VS	Bottle 0.5 L	[144]
Harvested hay	Steam explosion 175 °C; 10 min	243 mL/g VS	281 mL/g VS	Digester 0.25 L	[134]
Miscanthus lutarioriparius	Steam explosion 0.5 MPa; 153 °C; 5 min	181.2 mL/g VS	192.3 mL/g VS	Reactor 0.5 L	[136]
	Steam explosion 1.0 MPa; 180 °C; 5 min	181.2 mL/g VS	217.9 mL/g VS	Reactor 0.5 L	[136]
	Steam explosion 1.5 MPa; 198 °C; 3 min	181.2 mL/g VS	274.1 mL/g VS	Reactor 0.5 L	[136]
	Steam explosion 1.5 MPa; 198 °C; 5 min	181.2 mL/g VS	272.0 mL/g VS	Reactor 0.5 L	[136]
	Steam explosion 1.5 MPa; 198 °C; 10 min	181.2 mL/g VS	272.2 mL/g VS	Reactor 0.5 L	[136]
Reed biomass	Steam explosion 160 °C; 5 min	188 mL/g VS	226 mL/g VS	Digester 0.25 L	[147]
	Steam explosion 200 °C; 15 min	188 mL/g VS	355 mL/g VS	Digester 0.25 L	[147]
Rice straw	Hydrothermal 100 °C; 10 min	92 mL/g VS	280 mL/g VS	Glass Bottle 1 L	[143]
Safflower straw	Hydrothermal 120 °C; 60 min	96.5 mL/g VS	191.4 mL/g VS	Glass bottle 0.118 L	[142]
Sugarcane Bagasse	Ammonia 10% + ethanol 50%; 70 °C; 24 h	105.6 mL/g VS	299.3 mL/g VS	Bioreactor 0.118 L	[146]
Wheat straw	Hydrothermal 120 °C; 60 min	388.9 mL/g VS	483.3 mL/g VS	Serum bottle 0.3 L	[141]
	Hydrothermal 140 °C; 60 min	388.9 mL/g VS	511.2 mL/g VS	Serum bottle 0.3 L	[141]
	Hydrothermal 160 °C; 60 min	388.9 mL/g VS	552.7 mL/g VS	Serum bottle 0.3 L	[141]
	Hydrothermal 180 °C; 60 min	388.9 mL/g VS	611.7 mL/g VS	Serum bottle 0.3 L	[141]
	Ammonia 0.7%; 55 °C	407.8 mL/g VS	491.7 mL/g VS	Reactor 0.25 L	[145]
	Ammonia 0.7%; 105 °C	407.8 mL/g VS	538.1 mL/g VS	Reactor 0.25 L	[145]

3.4. Biological Pretreatments

Biological pretreatment offers an alternative choice to replace chemical pretreatment; however, its very slow reaction rate is unattractive from commercial point of view [148]. Different microorganisms are used for lignocellulose pretreatment, such as; white, brown, and soft rot fungi. Although, brown rot fungi participates in lignin decomposition [149], white rot fungi are well known and most effective microorganism for delignification process [150]. In general, most of the fungi degrading lignocellulose secrete accessory enzymes like cellobiose dehydrogenase, aryl alcohol oxidase, glyoxyl oxidase, copper oxidase, and hydrolytic enzymes, which cause the simultaneous or selective degradation of cellulose and hemicellulose along with lignin. These enzymes provide free radicals and intermediates that help in lignin and polysaccharide degradation [151]. Schroyen et al. [152] evaluated the impact of biological pretreatment with different enzymes such as laccase, manganese peroxidase, and versatile peroxidase, at different incubation times—0, 6, and 24 h. Pretreatments did not yield high concentrations of phenolic compounds, inhibitors of methane production. The laccase enzyme showed 25% increase in biomethane production after 24 h of incubation, meanwhile, pretreatment with peroxidase enzymes increased biomethane production by 17%, but aided in reducing the incubation time from 24 to 6 h. Liu et al. [153] evaluated fungal pretreatment of two forest residues, viz., hazel and acacia branches and as well as two agricultural residues, viz., barley straw and sugarcane bagasse using *Ceriporiopsis subvermispora* in order to improve the methane production. They reported that manganese peroxidase and laccase produced by the fungus increased the methane yield by 100% in comparison to untreated hazel. Although fungal pretreatments of lignocellulosic biomass are environmentally and economically friendly, it is a relatively time-consuming process and Liu et al. [153] observed that a special bioreactor should be designed to create aerobic and aseptic conditions for fungal pretreatments. It is also reported that selection of substrate specific and as well as effective strain and culture conditions can reduce treatment time and carbohydrate losses [154]. Dollhofer et al. [155] reported that biogas production from lignocellulosic biomass can be accelerated after pretreatment with *Neocallimastix frontalis*, an anaerobic fungus isolated from rumen fluid of a cow and of a chamois.

The byproducts formed after biological pretreatments normally do not inhibit subsequent hydrolysis, since the pretreatment is carried out at moderate reaction conditions, and also has other environmental benefits like low inputs (energy, chemicals) and outputs (inhibitors and wastes) [156]. However, as mentioned earlier, its process time is long compared to the others [148,157]. As a result, several studies select different microbial populations for hydrolysis of their specific substrates [158]. Shah and Ullah [39] selected strains producing lignin peroxidases and laccases and decreased 48.2% of lignin from wheat straw within 7 days of batch incubation; and observed 407.1% increase of the methane yield in comparison with the untreated wheat straw.

Microbial pretreatment, when compared to enzymatic pretreatment, demonstrates much better outcome in anaerobic digestion process due to their higher functional diversity and tolerance to environmental factors—i.e., temperature and pH. Barua et al. [38] isolated bacterial strains from soil (*Bordetella muralis* VKVVG5), the gut of silverfish (*Citrobacter werkmanii* VKVVG4), and millipede (*Paenibacillus* sp. VKVVG1); and employed these bacterial strains to accelerate hydrolysis of water hyacinth. *Citrobacter werkmanii* VKVVG4 pretreatment of water hyacinth at a density of 109 CFU/mL for 4 days increased the solubilization by 33.3%. On the other hand, several others report the use of anaerobic lignocellulose degrading consortia. Kong et al. [159] reported that anaerobic consortium TC-5 resulted in 45.7% degradation of wheat straw and showed an increase of 22.2% and 36.6% of methane yield under mesophilic and thermophilic conditions, respectively. Rumen fluid was found to be effective in increasing the yield from rapeseed stems and leaves by 47% within 24 h of treatment time [160]. Yuan et al. [19] used a thermophilic microbial consortium MC1 to pretreat the cotton stalk. The results indicated that the concentrations of soluble chemical oxygen demand and volatile organic products increased significantly in the early stages of the pretreatment, reaching the maximum methane yield of 128 mL/g VS after 6 days of MC1 pretreatment. However, with 122 mL/g VS, the methane yield was not significantly different. Ali et al. [33] observed a 72.6% increased higher methane yield over control from saw dust treated with a microbial consortium LCDC isolated from rotten sawdust. They also reported a significant reduction of lignocellulose compounds within 10 days of pretreatment.

Enzymatic pretreatment by using oxidative and hydrolytic enzymes produced by bacteria and fungi such as exo-, endo- glucanases, cellobiase, xylanase, pectinase; ligninolytic enzymes such as laccase, manganese, and versatile peroxidases; as well as α-amylases and protease is advantageous in comparison to bacterial/fungal pretreatments [161]. The enzymatic pretreatment method is gaining more interest due to the relatively short reaction time and the low nutrition requirement for the enzymatic reactions. In addition, the enzymatic pretreatment does not require expensive equipment for the processing. However, the activity of the enzymes and the efficiency of their reactions depends on several factors including the composition of the lignocellulosic substrate, temperature, pH, incubation time, and bioreactor configuration [157]. Additionally, the high enzyme cost remains a challenge facing the economic feasibility of this pretreatment methods for improved biogas production at an industrial scale [157,161]. However, the efficiency is well documented by Brémond et al. [162], who reported an increase of the methane yield by 111% from mix of corn stover after 18 h of reaction time using a blend of cellulase enzymes.

The conditions and effect of biological pretreatment on methane yield from different lignocellulosic substrates is presented in Table 6. In general, biological pretreatment shows an increase up to 50% in comparison with untreated lignocellulosic biomass.

Table 6. Effect of different biological pretreatments on methane yield from different lignocellulosic biomass.

Substrates	Pretreatment Conditions	Methane Yield (Untreated)	Methane Yield (After Treatment)	Reaction System	References
Corn stover	Fungi, *Pleurotus eryngii*, 30 days	301.5 mL/g VS	360.4 mL/g VS	Bottle 0.25 L	[148]
	Fungi, *Pleurotus ostreatus*, 30 days	294.3 mL/g VS	300.6 mL/g VS	Bottle 0.25 L	[148]
	Fungi, *Trametes versicolor*, 30 days	362.2 mL/g VS	222.83 mL/g VS	Bottle 0.25 L	[148]
Rapeseed stems and leaves	Rumen fluid, 24 h	485.5 mL/g VS	507.9 mL/g VS	Reactor 1 L	[160]
Sawdust	Microbial consortium, 10 days	89.9 mL/g VS	155.2 mL/g VS	Digester 5 L	[33]
Water Hyacinth	Microbial, *Citrobacter werkmanii* VKVVG4 109 cfu/mL; 4 days	338 mL/g VS	373 mL/g VS	Bottle 1 L	[38]
Wheat straw	Fungus culture, 0.3 OD; 7 days	78.1 mL/g VS	396.1 mL/g VS	-	[118]
	Microbial consortium, TC-5, 12 days	229.8 mL/g VS	314 mL/g VS	Reactor 0.18 L	[159]
Willow sawdust	Fungal, *Leiotrametes menziesii*, 30 days	95.5 mL/g VS	62.4 mL/g VS	Serum bottle 0.16 L	[40]
	Fungal, *Abortiporus biennis*, 30 days	95.5 mL/g VS	136.7 mL/g VS	Serum bottle 0.16 L	[40]

3.5. Other Pretreatments

Several other pretreatments such as ozonolysis, microwave irradiation, and high voltage pulse discharge have also been reported recently to increase the methane yield from lignocellulosic biomass. Ozonolysis is one of the most promising chemical oxidative pretreatments for lignin degradation with minimal losses of hemicellulose and cellulose contents [163], but most of the reported literature is on its application related to pulp. Dománski et al. [164] reported a specific methane production of 291 LCH_4/kg VS using ozonolysis pretreatment of rye straw with an ozone dose of 100 g O_3/m^3 for 60 min.

Microwave irradiation is known as an efficient technique in disrupting complex structures such as sludge and thereby increases the accessibility and bioavailability to enhance methane production [165]. Zhao et al. [166] reported a methane yield of 221 mL/g of water hyacinth after a pretreatment of 500 W for 14.6 min, which was about 38.3% more than the substrate pretreated with hot water. However, after an economic analysis, they reported that it is not economically viable. Previously, Pellera and Gidarakos [167] observed that pretreatment of winery waste, cotton gin waste, olive pomace, and juice industry waste by microwave did not show an increase in methane production. However, they observed that structural changes in cotton gin waste and olive pomace, whereas increased solubilization in the case of winery and juice industry waste. Kainthola et al. [168] evaluated the microwave pretreatment at different temperatures and exposure time, the ranges were 130–230 °C and 2–5 min, respectively. They obtained a specific methane yield of 325.7 mL/g VS of microwave pretreated rice straw, which was equivalent to a total net energy gain of 2388.5 J/g VS. They further reported that high heating rate with less residence time reduced the formation of inhibitory compounds.

High voltage pulse discharge has been a promising strategy for pretreatment of wastes and can shorten the processing time, from 24 h to 30 min, and the operation is more convenient than acid and alkaline pretreatments in the sense that there is no need to adjust pH after pretreatment [169]. Zou et al. [169] showed that methane production was higher by 50% from on food wastes in comparison with acid and alkaline pretreatments. Baijuan et al. [170] reported a gas production of 9587 mL, 27% higher than the untreated hybrid *Pennisetum* by applying a high voltage plus discharge for 60 min.

3.6. Combined Pretreatments

An appropriate combined pretreatment would not only improve the digestibility of lignocellulosic materials at a relatively low operating cost but also maximize the utilization of the lignocellulosic components [171]. A number of studies have shown that a combination of two pretreatments such as biological pretreatment with chemical or physical methods is more effective when compared to pretreatments with a chemical or biological method alone [172]. Although fungal pretreatment can provide a cost-effective feasibility for methane production, the low rate of hydrolysis and the reaction time of the process have made this method commercially inapplicable, but in combination with other pretreatment methods like mechanical or chemical there is an improvement to overall process productivity [173]. A study conducted by Mustafa et al. [174] used a combination of milling and biological pretreatment of rice straw with *Pleurotus ostreatus*, resulted in 30.4% of lignin removal and reported a methane yield of 258 mL/g VS, which is equivalent to 165% increase in methane yield compared to untreated rice straw. Gomez-Tovar et al. [27] used mild acid pretreatment (2% of HCl) in combination with an enzymatic pretreatment, which solubilized 96.8% of hemicellulose, 77.2% of cellulose, and 42.2% of lignin from oat straw; additionally, this combination produced a variety of sugars that could serve as substrates in the biomethanation process.

In general, chemical, physicochemical, or biological pretreatment is followed after mechanical pretreatment to improve the efficiency [75]. However, the combination is not restricted to different types of pretreatment, it can be even the use of two different chemical treatments; i.e., pretreatment with dilute acid for hemicellulose hydrolysis and then a pretreatment with dilute alkali for lignin removal; or alkaline hydrogen peroxide pretreatment is the combination of alkali reagent plus oxidant reagent [60]. Chen et al. [175] suggested combining dilute acid treatment of lignocellulose with dilute alkali to eliminates the need for acid/ base neutralization of lignocellulose biomass, which could also reduce the inhibitory effect of degradation products. The most common pretreatment combinations are physicochemical and thermochemical [176]. Sharma et al. [177] compared dilute acid and alkaline pretreatment, both at temperature of 150 °C and reported the maximum yield of 38 g/L of reducing sugars at 0.8 M sodium hydroxide after 30 min of incubation time, compared to 31 g/L of reducing sugars using dilute acid.

In some cases, even a combination of three pretreatments can be efficient. Moset et al. [97] combined mechanical, thermal, and chemical pretreatments for wheat straw. The advantage of the combination of three pretreatments was that there is no need for vigorous mechanical pretreatment with regard to particle size, reduced the concentration of chemicals required and low temperature process is sufficient to achieve greater efficiency. It clearly suggests that combination of right pretreatment conditions help to improve the energy and the economic balance of pretreatment step.

Table 7 summarizes the pretreatments addressed in which the mechanism of action on lignocellulosic biomass is pointed out, as well as the percentage of improvement for each. It can be seen that mechanical pretreatment improves up to 83% methane yield and can be seen as a pretreatment to be carried out in the first instance, to subsequently combine with other pretreatment to achieve a greater improvement.

Table 7. A brief summary on the effect of pretreatment methods on methane yield from lignocellulosic biomass.

Pretreatment Method		Mechanism of Action	Methane Yield Increase (%) (Substrate)	References
Mechanical	Milling, chipping, grinding	Increases accessibility and surface area. Reduces the crystallinity of cellulose	Up to 83 (Wheat straw)	[73]
Chemical	Acid	Hydrolyze hemicellulose	Up to 74 (Corn straw)	[18]
	Alkaline	Lignin removal, decrease of polymerization degree, and crystallinity of cellulose	Up to 105 (Corn straw)	[18]
	Organosolv	Allow isolation of each lignocellulose component	Up to 400 (Softwood spruce, rice straw and triticale straw)	[107]
	Ionic liquids	Disrupt lignin and decrystallize the cellulose	Up to 97 (Water hyacinth)	[35]
	Alkaline hydrogen peroxide	Lignin removal and improves enzymatic digestibility	Up to 115 (Corn straw)	[18]
Physiochemical	Steam explosion	Hydrolyze hemicellulose and breakage the lignocellulos structure	Up to 88 (Reed biomass)	[147]
	Hydrothermal	Ligning and hemicellulose removal	Up to 204 (Rice straw)	[143]
	AFEX	Lignin removal, decrystallize cellulosa, and hydrolyze hemicellulose	Up to 183 (Sugarcane bagasse)	[146]
Biological	Fungal, microbial, enzymatic	Delignification process and lignin decomposition	Up to 400 (Wheat straw)	[39]

3.7. Limitations

It can be observed that the efficiency of pretreatment methods depends on the feedstock and its composition. It is a complex issue to determine the optimal pretreatment method(s) of lignocellulosic biomass. Regardless of the pretreatment method to be employed, the particle size reduction of lignocellulosic biomass is the most common primary pretreatment method employed. Though, in general, the mechanical pretreatment is significantly related to increase methane yield, one of the disadvantages of mechanical process is its inability to remove the lignin, a critical barrier with regard to bioavailability of carbohydrates for biomethanation. Neumann et al. [176] suggested that lignocellulosic biomass must be reduced to 1–2 mm to eliminate limitations during hydrolysis; however, size reduction is a very expensive operation that consumes about 33% of the total electricity demand for the whole process. Considering the high energy requirement of mechanical pretreatment and the rise of energy cost, it is not economically sustainable. Thus, reducing the energy requirements and increasing efficiency of grinding and milling of biomass would help to improve the economics of the whole process.

In the case of chemical pretreatment, cost of reagents, and operation such as the extra step for neutralization and the requirement for corrosion resistant reactors are the known limitations [178]. Additionally, the formation of inhibitory compounds is a point to be considered since it can inhibit or significantly reduce the conversion efficiency of hydrolyzates of lignocellulosic biomass to methane. Thus, achieving higher efficiency and lowering the formation of inhibitory compounds at by combining lower concentration of chemical reagents with other pretreatments could aid in cost reduction. In the case of alkali pretreatment, it is effective in solubilization of lignin and the presence of a small amount of residual alkali in the treated biomass aids in neutralizing the pH reduction during the acidogenesis phase of biomethanation process. Therefore, alkaline pretreatment is more compatible with subsequent anaerobic digestion when compared to acid pretreatment [179].

On the other hand, organic solvents permit the recovery of individual components of lignocellulosic biomass, which make an improvement in capital cost in a biorefinery concept. However, large amounts of downstream wastes and specialized equipment are limitations [102]. Ionic liquids are superior to organic solvents for use in a wide range of applications and in extreme conditions [100], however, the cost and the need for recycling need is a limitation.

In spite of the advantages, biological pretreatments also have several major drawbacks such as specific growth conditions, larger space, longer treatment time, and loss of carbohydrates [178]. Biological pretreatment presents low formation of inhibitory compounds and in general its inhibition effect in the subsequent step of anaerobic digestion is low, when compared with chemical and physiochemical pretreatments.

3.8. Modeling and Parameters Optimization

Anaerobic digestion is a multi-step process that can be affected by many process conditions such as pH, temperature, substrate composition, pretreatment method, and digestion time. The optimization of these parameters may ensure high yields of biogas production [180]. Serious consideration should be given to the type of the substrate selected for biogas production because it can affect the amount and quality of the biogas obtained [181].

Although, the lignocellulosic material preparation procedures and pretreatment processes prior to anaerobic digestion are widely investigated, it is not possible to precisely predict the effectiveness of a pretreatment method on certain lignocellulosic substrate as many parameters control the process. However, the use of mathematical models can help to know the interrelationship between the parameters involved and predict the phenomena occurring in the process [182]. The kinetic of the anaerobic digestion provides information about the effect of the inhibitory compounds generated by the pretreatment method on the biodegradability. There are several models of the kinetic analysis of biogas production process being the Gompertz model, a model well known to study the kinetic behavior considering the inhibition parameter in the anaerobic digestion process. The Gompertz equation is used to estimate the biogas yield potential, duration of the lag phase, and maximum biogas production rate. However, when the hydrolysis reaction is the rate-limiting step of the overall process, as in the anaerobic degradation of some lignocellulosic substrates, the first order model is commonly used to estimate the extent of the reaction, and the hydrolysis constant [183].

Dahunsi [64] assessed the possible effect of mechanical pretreatment on the kinetics of the degradability of elephant grass, wild Mexican sunflower, and siam weed during the digestion process using a modified Gompertz equation and reported that the pretreatment reduced the lag period and increased the methane yield by around 22%. They positively correlated the chemical composition of lignocellulosic substrates to their methane potentials by employing single and multiple linear regressions models. Mancini et al. [130] also employed the first order model and a model based on modified Gompertz equation to evaluate three different chemical pretreatments and their effect on methane yield from wheat straw. They reported that pretreatment using organic solvent N-methylmorpholine N-oxide (NMMO) showed a significant biomethane production than untreated wheat straw and observed that it could be due to the increase in available surface area due to the pretreatment. Bolado-Rodríguez et al. [183] evaluated different physico-chemical pretreatments on the production of biogas from sugarcane bagasse and wheat straw, and reported that a combination of first order and a modified Gompertz equation provided valid information on the availability of substrate, hydrolysis, and effect of inhibitory compounds. Łukajtis et al. [184] proposed a quadratic equation to optimize process parameters such as concentration of chemicals, temperature, reaction time, and particle size in alkaline pretreatment of *Salix viminalis* L, based on the results obtained.

Although the biomass type is one of the most important factors in biogas production, it is important to identify the optimal levels of operational parameters of the reactor to increase the methane yield [185]. In this regard, a widely accepted model of the anaerobic process is the ADM1. It describes the process by analyzing 28 parameters and using 19 ordinary differential equations. Unfortunately, it may consume lots of time, and sometimes not possible due to economic and other factors. Rieke et al. [186] developed an alternate model based on four parameters, instead of the 28 of the ADM1 model and only two additive differential equations of first order and reported that the difficulties associated with ADM1 model in switching to different substrate due to the large number of parameters and equations could be overcome in this new model, and further reduced the simulation time needed.

3.9. Techno-Economic and Environmental Aspects

Techno-economic feasibility is one of the most important parameters for the commercial viability of biofuels production from lignocellulosic substrates [187]. It is expected that biomass pretreatment cost represents between 19% and 22% in a biofuel production chain [118]. Several authors have established techno-economic analysis of different pretreatments to elucidate the profitable conditions under the concept of a biorefinery. Vasco-Correa and Shah [188] compared the techno-economic feasibility of commercial scale fungal pretreatment of four feedstocks—perennial grasses, corn stover, agricultural residues other than corn stover, and hardwood—and observed that the sugar production costs were 1.7, 1.6, 2.0, and 2.8 $/kg for perennial grasses, corn stover, agricultural residues other than corn stover, and hardwood, respectively. Regarding to chemical pretreatment, Baral and Shah [118] evaluated the production of fermentable sugars from corn stover, switchgrass, and poplar to access the process improvement that would enable the commercialization of ionic liquids pretreatment. Unit sugar production costs ($/kg) for ionic liquids pretreatment of corn stover, switchgrass, and poplar were estimated to be 2.7, 3.2, and 3.0, respectively. In addition, other study assessed the economic and technical benefits of steam explosion, dilute sulfuric acid, ammonia fiber explosion, and biological (enzymatic hydrolysis) pretreatments in corn stover [189]. The results showed an estimated sugar production costs of ($/kg) 0.43, 0.42, 0.65, and 1.41 for steam explosion, dilute sulfuric acid, ammonia fiber explosion, and biological (enzymatic hydrolysis) pretreatments, respectively. On the other hand, Brandt et al. [190] provided economic analysis for converting softwood biomass into sugars using a fully mechanical pretreatment, using a three-stage milling process. The concentrated sugar syrup production cost was estimated in $0.496/kg of sugar. However, the environmental impact and sustainability of the pretreatment processes are not widely reported. Prasad et al. [191] carried out life cycle assessment analysis on four different pretreatments—viz., liquid hot water, steam explosion, dilute acid, and organosolv—using GaBi software and reported that dilute acid pretreatment performed the worst in all impact categories. Although, other three pretreatments performed similarly, liquid hot water significantly reduced the CO_2 emissions and the total sugars production was two times greater than organosolv and steam explosion pretreatments. However, Safarian and Unnthorsson [192] reported that steam explosion pretreatment is the most promising and beneficial technology in terms of economical, environmental and energy points of views and was followed by hot water method for production of bioethanol from paper and timber wastes. Though in terms of technical performance, dilute acid pretreatment was ranked highly, it is not preferable due to low energy performance, high production cost and high emission of greenhouse gases. Similarly, dilute alkali treatment also ranked highly on the technical side of bioethanol production, is not preferable due to high emission of greenhouse gases and other environmental impacts. Recently, Soam et al. [193] reported that an addition of low dosage alkali pretreatment to enzymatic pretreatment reduced the enzyme dosage by 23–39%, but observed a negative environmental impact. This clearly suggest that further studies are needed to select right combination of pretreatments and operation conditions of these combinations need to be standardized in order to reduce environmental impacts and costs involved in the pretreatment process. Recently, Rajendran and Murthy [194] also observed that it is important to do a realistic techno-economic and environmental impact analysis of anaerobic digestion processes as there is a need to include process mapping, create a database, and identify profitability indicators and geographic locations of the biogas plants.

4. Conclusions

Lignocellulose substrates are the main feedstocks for biogas production and this anaerobic digestion process help in sustainable management of wastes and as well as is one of the main sources of renewable bioenergy. However, the structure of lignocellulosic residues still present technological barriers due to their limited bioavailability and pretreatment of this recalcitrant substrates is a key to improve the performance of anaerobic digestion technology. However, as can be seen, the major limitations are the cost of energy, operational cost, and formation of inhibitory compounds, which significantly affect the downstream bioprocess of generating biogas or other value-added products. Though, combination of different pretreatments offers possible solution, it needs to be specific for the type of substrates and the type of downstream bioprocess to produce bioenergy and other products. To achieve a techno-economic viability of the use of lignocellulose substrates, the idea of integrated biorefineries in which two or more bioproducts are generated in the same platform could be a promising concept.

Author Contributions: Conceptualization, N.B. and J.U.H-B.; Methodology, J.U.H.-B., F.H.-T. and N.B.; Validation, J.U.H.-B., I.O.H.-D.L. and N.B.; formal analyses, J.U.H.-B., I.O.H.-D.L. and M.M.C.-S., Investigation, J.U.H.-B., I.O.H.-D.L., M.M.C.-S. and A.S.-L.; Supervision, N.B.; Writing—original draft preparation, J.U.H.-B.; Writing—review and editing, J.U.H.-B., I.O.H.-D.L., F.H.-T. and N.B.; supervision, N.B.; project administration, N.B.; funding acquisition, N.B.

Funding: This research was funded by Programa Fortalecimiento de la Investigación para el Desarrollo de la Educación y la sociedad (PROFIDES) convocatoria 2017, Secretary of Public Education, Government of Mexico for UACOAH-CA-92 Ciencias Biológicas Aplicadas.

Conflicts of Interest: The authors declare no conflict of interest.

References

1. Correa, D.F.; Beyer, H.L.; Fargione, J.E.; Hill, J.D.; Possingham, H.P.; Thomas-Hall, S.R.; Schenk, P.M. Towards the Implementation of Sustainable Biofuel Production Systems. *Renew. Sustain. Energy Rev.* **2019**, *107*, 250–263. [CrossRef]
2. Andrée, B.P.J.; Diogo, V.; Koomen, E. Efficiency of Second-Generation Biofuel Crop Subsidy Schemes: Spatial Heterogeneity and Policy Design. *Renew. Sustain. Energy Rev.* **2017**, *67*, 848–862. [CrossRef]
3. Perin, G.; Jones, P.R. Economic Feasibility and Long-Term Sustainability Criteria on the Path to Enable a Transition from Fossil Fuels to Biofuels. *Curr. Opin. Biotechnol.* **2019**, *57*, 175–182. [CrossRef]
4. Li, P.; Sakuragi, K.; Makino, H. Extraction Techniques in Sustainable Biofuel Production: A Concise Review. *Fuel Process. Technol.* **2019**, *193*, 295–303. [CrossRef]
5. Alalwan, H.A.; Alminshid, A.H.; Aljaafari, H.A.S. Promising Evolution of Biofuel Generations. Subject Review. *Renew. Energy Focus* **2019**, *28*, 127–139. [CrossRef]
6. Scaramuzzino, C.; Garegnani, G.; Zambelli, P. Integrated Approach for the Identification of Spatial Patterns Related to Renewable Energy Potential in European Territories. *Renew. Sustain. Energy Rev.* **2019**, *101*, 1–13. [CrossRef]

7. OECD. Renewable Energy (Indicator). Available online: https://www.oecd-ilibrary.org/energy/renewable-energy/indicator/english_aac7c3f1-en (accessed on 28 August 2019). [CrossRef]
8. Azevedo, S.G.; Sequeira, T.; Santos, M.; Mendes, L. Biomass-Related Sustainability: A Review of the Literature and Interpretive Structural Modeling. *Energy* **2019**, *171*, 1107–1125. [CrossRef]
9. Ibarra-Gonzalez, P.; Rong, B.G. A Review of the Current State of Biofuels Production from Lignocellulosic Biomass Using Thermochemical Conversion Routes. *Chin. J. Chem. Eng.* **2018**. [CrossRef]
10. SIAP. Servicio de Información Agroalimentaria y Pesquera. Available online: https://nube.siap.gob.mx/cierreagricola/ (accessed on 8 August 2019).
11. Delsart, C. Plant Cell Wall: Description, Role in Transport, and Effect of Electroporation. In *Handbook of Electroporation*; Miklavčič, D., Ed.; Springer International Publishing AG: Cham, Switzerland, 2017; pp. 489–509.
12. Jiang, B.; Zhang, Y.; Guo, T.; Zhao, H.; Jin, Y. Structural Characterization of Lignin and Lignin-Carbohydrate Complex (LCC) from Ginkgo Shells (*Ginkgo biloba* L.) by Comprehensive NMR Spectroscopy. *Polymers* **2018**, *10*, 736. [CrossRef]
13. Chen, J.-H.; Xu, J.K.; Huang, P.-L.; Sun, R.-C. Effect of Alkaline Pretreatment on the Preparation of Regenerated Lignocellulose Fibers from Bamboo Stem. *Cellulose* **2016**, *23*, 2727–2739. [CrossRef]
14. Luo, M.; Tian, D.; Shen, F.; Hu, J.; Zhang, Y.; Yang, G.; Zeng, Y.; Deng, S.; Hu, Y. A Comparative Investigation of H2O2-Involved Pretreatments on Lignocellulosic Biomass for Enzymatic Hydrolysis. *Biomass Convers. Biorefinery* **2019**, *9*, 321–331. [CrossRef]
15. Pérez-Rodríguez, N.; García-Bernet, D.; Domínguez, J.M. Extrusion and Enzymatic Hydrolysis as Pretreatments on Corn Cob for Biogas Production. *Renew. Energy* **2017**, *107*, 597–603. [CrossRef]
16. Dong, C.; Chen, J.; Guan, R.; Li, X.; Xin, Y. Dual-Frequency Ultrasound Combined with Alkali Pretreatment of Corn Stalk for Enhanced Biogas Production. *Renew. Energy* **2018**, *127*, 444–451. [CrossRef]
17. Cher, Y.; Stevens, M.A.; Zhu, Y.; Holmes, J.; Xu, H. Understanding the Alkaline Pretreatments Parameters of Corn Stover Enzymatic Saccarification. *Biotechnol. Biofuels* **2013**, *6*, 8. [CrossRef]
18. Song, Z.; Yang, G.; Liu, X.; Yan, Z.; Yuan, Y.; Liao, Y. Comparison of Seven Chemical Pretreatments of Corn Straw for Improving Methane Yield by Anaerobic Digestion. *PLoS ONE* **2014**, *9*, e93801. [CrossRef]
19. Yuan, X.; Wen, B.; Cui, Z.; Ma, L.; Zhou, D.; Kuang, M.; Yang, W. Enhancing Anaerobic Digestion of Cotton Stalk by Pretreatment with a Microbial Consortium (MC1). *Bioresour. Technol.* **2016**, *207*, 293–301. [CrossRef]
20. Xu, J.; Liu, B.; Hou, H.; Hu, J. Pretreatment of Eucalyptus with Recycled Ionic Liquids for Low-Cost Biorefinery. *Bioresour. Technol.* **2017**, *234*, 406–414. [CrossRef]
21. Zawawi, A.Z.; Gaik, L.P.; Sebran, N.H.; Othman, J.; Hussain, A.S. An Optimisation Study on Biomass Delignification Process Using Alkaline Wash. *Biomass Convers. Biorefinery* **2018**, *8*, 59–68. [CrossRef]
22. Jiang, D.; Ge, X.; Zhang, Q.; Li, Y. Comparison of Liquid Hot Water and Alkaline Pretreatments of Giant Reed for Improved Enzymatic Digestibility and Biogas Energy Production. *Bioresour. Technol.* **2016**, *216*, 60–68. [CrossRef]
23. Merino, O.; Almazán, V.; Martínez-Palou, R.; Aburto, J. Screening of Ionic Liquids for Pretreatment of Taiwan Grass in Q-Tube Minireactors for Improving Bioethanol Production. *Waste and Biomass Valorization* **2017**, *8*, 733–742. [CrossRef]
24. Khatri, S.; Wu, S.; Kizito, S.; Zhang, W.; Li, J.; Dong, R. Synergistic Effect of Alkaline Pretreatment and Fe Dosing on Batch Anaerobic Digestion of Maize Straw. *Appl. Energy* **2015**, *158*, 55–64. [CrossRef]
25. Tsapekos, P.; Kougias, P.G.; Angelidaki, I. Mechanical Pretreatment for Increased Biogas Production from Lignocellulosic Biomass; Predicting the Methane Yield from Structural Plant Components. *Waste Manag.* **2018**, *78*, 903–910. [CrossRef]
26. Thomas, H.L.; Arnoult, S.; Brancourt-Hulmel, M.; Carrère, H. Methane Production Variability According to Miscanthus Genotype and Alkaline Pretreatments at High Solid Content. *Bioenergy Res.* **2019**, *12*, 325–337. [CrossRef]
27. Gomez-Tovar, F.; Celis, L.B.; Razo-Flores, E.; Alatriste-Mondragón, F. Chemical and Enzymatic Sequential Pretreatment of Oat Straw for Methane Production. *Bioresour. Technol.* **2012**, *116*, 372–378. [CrossRef]
28. Charnnok, B.; Sakdaronnarong, C.; Sinbuathong, N. Hydrothermal Pretreatment with Sulfonated Bentonite Catalyst Enhances Potassium Removal and Bioconversion of Oil Palm Empty Fruit Bunch to Sugar and Biohydrogen. *Biomass Convers. Biorefinery* **2019**, *9*, 389–399. [CrossRef]

29. Salehian, P.; Karimi, K.; Zilouei, H.; Jeihanipour, A. Improvement of Biogas Production from Pine Wood by Alkali Pretreatment. *Fuel* **2013**, *106*, 484–489. [CrossRef]
30. Cabrera, E.; Muñoz, M.J.; Martín, R.; Caro, I.; Curbelo, C.; Díaz, A.B. Alkaline and Alkaline Peroxide Pretreatments at Mild Temperature to Enhance Enzymatic Hydrolysis of Rice Hulls and Straw. *Bioresour. Technol.* **2014**, *167*, 1–7. [CrossRef]
31. Mustafa, A.M.; Poulsen, T.G.; Sheng, K. Fungal Pretreatment of Rice Straw with Pleurotus Ostreatus and Trichoderma Reesei to Enhance Methane Production under Solid-State Anaerobic Digestion. *Appl. Energy* **2016**, *180*, 661–671. [CrossRef]
32. Smuga-Kogut, M.; Zgórska, K.; Szymanowska-Powałowska, D. Influence of the Crystalline Structure of Cellulose on the Production of Ethanol from Lignocellulose Biomass. *Int. Agrophys.* **2016**, *30*, 83–88. [CrossRef]
33. Ali, S.S.; Abomohra, A.E.-F.; Sun, J. Effective Bio-Pretreatment of Sawdust Waste with a Novel Microbial Consortium for Enhanced Biomethanation Biofuels Institute, School of the Environment and Safety Engineering, Jiangsu. *Bioresour. Technol.* **2017**, *238*, 425–432. [CrossRef]
34. Hernández-Beltrán, J.U.; Hernández-Escoto, H. Enzymatic Hydrolysis of Biomass at High-Solids Loadings through Fed-Batch Operation. *Biomass Bioenergy* **2018**, *119*, 191–197. [CrossRef]
35. Gao, J.; Chen, L.; Yuan, K.; Huang, H.; Yan, Z. Ionic Liquid Pretreatment to Enhance the Anaerobic Digestion of Lignocellulosic Biomass. *Bioresour. Technol.* **2013**, *150*, 352–358. [CrossRef]
36. Nosratpour, M.J.; Karimi, K.; Sadeghi, M. Improvement of Ethanol and Biogas Production from Sugarcane Bagasse Using Sodium Alkaline Pretreatments. *J. Environ. Manag.* **2018**, *226*, 329–339. [CrossRef]
37. Monlau, F.; Barakat, A.; Steyer, J.P.; Carrere, H. Comparison of Seven Types of Thermo-Chemical Pretreatments on the Structural Features and Anaerobic Digestion of Sunflower Stalks. *Bioresour. Technol.* **2012**, *120*, 241–247. [CrossRef]
38. Barua, V.B.; Goud, V.V.; Kalamdhad, A.S. Microbial Pretreatment of Water Hyacinth for Enhanced Hydrolysis Followed by Biogas Production. *Renew. Energy* **2018**, *126*, 21–29. [CrossRef]
39. Shah, T.A.; Ullah, R. Pretreatment of Wheat Straw with Ligninolytic Fungi for Increased Biogas Productivity. *Int. J. Environ. Sci. Technol.* **2019**, 1–12. [CrossRef]
40. Alexandropoulou, M.; Antonopoulou, G.; Fragkou, E.; Ntaikou, I.; Lyberatos, G. Fungal Pretreatment of Willow Sawdust and Its Combination with Alkaline Treatment for Enhancing Biogas Production. *J. Environ. Manag.* **2017**, *203*, 704–713. [CrossRef]
41. Marriott, P.E.; Gómez, L.D.; McQueen-Mason, S.J. Unlocking the Potential of Lignocellulosic Biomass through Plant Science. *New Phytol.* **2016**, *209*, 1366–1381. [CrossRef]
42. Chundawat, S.P.S.; Bellesia, G.; Uppugundla, N.; Da Costa Sousa, L.; Gao, D.; Cheh, A.M.; Agarwal, U.P.; Bianchetti, C.M.; Phillips, G.N.; Langan, P.; et al. Restructuring the Crystalline Cellulose Hydrogen Bond Network Enhances Its Depolymerization Rate. *J. Am. Chem. Soc.* **2011**, *133*, 11163–11174. [CrossRef]
43. Schneider, R.; Persson, S. Another Brick in the Wall. *Science* **2015**, *350*, 156–158. [CrossRef]
44. Hachem, K.; Faugeron, C.; Kaid-Harche, M.; Gloaguen, V. Structural Investigation of Cell Wall Xylan Polysaccharides from the Leaves of Algerian *Argania spinosa*. *Molecules* **2016**, *21*, 1587. [CrossRef]
45. Chen, H. Chemical Composition and Structure of Natural Lignocellulose. In *Biotechnology of Lignocellulose: Theory and Practice*; Chemical Industry Press: Beijing, China; Springer Science+Business Media: Dordrecht, The Netherlands, 2014; pp. 1–511. [CrossRef]
46. Vanholme, R.; Demedts, B.; Morreel, K.; Ralph, J.; Boerjan, W. Lignin Biosynthesis and Structure. *Plant Physiol.* **2010**, *153*, 895–905. [CrossRef]
47. Uzal, E.N.; Gómez Ros, L.V.; Pomar, F.; Bernal, M.A.; Paradela, A.; Albar, J.P.; Ros Barceló, A. The Presence of Sinapyl Lignin in *Ginkgo biloba* Cell Cultures Changes Our Views of the Evolution of Lignin Biosynthesis. *Physiol. Plant.* **2009**, *135*, 196–213. [CrossRef]
48. Zhao, Q.; Wang, H.; Yin, Y.; Xu, Y.; Chen, F.; Dixon, R.A. Syringyl Lignin Biosynthesis Is Directly Regulated by a Secondary Cell Wall Master Switch. *Proc. Natl. Acad. Sci. USA* **2010**, *107*, 14496–14501. [CrossRef]
49. Bekchanov, M.; Mondal, M.A.H.; De Alwis, A.; Mirzabaev, A. Why Adoption Is Slow despite Promising Potential of Biogas Technology for Improving Energy Security and Mitigating Climate Change in Sri Lanka? *Renew. Sustain. Energy Rev.* **2019**, *105*, 378–390. [CrossRef]

50. Montañez-Hernández, L.E.; Hernández-De Lira, I.O.; Rafael-Galindo, G.; Madariaga, M.L.F.; Balagurusamy, N. Sustainable Production of Biogas from Renewable Sources: Global Overview, Scale Up Opportunities and Potential Market Trends. In *Sustainable Biotechnology-Enzymatic Resources of Renewable Energy*; Singh, O.V., Chandel, A.K., Eds.; Springer International Publishing AG: Cham, Switzerland, 2018; pp. 325–354. [CrossRef]
51. Curto, D.; Martín, M. Renewable Based Biogas Upgrading. *J. Clean. Prod.* **2019**, *224*, 50–59. [CrossRef]
52. Parsaee, M.; Kiani, M.K.D.; Karimi, K. A Review of Biogas Production from Sugarcane Vinasse. *Biomass Bioenergy* **2019**, *122*, 117–125. [CrossRef]
53. Barbera, E.; Menegon, S.; Banzato, D.; D'Alpaos, C.; Bertucco, A. From Biogas to Biomethane: A Process Simulation-Based Techno-Economic Comparison of Different Upgrading Technologies in the Italian Context. *Renew. Energy* **2019**, *135*, 663–673. [CrossRef]
54. Paul, S.; Dutta, A. Challenges and Opportunities of Lignocellulosic Biomass for Anaerobic Digestion. *Resour. Conserv. Recycl.* **2018**, *130*, 164–174. [CrossRef]
55. Achinas, S.; Achinas, V.; Euverink, G.J.W. A Technological Overview of Biogas Production from Biowaste. *Engineering* **2017**, *3*, 299–307. [CrossRef]
56. Rajendran, K.; Drielak, E.; Varma, V.S.; Muthusamy, S.; Kumar, G. Updates on the Pretreatment of Lignocellulosic Feedstocks for Bioenergy Production—A Review. *Biomass Convers. Biorefinery* **2018**, *8*, 471–483. [CrossRef]
57. Karimi, K.; Taherzadeh, M.J. A Critical Review on Analysis in Pretreatment of Lignocelluloses: Degree of Polymerization, Adsorption/Desorption, and Accessibility. *Bioresour. Technol.* **2016**, *203*, 348–356. [CrossRef]
58. Mupondwa, E.; Li, X.; Tabil, L.; Sokhansanj, S.; Adapa, P. Status of Canada's Lignocellulosic Ethanol: Part I: Pretreatment Technologies. *Renew. Sustain. Energy Rev.* **2017**, *72*, 178–190. [CrossRef]
59. Bhutto, A.W.; Qureshi, K.; Harijan, K.; Abro, R.; Abbas, T.; Bazmi, A.A.; Karim, S.; Yu, G. Insight into Progress in Pre-Treatment of Lignocellulosic Biomass. *Energy* **2017**, *122*, 724–745. [CrossRef]
60. Lindner, J.; Zielonka, S.; Oechsner, H.; Lemmer, A. Effects of Mechanical Treatment of Digestate after Anaerobic Digestion on the Degree of Degradation. *Bioresour. Technol.* **2015**, *178*, 194–200. [CrossRef]
61. Kumari, D.; Singh, R. Pretreatment of Lignocellulosic Wastes for Biofuel Production: A Critical Review. *Renew. Sustain. Energy Rev.* **2018**, *90*, 877–891. [CrossRef]
62. Veluchamy, C.; Kalamdhad, A.S. Influence of Pretreatment Techniques on Anaerobic Digestion of Pulp and Paper Mill Sludge: A Review. *Bioresour. Technol.* **2017**, *245*, 1206–1219. [CrossRef]
63. Rodriguez, C.; Alaswad, A.; El-Hassan, Z.; Olabi, A.G. Mechanical Pretreatment of Waste Paper for Biogas Production. *Waste Manag.* **2017**, *68*, 157–164. [CrossRef]
64. Dahunsi, S. Mechanical Pretreatment of Lignocelluloses for Enhanced Biogas Production: Methane Yield Prediction from Biomass Structural Components. *Bioresour. Technol.* **2019**, *280*, 18–26. [CrossRef]
65. Dumas, C.; Damasceno, G.S.G.; Abdellatif, B.; Carrère, H.; Steyer, J.P.; Rouau, X. Effects of Grinding Processes on Anaerobic Digestion of Wheat Straw. *Ind. Crops Prod.* **2015**, *74*, 450–456. [CrossRef]
66. Sharma, S.K.; Mishra, I.M.; Sharma, M.P.; Saini, J.S. Effect of Particle Size on Biogas Generation from Biomass Residues. *Biomass* **1988**, *17*, 251–263. [CrossRef]
67. De la Rubia, M.A.; Fernández-Cegrí, V.; Raposo, F.; Borja, R. Influence of Particle Size and Chemical Composition on the Performance and Kinetics of Anaerobic Digestion Process of Sunflower Oil Cake in Batch Mode. *Biochem. Eng. J.* **2011**, *58–59*, 162–167. [CrossRef]
68. Chandra, R.; Takeuchi, H.; Hasegawa, T.; Vijay, V.K. Experimental Evaluation of Substrate's Particle Size of Wheat and Rice Straw Biomass on Methane Production Yield. *Agric. Eng. Int. CIGR J.* **2015**, *17*, 93–104.
69. Tsapekos, P.; Kougias, P.G.; Angelidaki, I. Biogas Production from Ensiled Meadow Grass; Effect of Mechanical Pretreatments and Rapid Determination of Substrate Biodegradability via Physicochemical Methods. *Bioresour. Technol.* **2015**, *182*, 329–335. [CrossRef]
70. Herrmann, C.; Heiermann, M.; Idler, C.; Prochnow, A. Particle Size Reduction during Harvesting of Crop Feedstock for Biogas Production I: Effects on Ensiling Process and Methane Yields. *Bioenergy Res.* **2012**, *5*, 926–936. [CrossRef]
71. Krause, M.J.; Chickering, G.W.; Townsend, T.G.; Pullammanappallil, P. Effects of Temperature and Particle Size on the Biochemical Methane Potential of Municipal Solid Waste Components. *Waste Manag.* **2018**, *71*, 25–30. [CrossRef]

72. Tsapekos, P.; Kougias, P.G.; Egelund, H.; Larsen, U.; Pedersen, J.; Trénel, P.; Angelidaki, I. Mechanical Pretreatment at Harvesting Increases the Bioenergy Output from Marginal Land Grasses. *Renew. Energy* **2017**, *111*, 914–921. [CrossRef]
73. Menardo, S.; Airoldi, G.; Balsari, P. The Effect of Particle Size and Thermal Pre-Treatment on the Methane Yield of Four Agricultural by-Products. *Bioresour. Technol.* **2012**, *104*, 708–714. [CrossRef]
74. Dell'Omo, P.; La Froscia, S. Enhancing Anaerobic Digestion of Wheat Straw through Multistage Milling. *Model. Meas. Control C* **2018**, *79*, 127–132. [CrossRef]
75. Frigon, J.; Mehta, P.; Guiot, S.R. Impact of Mechanical, Chemical and Enzymatic Pre-Treatments on the Methane Yield from the Anaerobic Digestion of Switchgrass. *Biomass Bioenergy* **2012**, *36*, 1–11. [CrossRef]
76. Nalinga, Y.; Legonda, I. The Effect of Particles Size on Biogas. *Int. J. Innov. Res. Technol. Sci.* **2016**, *4*, 9–13.
77. Gupta, A.; Verma, J.P. Sustainable Bio-Ethanol Production from Agro-Residues: A Review. *Renew. Sustain. Energy Rev.* **2015**, *41*, 550–567. [CrossRef]
78. Pellera, F.-M.; Gidarakos, E. Chemical Pretreatment of Lignocellulosic Agroindustrial Waste for Methane Production. *Waste Manag.* **2018**, *71*, 689–703. [CrossRef]
79. Behera, S.; Arora, R.; Nandhagopal, N.; Kumar, S. Importance of Chemical Pretreatment for Bioconversion of Lignocellulosic Biomass. *Renew. Sustain. Energy Rev. J.* **2014**, *36*, 91–106. [CrossRef]
80. Dagnino, E.P.; Chamorro, E.R.; Romano, S.D.; Felissia, F.E.; Area, M.C. Optimization of the Acid Pretreatment of Rice Hulls to Obtain Fermentable Sugars for Bioethanol Production. *Ind. Crops Prod.* **2013**, *42*, 363–368. [CrossRef]
81. Badiei, M.; Asim, N.; Jahim, J.M.; Sopian, K. Comparison of Chemical Pretreatment Methods for Cellulosic Biomass. *APCBEE Procedia* **2014**, *9*, 170–174. [CrossRef]
82. Saha, B.C.; Iten, L.B.; Cotta, M.A.; Wu, Y.V. Dilute Acid Pretreatment, Enzymatic Saccharification and Fermentation of Wheat Straw to Ethanol. *Process Biochem.* **2005**, *40*, 3693–3700. [CrossRef]
83. Saha, B.C.; Cotta, M.A. Comparison of Pretreatment Strategies for Enzymatic Saccharification and Fermentation of Barley Straw to Ethanol. *New Biotechnol.* **2010**, *27*, 10–16. [CrossRef]
84. Chen, H.; Liu, J.; Chang, X.; Chen, D.; Xue, Y.; Liu, P.; Lin, H.; Han, S. A Review on the Pretreatment of Lignocellulose for High-Value Chemicals. *Fuel Process. Technol.* **2017**, *160*, 196–206. [CrossRef]
85. Kumar, S.; Paritosh, K.; Pareek, N.; Chawade, A.; Vivekanand, V. De-Construction of Major Indian Cereal Crop Residues through Chemical Pretreatment for Improved Biogas Production: An Overview. *Renew. Sustain. Energy Rev.* **2018**, *90*, 160–170. [CrossRef]
86. Syaichurrozi, I.; Villta, P.K.; Nabilah, N.; Rusdi, R. Effect of Sulfuric Acid Pretreatment on Biogas Production from *Salvinia molesta*. *J. Environ. Chem. Eng.* **2019**, *7*, 102857. [CrossRef]
87. Germec, M.; Bader, N.B.; Turhan, I. Dilute Acid and Alkaline Pretreatment of Spent Tea Leaves to Determine the Potential of Carbon Sources. *Biomass Convers. Biorefinery* **2018**, *8*, 529–544. [CrossRef]
88. Liu, Y.-Y.; Xu, J.-L.; Zhang, Y.; Liang, C.-Y.; He, M.-C.; Yuan, Z.; Xie, J. Reinforced Alkali-Pretreatment for Enhancing Enzymatic Hydrolysis of Sugarcane Bagasse. *Fuel Process. Technol.* **2016**, *143*, 1–6. [CrossRef]
89. Park, Y.C.; Kim, J.S. Comparison of Various Alkaline Pretreatment Methods of Lignocellulosic Biomass. *Energy* **2012**, *47*, 31–35. [CrossRef]
90. Kim, I.; Han, J.-I. Optimization of Alkaline Pretreatment Conditions for Enhancing Glucose Yield of Rice Straw by Response Surface Methodology. *Biomass Bioenergy* **2012**, *6*, 1–8. [CrossRef]
91. Liu, X.; Zicari, S.M.; Liu, G.; Li, Y. Improving the Bioenergy Production from Wheat Straw with Alkaline Pretreatment. *Biosyst. Eng.* **2015**, *140*, 59–66. [CrossRef]
92. Kim, J.S.; Lee, Y.Y.; Kim, T.H. A Review on Alkaline Pretreatment Technology for Bioconversion of Lignocellulosic Biomass. *Bioresour. Technol.* **2016**, *199*, 42–48. [CrossRef]
93. Ravindran, R.; Jaiswal, A.K. A Comprehensive Review on Pre-Treatment Strategy for Lignocellulosic Food Industry Waste: Challenges and Opportunities. *Bioresour. Technol.* **2016**, *199*, 92–102. [CrossRef]
94. Peng, F.; Peng, P.; Xu, F.; Sun, R.-C. Fractional Purification and Bioconversion of Hemicelluloses. *Biotechnol. Adv.* **2012**, *30*, 879–903. [CrossRef]
95. Xihui, K.; Yongming, S.; Lianhua, L.; Xiaoying, K.; Zhenhong, Y. Improving Methane Production from Anaerobic Digestion of Pennisetum Hybrid by Alkaline Pretreatment. *Bioresour. Technol.* **2018**, *255*, 205–212. [CrossRef]

96. Romero-Güiza, M.S.; Wahid, R.; Hernández, V.; Møller, H.; Fernández, B. Improvement of Wheat Straw Anaerobic Digestion through Alkali Pre-Treatment: Carbohydrates Bioavailability Evaluation and Economic Feasibility. *Sci. Total Environ.* **2017**, *595*, 651–659. [CrossRef]
97. Moset, V.; Xavier, C.D.A.N.; Feng, L.; Wahid, R.; Møller, H.B. Combined Low Thermal Alkali Addition and Mechanical Pre-Treatment to Improve Biogas Yield from Wheat Straw. *J. Clean. Prod.* **2018**, *172*, 1391–1398. [CrossRef]
98. Gu, Y.; Zhang, Y.; Zhou, X. Effect of $Ca(OH)_2$ Pretreatment on Extruded Rice Straw Anaerobic Digestion. *Bioresour. Technol.* **2015**, *196*, 116–122. [CrossRef]
99. da Silva, A.R.G.; Errico, M.; Rong, B.G. Evaluation of Organosolv Pretreatment for Bioethanol Production from Lignocellulosic Biomass: Solvent Recycle and Process Integration. *Biomass Convers. Biorefinery* **2018**, *8*, 397–411. [CrossRef]
100. Zhao, X.; Cheng, K.; Liu, D. Organosolv Pretreatment of Lignocellulosic Biomass for Enzymatic Hydrolysis. *Appl. Microbiol. Biotechnol.* **2009**, *82*, 815–827. [CrossRef]
101. Zhang, K.; Pei, Z.; Wang, D. Organic Solvent Pretreatment of Lignocellulosic Biomass for Biofuels and Biochemicals: A Review. *Bioresour. Technol.* **2016**, *199*, 21–33. [CrossRef]
102. Capecchi, L.; Galbe, M.; Barbanti, L.; Wallberg, O. Combined Ethanol and Methane Production Using Steam Pretreated Sugarcane Bagasse. *Ind. Crops Prod.* **2015**, *74*, 255–262. [CrossRef]
103. Chaturvedi, V.; Verma, P. An Overview of Key Pretreatment Processes Employed for Bioconversion of Lignocellulosic Biomass into Biofuels and Value Added Products. *3 Biotech* **2013**, *3*, 415–431. [CrossRef]
104. Agnihotri, S.; Johnsen, I.A.; Bøe, M.S.; Øyaas, K.; Moe, S. Ethanol Organosolv Pretreatment of Softwood (*Picea abies*) and Sugarcane Bagasse for Biofuel and Biorefinery Applications. *Wood Sci. Technol.* **2015**, *49*, 881–896. [CrossRef]
105. Xu, Z.; Huang, F. Pretreatment Methods for Bioethanol Production. *Appl. Biochem. Biotechnol.* **2014**, *174*, 43–62. [CrossRef]
106. Amnuaycheewa, P.; Hengaroonprasan, R.; Rattanaporn, K.; Kirdponpattara, S.; Cheenkachorn, K.; Sriariyanun, M. Enhancing Enzymatic Hydrolysis and Biogas Production from Rice Straw by Pretreatment with Organic Acids. *Ind. Crops Prod.* **2016**, *87*, 247–254. [CrossRef]
107. Teghammar, A.; Karimi, K.; Horváth, I.S.; Taherzadeh, M.J. Enhanced Biogas Production from Rice Straw, Triticale Straw and Softwood Spruce by NMMO Pretreatment. *Biomass Bioenergy* **2012**, *36*, 116–120. [CrossRef]
108. Kabir, M.M.; Niklasson, C.; Taherzadeh, M.J.; Horváth, I.S. Biogas Production from Lignocelluloses by N-Methylmorpholine-N-Oxide (NMMO) Pretreatment: Effects of Recovery and Reuse of NMMO. *Bioresour. Technol.* **2014**, *161*, 446–450. [CrossRef]
109. Ostovareh, S.; Karimi, K.; Zamani, A. Efficient Conversion of Sweet Sorghum Stalks to Biogas and Ethanol Using Organosolv Pretreatment. *Ind. Crops Prod.* **2015**, *66*, 170–177. [CrossRef]
110. da Silva, A.R.G.; Errico, M.; Rong, B.G. Techno-Economic Analysis of Organosolv Pretreatment Process from Lignocellulosic Biomass. *Clean Technol. Environ. Policy* **2018**, *20*, 1401–1412. [CrossRef]
111. Weerachanchai, P.; Leong, S.S.J.; Chang, M.W.; Ching, C.B.; Lee, J.M. Improvement of Biomass Properties by Pretreatment with Ionic Liquids for Bioconversion Process. *Bioresour. Technol.* **2012**, *111*, 453–459. [CrossRef]
112. Konda, N.M.; Shi, J.; Singh, S.; Blanch, H.W.; Simmons, B.A.; Klein-Marcuschamer, D. Understanding Cost Drivers and Economic Potential of Two Variants of Ionic Liquid Pretreatment for Cellulosic Biofuel Production. *Biotechnol. Biofuels* **2014**, *7*, 1–11. [CrossRef]
113. Trinh, L.T.P.; Lee, Y.J.; Lee, J.W.; Lee, H.J. Characterization of Ionic Liquid Pretreatment and the Bioconversion of Pretreated Mixed Softwood Biomass. *Biomass Bioenergy* **2015**, *81*, 1–8. [CrossRef]
114. Weerachanchai, P.; Lee, J.M. Recyclability of an Ionic Liquid for Biomass Pretreatment. *Bioresour. Technol.* **2014**, *169*, 336–343. [CrossRef]
115. Kim, H.; Ahn, Y.; Kwak, S.-Y. Comparing the Influence of Acetate and Chloride Anions on the Structure of Ionic Liquid Pretreated Lignocellulosic Biomass. *Biomass Bioenergy* **2016**, *93*, 243–253. [CrossRef]
116. Hu, X.; Cheng, L.; Gu, Z.; Hong, Y.; Li, Z.; Li, C. Effects of Ionic Liquid/Water Mixture Pretreatment on the Composition, the Structure and the Enzymatic Hydrolysis of Corn Stalk. *Ind. Crops Prod.* **2018**, *122*, 142–147. [CrossRef]
117. Zhang, Q.; Hu, J.; Lee, D.J. Pretreatment of Biomass Using Ionic Liquids: Research Updates. *Renew. Energy* **2017**, *111*, 77–84. [CrossRef]

118. Baral, N.R.; Shah, A. Techno-Economic Analysis of Cellulose Dissolving Ionic Liquid Pretreatment of Lignocellulosic Biomass for Fermentable Sugars Production. *Biofuels Bioprod. Biorefin.* **2016**, *10*, 70–88. [CrossRef]
119. Xie, Y.; Ma, C.; Lu, X.; Ji, X. Evaluation of Imidazolium-Based Ionic Liquids for Biogas Upgrading. *Appl. Energy* **2016**, *175*, 69–81. [CrossRef]
120. Dutra, E.D.; Santos, F.A.; Alencar, B.R.A.; Reis, A.L.S.; de Souza, R.D.F.R.; Aquino, K.A.D.S.; Morais, M.A.; Menezes, R.S.C. Alkaline Hydrogen Peroxide Pretreatment of Lignocellulosic Biomass: Status and Perspectives. *Biomass Convers. Biorefinery* **2018**, *8*, 225–234. [CrossRef]
121. Song, Z.L.; Yag, G.H.; Feng, Y.Z.; Ren, G.X.; Han, X.H. Pretreatment of Rice Straw by Hydrogen Peroxide for Enhanced Methane Yield. *J. Integr. Agric.* **2013**, *12*, 1258–1266. [CrossRef]
122. Yuan, Z.; Wen, Y.; Li, G. Production of Bioethanol and Value Added Compounds from Wheat Straw through Combined Alkaline/Alkaline-Peroxide Pretreatment. *Bioresour. Technol.* **2018**, *259*, 228–236. [CrossRef]
123. Su, Y.; Du, R.; Guo, H.; Cao, M.; Wu, Q.; Su, R.; Qi, W.; He, Z. Fractional Pretreatment of Lignocellulose by Alkaline Hydrogen Peroxide: Characterization of Its Major Components. *Food Bioprod. Process.* **2015**, *94*, 322–330. [CrossRef]
124. Rabelo, S.C.; Andrade, R.R.; Maciel Filho, R.; Costa, A.C. Alkaline Hydrogen Peroxide Pretreatment, Enzymatic Hydrolysis and Fermentation of Sugarcane Bagasse to Ethanol. *Fuel* **2014**, *136*, 349–357. [CrossRef]
125. Ayeni, A.O.; Oyekunle, D.T.; Shodipe, O.C.; Folayan, J.A. Data on the Enzymatic Conversion of Alkaline Peroxide Oxidative Pretreated Sugarcane Bagasse for the Production of Fermentable Sugars. *Data Brief* **2019**, *24*, 103867. [CrossRef]
126. Siciliano, A.; Stillitano, M.A.; De Rosa, S. Biogas Production from Wet Olive Mill Wastes Pretreated with Hydrogen Peroxide in Alkaline Conditions. *Renew. Energy* **2016**, *85*, 903–916. [CrossRef]
127. Katukuri, N.R.; Fu, S.; He, S.; Xu, X.; Yuan, X.; Yang, Z.; Guo, R.B. Enhanced Methane Production of Miscanthus Floridulus by Hydrogen Peroxide Pretreatment. *Fuel* **2017**, *199*, 562–566. [CrossRef]
128. Alencar, B.R.A.; Reis, A.L.S.; de Souza, R.D.F.R.; Morais, M.A.; Menezes, R.S.C.; Dutra, E.D. Recycling the Liquid Fraction of Alkaline Hydrogen Peroxide in the Pretreatment of Corn Stover. *Bioresour. Technol.* **2017**, *241*, 928–935. [CrossRef]
129. Sarto, S.; Hildayati, R.; Syaichurrozi, I. Effect of Chemical Pretreatment Using Sulfuric Acid on Biogas Production from Water Hyacinth and Kinetics. *Renew. Energy* **2019**, *132*, 335–350. [CrossRef]
130. Mancini, G.; Papirio, S.; Lens, P.N.L.; Esposito, G. Increased Biogas Production from Wheat Straw by Chemical Pretreatments. *Renew. Energy* **2018**, *119*, 608–614. [CrossRef]
131. Theuretzbacher, F.; Lizasoain, J.; Lefever, C.; Saylor, M.K.; Enguidanos, R.; Weran, N.; Gronauer, A.; Bauer, A. Steam Explosion Pretreatment of Wheat Straw to Improve Methane Yields: Investigation of the Degradation Kinetics of Structural Compounds during Anaerobic Digestion. *Bioresour. Technol.* **2015**, *179*, 299–305. [CrossRef]
132. Rabemanolontsoa, H.; Saka, S. Various Pretreatments of Lignocellulosics. *Bioresour. Technol.* **2016**, *199*, 83–91. [CrossRef]
133. Ullah, K.; Sharma, V.K.; Ahmad, M.; Lv, P.; Krahl, J.; Wang, Z. The Insight Views of Advanced Technologies and Its Application in Bio-Origin Fuel Synthesis from Lignocellulose Biomasses Waste, a Review. *Renew. Sustain. Energy Rev.* **2018**, *82*, 3992–4008. [CrossRef]
134. Bauer, A.; Lizasoain, J.; Theuretzbacher, F.; Agger, J.W.; Rincón, M.; Menardo, S.; Saylor, M.K.; Enguídanos, R.; Nielsen, P.J.; Potthast, A.; et al. Steam Explosion Pretreatment for Enhancing Biogas Production of Late Harvested Hay. *Bioresour. Technol.* **2014**, *166*, 403–410. [CrossRef]
135. Nges, I.A.; Li, C.; Wang, B.; Xiao, L.; Yi, Z.; Liu, J. Physio-Chemical Pretreatments for Improved Methane Potential of *Miscanthus lutarioriparius*. *Fuel* **2016**, *166*, 29–35. [CrossRef]
136. Li, C.; Liu, G.; Nges, I.A.; Liu, J. Enhanced Biomethane Production from *Miscanthus lutarioriparius* Using Steam Explosion Pretreatment. *Fuel* **2016**, *179*, 267–273. [CrossRef]
137. Chandra, R.; Takeuchi, H.; Hasegawa, T. Hydrothermal Pretreatment of Rice Straw Biomass: A Potential and Promising Method for Enhanced Methane Production. *Appl. Energy* **2012**, *94*, 129–140. [CrossRef]
138. Wang, D.; Shen, F.; Yang, G.; Zhang, Y.; Deng, S.; Zhang, J.; Zeng, Y.; Luo, T.; Mei, Z. Can Hydrothermal Pretreatment Improve Anaerobic Digestion for Biogas from Lignocellulosic Biomass? *Bioresour. Technol.* **2018**, *249*, 117–124. [CrossRef]

139. Zhou, X.; Li, Q.; Zhang, Y.; Gu, Y. Effect of Hydrothermal Pretreatment on Miscanthus Anaerobic Digestion. *Bioresour. Technol.* **2017**, *224*, 721–726. [CrossRef]
140. Phuttaro, C.; Sawatdeenarunat, C.; Surendra, K.C.; Boonsawang, P.; Chaiprapat, S.; Khanal, S.K. Anaerobic Digestion of Hydrothermally-Pretreated Lignocellulosic Biomass: Influence of Pretreatment Temperatures, Inhibitors and Soluble Organics on Methane Yield. *Bioresour. Technol.* **2019**, *284*, 128–138. [CrossRef]
141. Rajput, A.A.; Visvanathan, C. Effect of Thermal Pretreatment on Chemical Composition, Physical Structure and Biogas Production Kinetics of Wheat Straw. *J. Environ. Manag.* **2018**, *221*, 45–52. [CrossRef]
142. Hashemi, S.S.; Karimi, K.; Mirmohamadsadeghi, S. Hydrothermal Pretreatment of Safflower Straw to Enhance Biogas Production. *Energy* **2019**, *172*, 545–554. [CrossRef]
143. Luo, T.; Huang, H.; Mei, Z.; Shen, F.; Ge, Y.; Hu, G.; Meng, X. Hydrothermal Pretreatment of Rice Straw at Relatively Lower Temperature to Improve Biogas Production via Anaerobic Digestion. *Chin. Chem. Lett.* **2019**, *30*, 1219–1223. [CrossRef]
144. Antwi, E.; Engler, N.; Nelles, M.; Schüch, A. Anaerobic Digestion and the Effect of Hydrothermal Pretreatment on the Biogas Yield of Cocoa Pods Residues. *Waste Manag.* **2019**, *88*, 131–140. [CrossRef]
145. Wang, D.; Xin, Y.; Shi, H.; Ai, P.; Yu, L.; Li, X.; Chen, S. Closing Ammonia Loop in Efficient Biogas Production: Recycling Ammonia Pretreatment of Wheat Straw. *Biosyst. Eng.* **2019**, *180*, 182–190. [CrossRef]
146. Hashemi, S.S.; Karimi, K.; Majid Karimi, A. Ethanolic Ammonia Pretreatment for Efficient Biogas Production from Sugarcane Bagasse. *Fuel* **2019**, *248*, 196–204. [CrossRef]
147. Lizasoain, J.; Rincón, M.; Theuretzbacher, F.; Enguídanos, R.; Nielsen, P.J.; Potthast, A.; Zweckmair, T.; Gronauer, A.; Bauer, A. Biogas Production from Reed Biomass: Effect of Pretreatment Using Different Steam Explosion Conditions. *Biomass Bioenergy* **2016**, *95*, 84–91. [CrossRef]
148. Wyman, V.; Henríquez, J.; Palma, C.; Carvajal, A. Lignocellulosic Waste Valorisation Strategy through Enzyme and Biogas Production. *Bioresour. Technol.* **2018**, *247*, 402–411. [CrossRef]
149. Rouches, E.; Herpoël-Gimbert, I.; Steyer, J.P.; Carrere, H. Improvement of Anaerobic Degradation by White-Rot Fungi Pretreatment of Lignocellulosic Biomass: A Review. *Renew. Sustain. Energy Rev.* **2016**, *59*, 179–198. [CrossRef]
150. Raveendran, S.; Parameswaran, B.; Ashok, P. Biological Pretreatment of Lignocellulosic Biomass—An Overview. *Bioresour. Technol.* **2016**, *199*, 76–82. [CrossRef]
151. Rudakiya, D.M.; Gupte, A. Degradation of Hardwoods by Treatment of White Rot Fungi and Its Pyrolysis Kinetics Studies. *Int. Biodeterior. Biodegrad.* **2017**, *120*, 21–35. [CrossRef]
152. Schroyen, M.; Vervaeren, H.; Van Hulle, S.W.H.; Raes, K. Impact of Enzymatic Pretreatment on Corn Stover Degradation and Biogas Production. *Bioresour. Technol.* **2014**, *173*, 59–66. [CrossRef]
153. Liu, X.; Hiligsmann, S.; Gourdon, R.; Bayard, R. Anaerobic Digestion of Lignocellulosic Biomasses Pretreated with *Ceriporiopsis subvermispora*. *J. Environ. Manag.* **2017**, *193*, 154–162. [CrossRef]
154. Van Kuijk, S.J.A.; Sonnenberg, A.S.M.; Baars, J.J.P.; Hendriks, W.H.; Cone, J.W. Fungal Treated Lignocellulosic Biomass as Ruminant Feed Ingredient: A Review. *Biotechnol. Adv.* **2015**, *33*, 191–202. [CrossRef]
155. Dollhofer, V.; Dandikas, V.; Dorn-In, S.; Bauer, C.; Lebuhn, M.; Bauer, J. Accelerated Biogas Production from Lignocellulosic Biomass after Pre-Treatment with *Neocallimastix Frontalis*. *Bioresour. Technol.* **2018**, *264*, 219–227. [CrossRef]
156. Rouches, E.; Zhou, S.; Steyer, J.P.; Carrere, H. White-Rot Fungi Pretreatment of Lignocellulosic Biomass for Anaerobic Digestion: Impact of Glucose Supplementation. *Process Biochem.* **2016**, *51*, 1784–1792. [CrossRef]
157. Hosseini Koupaie, E.; Dahadha, S.; Bazyar Lakeh, A.A.; Azizi, A.; Elbeshbishy, E. Enzymatic Pretreatment of Lignocellulosic Biomass for Enhanced Biomethane Production-A Review. *J. Environ. Manag.* **2019**, *233*, 774–784. [CrossRef]
158. Shrestha, S.; Fonoll, X.; Khanal, S.K.; Raskin, L. Biological Strategies for Enhanced Hydrolysis of Lignocellulosic Biomass during Anaerobic Digestion: Current Status and Future Perspectives. *Bioresour. Technol.* **2017**, *245*, 1245–1257. [CrossRef]
159. Kong, X.; Du, J.; Ye, X.; Xi, Y.; Jin, H.; Zhang, M.; Guo, D. Enhanced Methane Production from Wheat Straw with the Assistance of Lignocellulolytic Microbial Consortium TC-5. *Bioresour. Technol.* **2018**, *263*, 33–39. [CrossRef]
160. Baba, Y.; Matsuki, Y.; Mori, Y.; Suyama, Y.; Tada, C.; Fukuda, Y.; Saito, M.; Nakai, Y. Pretreatment of Lignocellulosic Biomass by Cattle Rumen Fluid for Methane Production: Bacterial Flora and Enzyme Activity Analysis. *J. Biosci. Bioeng.* **2017**, *123*, 489–496. [CrossRef]

161. Carrere, H.; Antonopoulou, G.; Affes, R.; Passos, F.; Battimelli, A.; Lyberatos, G.; Ferrer, I. Review of Feedstock Pretreatment Strategies for Improved Anaerobic Digestion: From Lab-Scale Research to Full-Scale Application. *Bioresour. Technol.* **2016**, *199*, 386–397. [CrossRef]
162. Brémond, U.; de Buyer, R.; Steyer, J.P.; Bernet, N.; Carrere, H. Biological Pretreatments of Biomass for Improving Biogas Production: An Overview from Lab Scale to Full-Scale. *Renew. Sustain. Energy Rev.* **2018**, *90*, 583–604. [CrossRef]
163. Travaini, R.; Martín-Juárez, J.; Lorenzo-Hernando, A.; Bolado-Rodríguez, S. Ozonolysis: An Advantageous Pretreatment for Lignocellulosic Biomass Revisited. *Bioresour. Technol.* **2016**, *199*, 2–12. [CrossRef]
164. Domański, J.; Marchut-Mikołajczyk, O.; Polewczyk, A.; Januszewicz, B. Ozonolysis of Straw from *Secale cereale* L. for Anaerobic Digestion. *Bioresour. Technol.* **2017**, *245*, 394–400. [CrossRef]
165. Li, Y.; Chen, Y.; Wu, J. Enhancement of Methane Production in Anaerobic Digestion Process: A Review. *Appl. Energy* **2019**, *240*, 120–137. [CrossRef]
166. Zhao, B.-H.; Chen, J.; Yu, H.-Q.; Hu, Z.-H.; Yue, Z.-B.; Li, J. Optimization of Microwave Pretreatment of Lignocellulosic Waste for Enhancing Methane Production: Hyacinth as an Example. *Front. Environ. Sci. Eng.* **2017**, *11*, 1–9. [CrossRef]
167. Pellera, F.-M.; Gidarakos, E. Microwave Pretreatment of Lignocellulosic Agroindustrial Waste for Methane Production. *J. Environ. Chem. Eng.* **2016**, *5*, 352–365. [CrossRef]
168. Kainthola, J.; Shariq, M.; Kalamdhad, A.S.; Goud, V.V. Enhanced Methane Potential of Rice Straw with Microwave Assisted Pretreatment and Its Kinetic Analysis. *J. Environ. Manag.* **2019**, *232*, 188–196. [CrossRef]
169. Zou, L.; Ma, C.; Liu, J.; Li, M.; Ye, M.; Qian, G. Pretreatment of Food Waste with High Voltage Pulse Discharge towards Methane Production Enhancement. *Bioresour. Technol.* **2016**, *222*, 82–88. [CrossRef]
170. Baijuan, W.; Ting, C.; Xiangdong, Q.; Qi, W.; Yan, Z.; Shunxing, B.; Wen, P.; Bo, F. Effect of High-Voltage Pulsed Electric Field (HPEF) Pretreatment on Biogas Production Rates of Hybrid Pennisetum by Anaerobic Fermentation. *Nat. Gas Ind. B* **2018**, *5*, 48–53. [CrossRef]
171. De Bhowmick, G.; Sarmah, A.K.; Sen, R. Lignocellulosic Biorefinery as a Model for Sustainable Development of Biofuels and Value Added Products. *Bioresour. Technol.* **2018**, *247*, 1144–1154. [CrossRef]
172. Ponnusamy, V.K.; Nguyen, D.D.; Dharmaraja, J.; Shobana, S.; Banu, J.R.; Saratale, R.G.; Chang, S.W.; Kumar, G. A Review on Lignin Structure, Pretreatments, Fermentation Reactions and Biorefinery Potential. *Bioresour. Technol.* **2019**, *271*, 462–472. [CrossRef]
173. Shirkavand, E.; Baroutian, S.; Gapes, D.J.; Young, B.R. Combination of Fungal and Physicochemical Processes for Lignocellulosic Biomass Pretreatment—A Review. *Renew. Sustain. Energy Rev.* **2016**, *54*, 217–234. [CrossRef]
174. Mustafa, A.M.; Poulsen, T.G.; Xia, Y.; Sheng, K. Combinations of Fungal and Milling Pretreatments for Enhancing Rice Straw Biogas Production during Solid-State Anaerobic Digestion. *Bioresour. Technol.* **2017**, *224*, 174–182. [CrossRef]
175. Chen, X.; Zhai, R.; Shi, K.; Yuan, Y.; Dale, B.E.; Gao, Z.; Jin, M. Mixing Alkali Pretreated and Acid Pretreated Biomass for Cellulosic Ethanol Production Featuring Reduced Chemical Use and Decreased Inhibitory Effect. *Ind. Crops Prod.* **2018**, *124*, 719–725. [CrossRef]
176. Neumann, P.; Pesante, S.; Venegas, M.; Vidal, G. Developments in Pre-Treatment Methods to Improve Anaerobic Digestion of Sewage Sludge. *Rev. Environ. Sci. Biotechnol.* **2016**, *15*, 173–211. [CrossRef]
177. Sharma, S.; Sharma, V.; Kuila, A. Thermochemical Pretreatment of Corn Husk and Enzymatic Hydrolysis Using Mixture of Different Cellulases. *Biomass Convers. Biorefinery* **2018**, *8*, 179–188. [CrossRef]
178. Paudel, S.R.; Banjara, S.P.; Choi, O.K.; Park, K.Y.; Kim, Y.M.; Lee, J.W. Pretreatment of Agricultural Biomass for Anaerobic Digestion: Current State and Challenges. *Bioresour. Technol.* **2017**, *245*, 1194–1205. [CrossRef]
179. Taherdanak, M.; Zilouei, H. Improving Biogas Production from Wheat Plant Using Alkaline Pretreatment. *Fuel* **2014**, *115*, 714–719. [CrossRef]
180. Kucharska, K.; Hołowacz, I.; Konopacka-Łyskawa, D.; Rybarczyk, P.; Kamiński, M. Key Issues in Modeling and Optimization of Lignocellulosic Biomass Fermentative Conversion to Gaseous Biofuels. *Renew. Energy* **2018**, *129*, 384–408. [CrossRef]
181. Abdul Aziz, N.I.H.; Hanafiah, M.M.; Mohamed Ali, M.Y. Sustainable Biogas Production from Agrowaste and Effluents—A Promising Step for Small-Scale Industry Income. *Renew. Energy* **2019**, *132*, 363–369. [CrossRef]
182. Kucharska, K.; Rybarczyk, P.; Hołowacz, I.; Łukajtis, R.; Glinka, M.; Kamiński, M. Pretreatment of Lignocellulosic Materials as Substrates for Fermentation Processes. *Molecules* **2018**, *23*, 2937. [CrossRef]

183. Bolado-Rodríguez, S.; Toquero, C.; Martín-Juárez, J.; Travaini, R.; García-Encina, P.A. Effect of Thermal, Acid, Alkaline and Alkaline-Peroxide Pretreatments on the Biochemical Methane Potential and Kinetics of the Anaerobic Digestion of Wheat Straw and Sugarcane Bagasse. *Bioresour. Technol.* **2016**, *201*, 182–190. [CrossRef]
184. Łukajtis, R.; Rybarczyk, P.; Kucharska, K.; Konopacka-Łyskawa, D.; Słupek, E.; Wychodnik, K.; Kamiński, M. Optimization of Saccharification Conditions of Lignocellulosic Biomass under Alkaline Pre-Treatment and Enzymatic Hydrolysis. *Energies* **2018**, *11*, 886. [CrossRef]
185. Díaz-Trujillo, L.A.; Nápoles-Rivera, F. Optimization of Biogas Supply Chain in Mexico Considering Economic and Environmental Aspects. *Renew. Energy* **2019**, *139*, 1227–1240. [CrossRef]
186. Rieke, C.; Stollenwerk, D.; Dahmen, M.; Pieper, M. Modeling and Optimization of a Biogas Plant for a Demand-Driven Energy Supply. *Energy* **2018**, *145*, 657–664. [CrossRef]
187. Carlsson, M.; Lagerkvist, A.; Morgan-Sagastume, F. The Effects of Substrate Pre-Treatment on Anaerobic Digestion Systems: A Review. *Waste Manag.* **2012**, *32*, 1634–1650. [CrossRef]
188. Vasco-Correa, J.; Shah, A. Techno-Economic Bottlenecks of the Fungal Pretreatment of Lignocellulosic Biomass. *Fermentation* **2019**, *5*, 30. [CrossRef]
189. Baral, N.R.; Shah, A. Comparative Techno-Economic Analysis of Steam Explosion, Dilute Sulfuric Acid, Ammonia Fiber Explosion and Biological Pretreatments of Corn Stover. *Bioresour. Technol.* **2017**, *232*, 331–343. [CrossRef]
190. Brandt, K.L.; Gao, J.; Wang, J.; Wooley, R.J.; Wolcott, M. Techno-Economic Analysis of Forest Residue Conversion to Sugar Using Three-Stage Milling as Pretreatment. *Front. Energy Res.* **2018**, *6*, 1–11. [CrossRef]
191. Prasad, A.; Sotenko, M.; Blenkinsopp, T.; Coles, S.R. Life Cycle Assessment of Lignocellulosic Biomass Pretreatment Methods in Biofuel Production. *Int. J. Life Cycle Assess.* **2016**, *21*, 44–50. [CrossRef]
192. Safarian, S.; Unnthorsson, R. An Assessment of the Sustainability of Lignocellulosic Bioethanol Production from Wastes in Iceland. *Energies* **2018**, *11*, 1493. [CrossRef]
193. Soam, S.; Kapoor, M.; Kumar, R.; Gupta, R.P.; Puri, S.K.; Ramakumar, S.S.V. Life Cycle Assessment and Life Cycle Costing of Conventional and Modified Dilute Acid Pretreatment for Fuel Ethanol Production from Rice Straw in India. *J. Clean. Prod.* **2018**, *197*, 732–741. [CrossRef]
194. Rajendran, K.; Murthy, G.S. Techno-Economic and Life Cycle Assessments of Anaerobic Digestion—A Review. *Biocatal. Agric. Biotechnol.* **2019**, *20*, 101207. [CrossRef]

© 2019 by the authors. Licensee MDPI, Basel, Switzerland. This article is an open access article distributed under the terms and conditions of the Creative Commons Attribution (CC BY) license (http://creativecommons.org/licenses/by/4.0/).

applied sciences

MDPI

Article

Water-Soluble Sugars of Pedigreed Sorghum Mutant Stalks and Their Recovery after Pretreatment

Youjie Xu [1], Jun Li [1], Zhanguo Xin [2], Scott R. Bean [3], Michael Tilley [3] and Donghai Wang [1,*]

1 Department of Biological and Agricultural Engineering, Kansas State University, Manhattan, KS 66506, USA; youjiex@ksu.edu (Y.X.); lijun1989730@ksu.edu (J.L.)

2 Plant Stress and Germplasm Development Research Laboratory, USDA-ARS, Lubbock, TX 79415, USA; Zhanguo.Xin@ARS.USDA.GOV

3 Center for Grain and Animal Health Research, USDA-ARS, 1515 College Ave. Manhattan, KS 66502, USA; scott.bean@usda.gov (S.R.B.); michael.tilley@usda.gov (M.T.)

* Correspondence: dwang@ksu.edu; Tel.: +1-785-532-2919; Fax: +1-785-5325825

Received: 11 July 2020; Accepted: 5 August 2020; Published: 7 August 2020

Abstract: Chemical composition of biomass, especially carbohydrate content, is a critical indicator of a biomass source's potential for biofuel applications. This study characterized physico-chemical properties of stalks from 16 representative pedigreed sorghum mutant lines. The objectives of this study were to evaluate the recovery of sucrose and its hydrolysis products, glucose and fructose, during dilute sulfuric acid pretreatment at conditions typically used for lignocellulosic biomass, and to determine the relationship between water-extractive contents and sugar recovery after pretreatment. Dilute acid-pretreated sorghum stalks had enzymatic saccharification of >82.4% glucose yield for all treated samples with more than 82.3% cellulose recovery and 85% hemicellulose removal. A single-step, one-pot process was recommended for sorghum mutant stalks with low water-extractive content (<35%, *w/w*) to reduce processing cost and minimize wastewater disposal since the majority of sugars will be recovered after dilute acid pretreatment with minimal degradation products. However, for sorghum mutant stalks with high water-extractive content (>35%, *w/w*), a pre-washing step is beneficial to recover the water-soluble sugars before subjecting to the pretreatment process in order to avoid sugar losses during the pretreatment stage. Thus, different processing technologies should be applied to lignocellulosic biomass with various water-extractive contents and water-soluble sugar concentrations.

Keywords: sorghum mutant; biomass; soluble sugars; dilute acid pretreatment; one-pot process

1. Introduction

Bioethanol is currently used as an alternative liquid fuel for transportation in the United States of America (USA) and Brazil. Ethanol with an octane rating of 113, can function as an octane booster to reduce engine knocking and contribute to national energy security by reducing oil imports. Bioethanol is a sustainable alternative fuel, which can be derived from various sustainable feedstocks, including sugar-based crops, starch-based crops, and cellulosic biomass [1–3]. Cellulosic biofuels provide environmental benefits not available from grain- or sugar-based biofuels and are considered as a solid foundation to meet the needs for transportation fuels in a low-carbon economy, albeit with electrified vehicles and other technical advances [4].

Sorghum (*Sorghum bicolor* L. Moench) is a C4 photosynthetic species with high productivity and drought tolerance, ranking the fifth among widely grown cereal crops globally [5]. Unique characteristics of sorghum position the crop as a viable bioenergy source, including high biomass yield and high sugar content, high moisture-use efficiency and strong drought tolerance, a well-established management system, and breeding potential for genetic improvement [6,7]. Sweet sorghum contains high fermentable sugars such as sucrose, glucose, and fructose in the stalk which can be extracted easily and directly fermented to bioethanol; in contrast, hybrids of grain sorghum provide starch for bioethanol production. Sweet sorghum, previously used as a syrup, is now being researched for use in sugarcane processing facilities to produce ethanol (1.5 generation biofuel), while biomass sorghums contain a significant amount of structural carbohydrates which can be hydrolyzed and fermented to biofuel [7,8]. After pressing sorghum stalks, approximately 50% of water-soluble sugars and 100% water-insoluble structural carbohydrates remain in the bagasse; thus, recovery of all the residual sugars can significantly improve the process economics and increase the overall productivity [9,10].

Lignocellulosic biomass is the most abundant renewable source for biofuel production, mainly consisted of cellulose, hemicellulose, and lignin. Cellulose-based biomass such as sorghum stalk must be pretreated in order to open its complex structure for enzymatic hydrolysis and subsequent fermentation [11–15]. Currently, pretreatment methods mainly include physical methods, chemicals methods, biological methods, and a combination of methods. Numerous pretreatment methods were developed to overcome the recalcitrant structure of sorghum biomass, such as ball milling, steam explosion, liquid hot water, dilute acid, lime, ammonia, organic solvent, and ionic liquid pretreatments [16–19]. Challenges of using the current pretreatment processes include incomplete separation of cellulose and lignin, which results in reduced subsequent enzymatic saccharification efficiency, formation of inhibitors (such as acetic acid, furans, or phenolic compounds) that affect ethanol fermentation, and increased usage of chemicals and energy-intensive processes, as well as high cost of waste disposal [20–22].

A two-step process, consisting of prewashing and a subsequent pretreatment process was recommended to extract considerable amounts of water-soluble sugars to avoid sugar losses or degradation during the pretreatment stage [23]. However, the economic viability of adding an extraction step is questionable due to the increased process cost due to high water usage [24]. Instead, a single pretreatment step without the pre-extraction processing, i.e., a "one-pot" process, was proposed to recover the water-soluble sugars and improve the enzymatic digestibility of cellulosic biomass. The improved digestibility was based on the possibility that sucrose or its immediate hydrolysis products, glucose and fructose, will survive the pretreatment step and will be available to fermentative microorganisms [25].

The objective of this research was to evaluate the recovery of sucrose and its hydrolysis products, glucose and fructose, during dilute sulfuric acid pretreatment process at conditions typically used to pretreat lignocellulosic biomass. Sixteen representative stalk samples (sorghum stalks without panicle and leaves) from pedigreed mutant sorghum lines with various water-extractive contents (18 to 52%) were selected from a pedigreed sorghum mutant library consisting of 6000 individually mutagenized M_4 seed pools for the proposed one-pot process test and determination of the relationship between water extractive contents and sugar recovery after pretreatment.

2. Materials and Methods

2.1. Materials

Stalk samples from 16 pedigreed sorghum mutant lines, including wild-type sorghum used as control, were selected from a large sample pool provided by the Plant Stress and Germplasm Development Unit of the US Department of Agriculture Agricultural Research Services (Lubbock, TX, USA). Sorghum stalks were manually harvested. Sorghum stalks were ground (<1 mm) using a cutting mill (SM 2000, Retsch Inc., Newton, PA, USA) and placed into a plastic bag kept at room temperature. The moisture content was measured according to the convection oven drying method described in NREL LAP Determination of Total Solids in Biomass [26]. One gram of sorghum biomass was placed in a pre-weighed aluminum weighing dish and dried to a constant weight at 105 °C. Total solid percentage was defined as dry weight of sample per sample weight as received [27].

All chemicals, such as 72% (*w*/*w*) sulfuric acid, pure ethanol 200 proof, and HPLC grade water, used for this research were purchased from Sigma Chemical Co. (St. Louis, MO, USA).

2.2. Compositional Analysis

2.2.1. Extractives

Water and ethanol extraction was used to determine extractive content of the sorghum samples. Five grams of sorghum biomass samples were put into a Soxhlet thimble and washed using distilled water for 8 h according to the NREL Laboratory Analytical Procedure (LAP) Determination of Extractives in Biomass [26]. Then, ethanol was added to continue extraction for another 16 h. Extractive percentage was defined as dry weight of sample per sample weight as received. Water-soluble extractives contain inorganic materials, non-structural sugars, and nitrogenous materials [27].

2.2.2. Structural Carbohydrates, Lignin, and Ash

The chemical composition of samples was determined according to the National Renewable Energy Laboratory (NREL) procedure [26]. Extraction-free samples were placed into pressure tubes and through a two-step acid hydrolysis with 72% (*w*/*w*) sulfuric acid at 30 °C for 60 min, followed by 4% (*w*/*w*) dilute sulfuric acid at 121 °C for another 60 min. Carbohydrates such as cellulose and hemicellulose were converted to monosaccharides glucose, fructose, xylose, and arabinose, which were determined by high-performance liquid chromatography (HPLC) equipped with an RCM monosaccharide column (300 × 7.8 mm) (Phenomenex, Torrance, CA, USA) and a refractive index detector, under a mobile phase of 0.6 mL·min^{-1} water and a column temperature of 80 °C. Acid-soluble lignin in the hydrolysis liquor was determined using an ultraviolet (UV)–visible spectrophotometer at 320 nm wavelength blanked against distilled water [26]. Acid-insoluble lignin was weighed from the residue solids after oven heating overnight at 105 °C and then at 575 °C using a muffle furnace for at least 6 h to measure the ash content [27].

2.3. Dilute Acid Pretreatment

For dilute acid pretreatment, sulfuric acid (2%, *w*/*v*) was applied as a reaction medium, and reaction temperature was kept at 120 °C for 60 min (Figure 1). Once the treatment was complete, the reactor (Parr Instrument Co., Moline, IL, USA) was removed from the electric heater and placed into tap water to cool down within 50 °C in 5 min. Then, the slurry was vacuum-filtered using Whatman Paper (No. 4). Treated biomass was washed thoroughly with water and collected for composition analysis and enzymatic hydrolysis to evaluate the effect pretreatment on sugar yields and other hydrolytes.

$$\textit{Mass recovery}\ (\%) = \frac{M_{tre}}{M_{ori}} \times 100\%, \tag{1}$$

$$\textit{Cellulose recovery}\ (\%) = \frac{M_{tre} \times C_{tre}}{M_{ori} \times C_{ori}} \times 100\%, \tag{2}$$

$$\textit{Hemicellulose recovery}\ (\%) = \left(1 - \frac{M_{tre} \times H_{tre}}{M_{ori} \times H_{ori}}\right) \times 100\%, \tag{3}$$

where M_{tre} is dry mass weight after pretreatment, M_{ori} is original dry mass weight, C_{tre} is the cellulose percentage of solid biomass after pretreatment, C_{ori} is the cellulose percentage in raw material, H_{tre} is the hemicellulose percentage of solid biomass after pretreatment, and H_{ori} is the hemicellulose percentage in raw material.

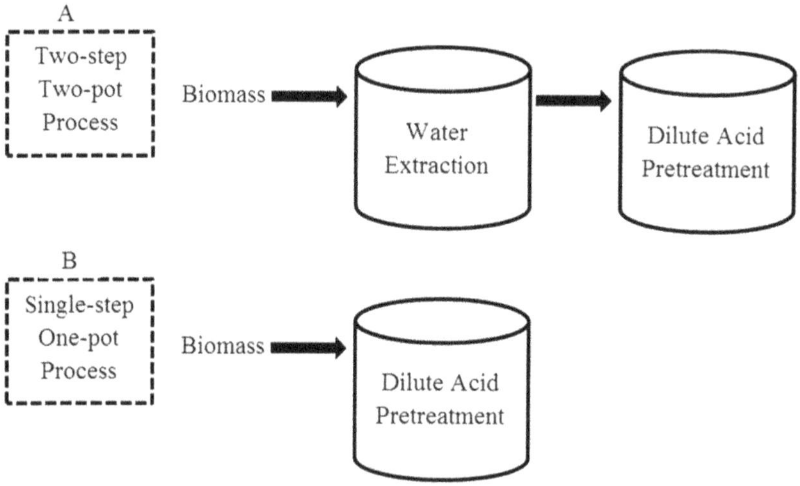

Figure 1. Flowchart of process designs: (**A**) two-step, two-pot process; (**B**): single-step, one-pot process.

The National Renewable Energy Laboratory (NREL) method for the determination of degradation products in pretreatment hydrolysate was applied to measure the furfural and hydromethylfurfural (HMF) concentration in biomass hydrolysates using HPLC. The injection volume was 20 μL; the solvent was water containing 0.005 M sulfuric acid at a flow rate of 0.6 mL/min. The column and refractive index detector (RID) temperatures were set at 60 and 45 °C.

2.4. Enzymatic Saccharification

After pretreatment, biomass (4%, *w*/*v*) was added to flasks to perform enzymatic hydrolysis using Accellerase 1500, generously provided by Genencor Dupont, at the recommended dosage (0.5 mL/g cellulose). Flasks were incubated at 50 °C in a rotary shaker (Model I2400, New Brunswick Scientific Inc., Edison, NJ, USA) with a speed of 150 rpm. After 72 h of enzymatic hydrolysis, supernatants were extracted by filtration and stored at 4 °C for further experiments. Glucose concentration was measured by HPLC. All reactions were performed in duplicate. Glucose yield was calculated as follows:

$$Glucose\ yield\ (\%) = \frac{Released\ sugar\ amount}{Theoretical\ sugar\ amount\ in\ raw\ material} \times 100\%. \quad (4)$$

2.5. Elemental Analysis

The elemental composition of sorghum stalks was determined using a CHNS/O Elemental Analyzer (PerkinElmer 2400 Series II, PerkinElmer Inc., Waltham, MA, USA). Approximately 2 mg of ground sorghum stalk packed with foil was burned under a pure oxygen atmosphere. The gases including CO_2, N_2, SO_2, and H_2O were burned and separated in a quartz column containing copper wires detected by a thermo-conductometer detector and reported as a percentage of initial dry weight (*w*/*w*, db) [27].

2.6. Heating Value

The energy content of pedigreed sorghum mutant stalks was determined using a calorimeter (IKA-Calorimeter C 200, IKA-Werke GmbH and Co. KG, Staufen, Germany) with a benzoic acid standard. A crucible with approximately 1.00 g of biomass sample was placed into an adiabatic bomb. After sealing the bomb, it was filled with oxygen and placed into the calorimetric equipment filled with water. The sample was ignited electrically through the cotton line. Powder sorghum stalks were pressed into pellets in order to reduce experimental error caused by incomplete combustion resulting from loose samples blown away during burning. The resultant water temperature increase allows the calculation of high heating value of the sample [27]. The heat capacity of the calorimeter was measured using benzoic acid as a reference standard.

2.7. Statistics

Analysis of variance and pairwise comparisons for the means using the Tukey adjustment at the confidence level of 95% were performed with SAS (SAS Institute, Inc., Cary, NC, USA). Mean values and standard deviations from the duplicated experiments are reported.

3. Results and Discussion

3.1. Physical and Chemical Properties of Pedigreed Sorghum Mutant Stalks

Sixteen sorghum mutant stalks with evenly distributed water-extractive contents ranging from the lowest, 18%, to the highest, 52%, were selected for this study (Table 1). The carbon and oxygen contents of selected samples ranged from 37.4% to 41.4% and 49.2% to 55.3%, respectively. The hydrogen content was approximately 5% to 6%, and nitrogen and sulfur contents were typically less than 2%. The wild-type sorghum samples (E25a and E25b, referred to as the control) contained relatively low water-extractive content of 25.8% with an average of 39.4% carbon, 52.5% oxygen, 6.4% hydrogen, 0.8% nitrogen, and 1.3% sulfur contents (Table 1). The selected sorghum mutants with higher water-extractive contents were expected to contain significantly larger portions of non-structural carbohydrates, including sucrose, glucose, and fructose, as compared to the control wild-type sorghum biomass (Table 2).

Table 1. Physico-chemical characteristics of sorghum stalks.

Sample Identity	Water-Soluble Extractive (%)	Carbon (%)	Hydrogen (%)	Nitrogen (%)	Sulfur (%)	Oxygen (%)	Heating Value (J/g)
E18A	17.6 ± 0.6a [1]	40.3 ± 0.8a	6.4 ± 0.2a	0.96 ± 0.3a	1.3 ± 0.3a	51.1 ± 0.6a	16,419.5 ± 21a
E18B	18.7 ± 0.8a	41.4 ± 0.9a	6.5 ± 0.3a	0.91 ± 0.2a	1.3 ± 0.2a	49.9 ± 0.7a	17,047.4 ± 31a
E20A	21.2 ± 0.9b	38.9 ± 1.3b	5.6 ± 0.2b	0.64 ± 0.3b	1.2 ± 0.2a	53.6 ± 0.8a	15,935.1 ± 26b
E20B	21.0 ± 0.8b	37.4 ± 0.7b	5.5 ± 0.4b	0.73 ± 0.1b	1.1 ± 0.1a	55.3 ± 0.4b	14,748.6 ± 19b
E25A	25.4 ± 1.3b	39.4 ± 1.6a	6.4 ± 0.1a	0.77 ± 0.2b	1.2 ± 0.1a	52.2 ± 0.5a	16,039.3 ± 21b
E25B	26.3 ± 1.6b	39.4 ± 0.7a	6.4 ± 0.2a	0.73 ± 0.3b	1.3 ± 0.2a	52.2 ± 0.6a	15,953.4 ± 18b
E30A	30.3 ± 1.8c	39.5 ± 0.6a	6.4 ± 0.3a	1.10 ± 0.2c	1.5 ± 0.2b	51.5 ± 0.7a	16,651.4 ± 25a
E30B	30.3 ± 1.6c	41.1 ± 1.5a	6.5 ± 0.4a	1.83 ± 0.3c	1.5 ± 0.2b	49.2 ± 0.6c	17,327.4 ± 24a
E35A	35.3 ± 2.2d	38.9 ± 1.8b	5.8 ± 0.5b	0.72 ± 0.1b	1.2 ± 0.1a	53.4 ± 0.8a	16,071.9 ± 29b
E35B	35.3 ± 2.1d	39.9 ± 0.9a	5.2 ± 0.2b	0.92 ± 0.2a	0.9 ± 0.1c	53.0 ± 0.9a	16,510.8 ± 28a
E40A	40.3 ± 2.3d	38.6 ± 0.8a	5.6 ± 0.3b	1.07 ± 0.1a	1.1 ± 0.2a	53.6 ± 0.6a	16,085.9 ± 28b
E40B	40.8 ± 1.9d	38.3 ± 0.7a	5.9 ± 0.5a	0.68 ± 0.1b	1.2 ± 0.1a	53.9 ± 1.1d	16,144.3 ± 22b
E45A	45.1 ± 2.3e	37.6 ± 0.5a	5.8 ± 0.4b	1.01 ± 0.2a	1.2 ± 0.2a	54.4 ± 1.2d	16,035.2 ± 17b
E45B	46.4 ± 2.1e	39.0 ± 0.6a	5.9 ± 0.6a	1.00 ± 0.2a	1.2 ± 0.1a	52.9 ± 0.8a	16,461.8 ± 26a
E52A	52.7 ± 2.5e	39.7 ± 0.8b	5.5 ± 0.3b	0.78 ± 0.1b	1.1 ± 0.1a	52.9 ± 0.5a	16,594.9 ± 25a
E52B	53.4 ± 2.6e	40.0 ± 1.3b	5.8 ± 0.4b	0.70 ± 0.1b	1.8 ± 0.2d	51.8 ± 0.7a	16,436.9 ± 21a

Note: Samples E25A and E25B are wild-type sorghum used as control; the remaining samples are pedigreed sorghum mutant stalks. [1] Column means with different letters than the control are significantly different at the 0.05 level.

Table 2. Water-soluble sugars in the raw biomass and recovery of sugars after pretreatment.

Sample Identity	Raw Biomass			Pretreatment Hydrolysate						Glucose Recovery (%)	Fructose Recovery (%)
	Water-Soluble Sugars (g/L)			Water-Soluble Sugars (g/L)				Sugar Degradation (g/L)			
	Sucrose	Glucose	Fructose	Glucose	Fructose	Xylose	Arabinose	HMF	Furfural		
E18A	0.13a [1]	0.1a	0.08a	0.16a	0.14a	4.3a	0.56a	0a	0.07a	97.4a	97.0a
E18B	0.16a	0.16a	0.07a	0.23a	0.14a	4.42a	0.51a	0.04a	0.06a	96.2a	93.9a
E20A	0.12a	0.08a	0.07a	0.14a	0.12a	4.28a	0.59a	0.02a	0.06a	96.9a	92.8a
E20B	0.26a	0.11a	0.09a	0.23a	0.20a	3.9a	0.52a	0.03a	0.05a	96.4a	91.5a
E25A	0.27a	0.15a	0.16a	0.28a	0.28a	4.07a	0.55a	0.04a	0.05a	98.8a	95.4a
E25B	0.28a	0.17a	0.19a	0.3a	0.31a	3.99a	0.81a	0.07a	0.06a	97.3a	94.4a
E30A	0.57b	0.49a	0.47a	0.77b	0.69b	3.52b	0.66a	0.24a	0.06a	99.8a	91.8a
E30B	0.77b	0.4a	0.42a	0.73b	0.68b	4.12a	0.75a	0.23a	0.06a	93.5a	84.9b
E35A	1.42b	0.78b	0.89b	1.39b	1.33b	3.51b	0.53a	0.39b	0.05a	93.8a	83.5b
E35B	0.82b	0.83b	1.07b	1.21b	1.29b	3.49b	0.54a	0.38b	0.05a	97.9a	87.4b
E40A	1.78b	0.78b	0.95b	1.48b	1.51b	3.17b	0.53a	0.45b	0.04a	89.1b	82.5b
E40B	1.63b	1.23b	1.44b	1.82b	1.52b	3.11b	0.44b	0.48b	0.04a	89.4b	67.7b
E45A	3.36b	1.32b	1.29b	2.55b	1.68b	2.66b	0.49b	0.55b	0.05a	85.5b	56.9b
E45B	3.05b	1.24b	1.40b	2.39b	1.78b	2.52b	0.40b	0.66b	0.04a	87.0b	61.2b
E52A	5.20b	1.15b	1.10b	3.20b	2.06b	2.28b	0.37b	0.79b	0.03b	86.0b	56.1b
E52B	4.75b	1.17b	1.08b	2.85b	2.06b	2.39b	0.34b	0.72b	0.04a	81.0b	60.1b

Note: Samples E25A and E25B are wild-type sorghum used as control; the remaining samples are pedigreed sorghum mutant stalks. [1] Column means with different letters than the control are significantly different at the 0.05 level.

Biomass samples with high carbon content are generally predicted to exhibit high caloric energy. The control samples had a mean heating value of 15,996.4 J/g, relatively lower than other selected sorghum mutant samples, especially those with higher water-extractive content. The heating value of selected sorghum mutant samples was positively correlated to the carbon content ($R^2 = 0.66$), while negatively correlated with the oxygen content ($R^2 = 0.69$). The heating value is critical for biomass pyrolysis evaluation and provided as comprehensive information for alternative use.

3.2. Sugar Recovery after Dilute Acid Pretreatment

The control sample contained minimal amounts of water-soluble sugars with approximately 0.28 g/L sucrose, 0.16 g/L glucose, and 0.18 g/L fructose (totally <3%, *w/w* of dry biomass) (Table 2). The content of water-soluble sugars in sorghum stalks increased as water-extractive content increased. Significantly larger amounts of water-soluble sugars were detected in sorghum mutant stalks with higher water-extractive content, as high as 4.98 g/L sucrose, 1.16 g/L glucose, and 1.09 g/L fructose (approximately 36%, *w/w* of dry biomass).

Sucrose present in sorghum stalks was completely hydrolyzed during the dilute acid pretreatment since no sucrose remained after treatment. Sucrose is a non-reducing disaccharide composed of β-D-fructose and α-D-glucose. In commercial industrial production, sucrose is hydrolyzed to monomers under typically low temperature (<120 °C) to minimize fructose and glucose degradation [24]. Both glucose and fructose are usually degraded when temperatures are higher than 106 °C and pH is less than 2.0, which is similar to the conditions present during dilute acid pretreatment of lignocellulosic biomass. The primary degradation pathway is dehydration of the sugars, glucose and fructose, to 5-hydroxymethylfurfural (5-HMF), which hydrolyzes and further degrades to levulinic and formic acid. Appreciable concentrations of sucrose hydrolysis products, glucose and fructose, were detected after dilute acid pretreatment (Table 2). Glucose and fructose levels were as high as 3.2 g/L and 2.06 g/L, respectively, in sample E52A, which were significantly higher than the control E25A and E25B. However, higher amounts of the sugar degradation product (approximately 0.79 g/L), 5-HMF, were detected in the pretreatment hydrolysate. A significant concentration of the hemicellulose degradation product, xylose, was also detected in the pretreatment hydrolysate, ranging from 2.28 to 4.42 g/L. Relatively low amounts of xylose degradation product, furfural, were detected in all the selected sorghum samples (Table 2).

For wild-type sorghum, approximately 98% and 95% recovery was achieved for glucose and fructose, respectively. However, as the water-soluble sugar content in the sorghum stalks increased, significant amounts of sugars were lost during the dilute acid pretreatment, especially for fructose. When the water-extractive content was greater than 35% (*w*/*w*), fructose recovery was usually less than 60% and 5-HMF concentration was significantly higher (>0.7 g/L). In contrast, the glucose recoveries of all the selected sorghum samples were higher than 80%, which further proved that fructose was easier to degrade than glucose, as found in other studies [24,25].

For sorghum mutants with high water-extractive content (>35%, *w*/*w*), a pre-washing step was beneficial to recover the present water-soluble sugars before performing the pretreatment process in order to avoid sugar losses or degradation during the pretreatment stage [23,28]. However, for those sorghum mutants with less water-extractive content, a single-step, one-pot process was recommended to reduce processing cost and minimize wastewater disposal since a majority of sugars can be recovered after dilute acid pretreatment with minimal amount of degradation products.

For high water-extractive biomass, an integrated bioprocess of combined advanced solid-state fermentation technology (ASSF) and alkaline pretreatment can be used to convert all the present sugars, including sucrose and structural carbohydrate. This process allows soluble sugars in the biomass (e.g., sweet sorghum stalks) to be converted into ethanol by ASSF using crushed stalks directly, and then ethanol distillation combined with alkaline pretreatment is simultaneously performed using a single distillation reactor. This integrated one-pot process can significantly reduce the production cost and minimize the wastewater pollution [29].

3.3. Chemical Composition of Pretreated Solid Biomass

For raw biomass, the maximum cellulose content of the selected 16 sorghum mutants was 31.3%, whereas the minimum was only 18.1%. The average cellulose content was 24.6% with a standard deviation of 4.8% (Table 3). Large variations of hemicellulose content were also observed from 14.7% to 26.6%, with an average of 20.4% and standard deviation of 4.20%. Relatively low amounts of lignin content were observed in this study, ranging from 6.6% to 13.9% with a mean value of 10.4% and standard deviation of 2.3%, as compared to corn stover and herbaceous grass [3,12].

Significant amounts of hemicellulose were hydrolyzed into xylose and arabinose after the dilute acid pretreatment. Hemicellulose content after the dilute acid pretreatment was reduced to approximately 6%, which resulted in significantly higher cellulose and lignin content. The majority of hemicellulose (>85%) was hydrolyzed during the pretreatment, which could increase the pore size and provide greater access for enzymes to digest cellulose [7,23]. Lignin relocation and removal were reported to enhance the enzymatic hydrolysis of treated sweet sorghum bagasse [30,31]. Kim and Day reported that sweet sorghum bagasse contained 45% cellulose, 27% hemicellulose, and 21% lignin [32]. The maximum cellulose content after dilute acid pretreatment was 55.8%, whereas the minimum was still as high as 47.0% with an average value of 50.7% and standard deviation of 2.2% (Table 3), which were higher than the cellulose content of acid-treated sweet sorghum bagasse reported by Yu et al. [33]. The increased cellulose content contributed to higher fermentable sugar concentration after enzymatic hydrolysis as compared to the untreated biomass [34,35]. A relatively high proportion of solid biomass was lost after dilute acid pretreatment, especially for those samples with high water-soluble sugar percentages. For pretreated pedigreed sorghum mutant stalks, the maximum solid recovery was 53.7%, and the minimum solid recovery was only 29.3% (E52A) (Table 3). However, it was more important to preserve as much of the cellulose present in the pretreated biomass. Cellulose recovery was defined as the cellulose ratio before and after pretreatment multiplied by the mass recovery. All the sorghum samples obtained greater than 82.3% cellulose recovery with a maximum recovery of 92.5%.

Table 3. Chemical composition of sorghum stalks before and after pretreatment.

Sample Identity	Raw Sorghum Stalk			Pretreated Sorghum Stalk			Mass Recovery (%)	Cellulose Recovery (%)	Hemicellulose Removal (%)
	Cellulose (%)	Hemicellulose (%)	Lignin (%)	Cellulose (%)	Hemicellulose (%)	Lignin (%)			
E18A	31.3 ± 0.2a [1]	26.5 ± 0.6a	13.8 ± 0.2a	53.4 ± 0.3a	6.4 ± 0.1a	29.1 ± 0.6a	53.7a	91.6a	87.0a
E18B	29.3 ± 0.3b	26.6 ± 0.7a	13.9 ± 0.3a	52.0 ± 0.2a	7.6 ± 0.2a	29.6 ± 0.5a	52.1a	92.5a	85.2b
E20A	28.0 ± 0.1b	24.6 ± 0.9b	12.8 ± 0.2a	50.0 ± 0.4a	5.9 ± 0.3a	28.3 ± 0.4a	50.8b	90.9a	87.9a
E20B	29.8 ± 0.2b	22.6 ± 1.2b	12.2 ± 0.1a	52.0 ± 0.9a	5.8 ± 0.2a	25.7 ± 0.6b	53.0a	92.3a	86.3a
E25A	30.2 ± 0.5b	23.9 ± 0.6b	11.9 ± 0.1b	55.8 ± 1.4b	6.9 ± 0.4a	27.2 ± 1.1c	48.8c	90.4a	85.9b
E25B	28.0 ± 0.7c	24.0 ± 0.5b	11.8 ± 0.2b	50.3 ± 0.6a	5.8 ± 0.2a	27.4 ± 0.8c	48.5c	86.9b	88.3a
E30A	27.7 ± 0.8c	21.0 ± 0.3c	11.0 ± 0.5b	52.2 ± 0.7a	5.0 ± 0.3b	27.5 ± 0.9c	43.6d	82.3c	89.6a
E30B	25.7 ± 0.2d	23.2 ± 0.2b	11.6 ± 0.6b	50.9 ± 0.5a	6.9 ± 0.6a	30.6 ± 0.8a	42.4d	84.0c	87.4a
E35A	24.2 ± 0.4e	19.3 ± 0.5d	9.6 ± 0.6c	49.6 ± 0.5a	5.4 ± 0.2b	26.9 ± 0.6d	42.5d	87.3b	88.0a
E35B	24.3 ± 0.3e	19.5 ± 0.1d	10.0 ± 0.5c	49.5 ± 0.6a	4.8 ± 0.3c	29.1 ± 0.7a	40.7f	83.1c	90.0a
E40A	19.8 ± 0.2f	18.0 ± 0.2d	9.5 ± 0.4c	47.0 ± 0.8c	5.2 ± 0.2b	29.5 ± 0.5a	36.8e	87.4b	89.3a
E40B	21.2 ± 0.5f	18.1 ± 0.3d	8.9 ± 0.3c	50.3 ± 0.3a	5.3 ± 0.2b	27.0 ± 0.6c	36.9e	87.5b	89.2a
E45A	18.6 ± 0.8f	15.2 ± 0.5e	8.5 ± 0.2c	47.1 ± 0.2c	5.3 ± 0.5b	28.8 ± 0.5a	35.5e	89.7a	87.6a
E45B	19.2 ± 0.6f	14.9 ± 0.4e	8.1 ± 0.4d	48.5 ± 0.4c	4.3 ± 0.3c	28.9 ± 0.8a	32.9g	83.1c	90.5a
E52A	18.3 ± 0.4f	14.9 ± 0.3e	6.6 ± 0.2d	51.5 ± 0.5a	5.5 ± 0.4b	27.2 ± 0.3c	29.3h	82.3c	89.2a
E52B	18.1 ± 0.5f	14.7 ± 0.2e	6.7 ± 0.3d	51.3 ± 0.7a	5.3 ± 0.2b	28.1 ± 0.4c	29.9h	84.5c	89.3a

Note: Samples E25A and E25B are wild-type sorghum and used as control; the remaining samples are pedigreed sorghum mutant stalks. [1] Column means with different letters than the control are significantly different at the 0.05 level.

3.4. Glucose Yield from Enzymatic Hydrolysis of Dilute Acid-Treated Solid Biomass

The reaction temperature in dilute acid pretreatment ranged from 120 to 160 °C, while sulfuric acid levels were in the range of 0.5 to 2% (*w*/*w*) [12,35]. The yield of reducing sugars through enzymatic hydrolysis was mainly influenced by acid concentration, reaction temperature, and reaction time. Acid hydrolysis releases oligosaccharides and monosaccharides but also results in the formation of degradation products such as aldehydes. In order to reduce the decomposition of the sugars, a hydrolysis time of 60 min was used in this study. Dilute acid-pretreated sorghum stalks demonstrated good enzymatic digestibility with more than 82.4% glucose yield for all the treated samples (Figure 2), which showed that the dilute acid pretreatment condition of reaction temperature 120 °C for 60 min with 2% (*w*/*v*) sulfuric acid was sufficient to disrupt the rigid cell-wall structure of sorghum biomass and significantly improve the enzymatic hydrolysis of treated sorghum biomass. Half of the selected sorghum mutants achieved greater than 90% glucose yields with a maximum glucose yield of 98.2% for sample E40B.

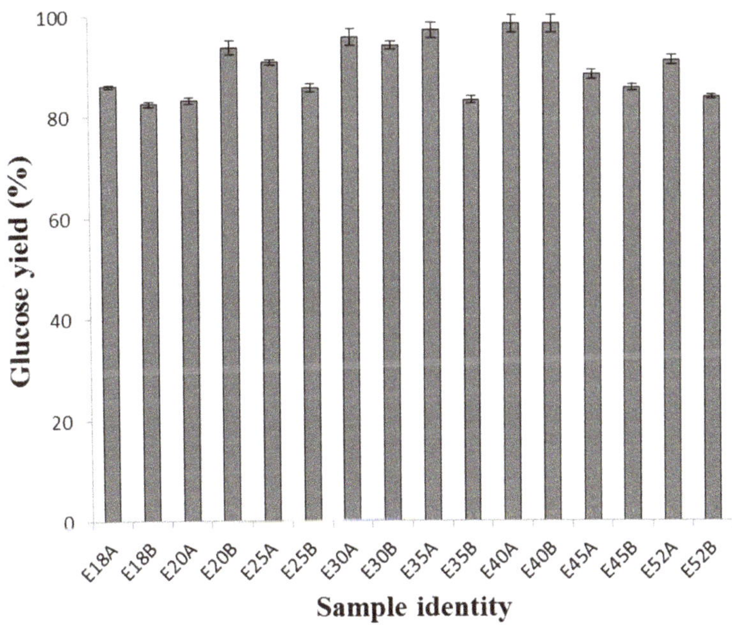

Figure 2. Glucose yield of dilute acid-pretreated sorghum stalks.

4. Conclusions

The content of water-soluble sugars in pedigreed sorghum mutant stalks increased as water-extractive content increased. The recovery of sugars including sucrose, glucose, and fructose during dilute sulfuric acid pretreatment also increased as water-extractive content increased. More than 82.4% of enzymatic hydrolysis yield and more than 82.3% cellulose recovery and 85% hemicellulose removal were achieved for all pretreated sorghum samples. To consider the effect of water-soluble content on the bioconversion process, a single-step, one-pot process is recommended for sorghum mutants with water-extractive content of equal or less than 35% (*w*/*w*), which can reduce the production cost and minimize wastewater generation, since the majority of sugars can be recovered after pretreatment with a minimum amount of degradation products. However, for sorghum mutants with higher water-extractive content (>35%, *w*/*w*), a pre-washing step is recommended to extract the exiting water-soluble sugars before subjecting to the pretreatment process in order to avoid sugar losses or degradation during the pretreatment stage. Thus, different processing strategies should be considered for the lignocellulosic biomass with various water-extractive contents and various water-soluble sugar levels.

Author Contributions: Conceptualization, Y.X. and J.L.; methodology, Y.X.; software, Y.X.; validation, Y.X., J.L. and D.W.; formal analysis, Y.X.; investigation, Y.X.; resources, Z.X.; data curation, Y.X.; writing—original draft preparation, Y.X.; writing—review and editing, Y.X., D.W.; visualization, Y.X.; supervision, D.W., Z.X., S.R.B., M.T.; project administration, X.Y.; funding acquisition, D.W. All authors have read and agreed to the published version of the manuscript.

Funding: This work was supported by the United Sorghum Checkoff Program with grant number RN002-16.

Acknowledgments: This publication is contribution no. 18-333-J from the Kansas Agricultural Experiment Station, Manhattan, KS. Names are necessary to report factually on available data; however, the US Department of Agriculture neither guarantees nor warrants the standard of the product, and use of the name by the US Department of Agriculture implies no approval of the product to the exclusion of others that may also be suitable. USDA is an equal opportunity provider and employer.

Conflicts of Interest: The authors declare no conflict of interest.

References

1. Agblevor, F.; Murden, A.; Hames, B. Improved method of analysis of biomass sugars using high-performance liquid chromatography. *Biotechnol. Lett.* **2014**, *26*, 1207–1211. [CrossRef] [PubMed]
2. Vassilev, S.V.; Vassileva, C.G. Water-soluble fractions of biomass and biomass ash and their significance for biofuel application. *Energy Fuels* **2019**, *33*, 2763–2777. [CrossRef]
3. Bower, S.; Wickramasinghe, R.; Nagle, N.J.; Schell, D.J. Modeling sucrose hydrolysis in dilute sulfuric acid solutions at pretreatment conditions for lignocellulosic biomass. *Bioresour. Technol.* **2008**, *99*, 7354–7362. [CrossRef]
4. Cao, W.; Sun, C.; Liu, R.; Yin, R.; Wu, X. Comparison of the effects of five pretreatment methods on enhancing the enzymatic digestibility and ethanol production from sweet sorghum bagasse. *Bioresour. Technol.* **2012**, *111*, 215–221. [CrossRef] [PubMed]
5. Cybulska, I.; Brudecki, G.P.; Zembrzuska, J.; Schmidt, J.E.; Lopez, C.G.; Thomsen, M.H. Organosolv delignification of agricultural residues (date palm fronds, *Phoenix dactylifera* L.) of the United Arab Emirates. *Appl. Energy* **2017**, *185*, 1040–1050. [CrossRef]
6. Xu, Y.; Li, J.; Zhang, M.; Wang, D. Modified simultaneous saccharification and fermentation to enhance bioethanol titers and yields. *Fuel* **2018**, *215*, 647–654. [CrossRef]
7. Xu, Y.; Zhang, K.; Wang, D. High gravity enzymatic hydrolysis of hydrothermal and ultrasonic pretreated big bluestem with recycling prehydrolysate water. *Renew. Energy* **2017**, *114*, 351–356. [CrossRef]
8. Raj, T.; Kapoor, M.; Gaur, R.; Christopher, J.; Lamba, B.; Tuli, D.K.; Kumar, R. Physical and chemical characterization of various Indian agriculture residues for biofuels production. *Energy Fuels* **2015**, *29*, 3111–3118. [CrossRef]
9. Wang, Z.; Xu, J.; Pandey, P.; Cheng, J.J.; Li, R.; Qu, R. Improvement of sugar production from transgenic switchgrass with low-temperature alkali pretreatment. *Energy Fuels* **2012**, *26*, 3054–3061. [CrossRef]
10. Godin, B.; Lamaudière, S.; Agneessens, R.; Schmit, T.; Goffart, J.P.; Stilmant, D.; Gerin, P.A.; Delcarte, J. Chemical composition and biofuel potentials of a wide diversity of plant biomasses. *Energy Fuels* **2013**, *27*, 2588–2598. [CrossRef]
11. Jin, M.; Gunawan, C.; Uppugundla, N.; Balan, V.; Dale, B.E. A novel integrated biological process for cellulosic ethanol production featuring high ethanol productivity, enzyme recycling and yeast cells reuse. *Energy Environ. Sci.* **2012**, *5*, 7168–7175. [CrossRef]
12. Yang, F.; Liu, Z.; Afzal, W.; Liu, Z.; Bell, A.T.; Prausnitz, J.M. Pretreatment of Miscanthus giganteus with Lime and Oxidants for Biofuels. *Energy Fuels* **2015**, *29*, 1743–1750. [CrossRef]
13. Anitescu, G.; Bruno, T.J. Liquid biofuels: Fluid properties to optimize feedstock selection, processing, refining/blending, storage/transportation, and combustion. *Energy Fuels* **2012**, *26*, 324–348. [CrossRef]
14. Zhao, X.; Wen, J.; Chen, H.; Liu, D. The fate of lignin during atmospheric acetic acid pretreatment of sugarcane bagasse and the impacts on cellulose enzymatic hydrolyzability for bioethanol production. *Renew. Energy* **2018**, *128*, 200–209. [CrossRef]

15. Ramadoss, G.; Muthukumar, K. Ultrasound assisted metal chloride treatment of sugarcane bagasse for bioethanol production. *Renew. Energy* **2016**, *99*, 1092–1102. [CrossRef]
16. Ju, X.; Engelhard, M.; Zhang, X. An advanced understanding of the specific effects of xylan and surface lignin contents on enzymatic hydrolysis of lignocellulosic biomass. *Bioresour. Technol.* **2013**, *132*, 137–145. [CrossRef]
17. Kim, M.; Day, D.F. Composition of sugar cane, energy cane, and sweet sorghum suitable for ethanol production at Louisiana sugar mills. *J. Ind. Microbiol. Biotechnol.* **2011**, *38*, 803–807. [CrossRef]
18. Li, J.; Li, S.; Han, B.; Yu, M.; Li, G.; Jiang, Y. A novel cost-effective technology to convert sucrose and homocelluloses in sweet sorghum stalks into ethanol. *Biotechnol. Biofuels* **2013**, *6*, 174. [CrossRef]
19. Li, M.; Feng, S.; Wu, L.; Li, Y.; Fan, C.; Zhang, R. Sugar-rich sweet sorghum is distinctively affected by wall polymer features for biomass digestibility and ethanol fermentation in bagasse. *Bioresour. Technol.* **2014**, *167*, 14–23. [CrossRef]
20. Mesa, L.; Martínez, Y.; Barrio, E.; González, E. Desirability function for optimization of Dilute Acid pretreatment of sugarcane straw for ethanol production and preliminary economic analysis based in three fermentation configurations. *Appl. Energy* **2017**, *198*, 299–311. [CrossRef]
21. Robertson, G.P.; Hamilton, S.K.; Barham, B.L.; Dale, B.E.; Izaurralde, R.C.; Jackson, R.D. Cellulosic biofuel contributions to a sustainable energy future: Choices and outcomes. *Science* **2017**, *356*, 6345. [CrossRef] [PubMed]
22. Rohowsky, B.; Häßler, T.; Gladis, A.; Remmele, E.; Schieder, D.; Faulstich, M. Feasibility of simultaneous saccharification and juice co-fermentation on hydrothermal pretreated sweet sorghum bagasse for ethanol production. *Appl. Energy* **2013**, *102*, 211–219. [CrossRef]
23. Rooney, W.L.; Blumenthal, J.; Bean, B.; Mullet, J.E. Designing sorghum as a dedicated bioenergy feedstock. Biofuel. *Bioprod. Biorefin.* **2007**, *1*, 147–157. [CrossRef]
24. Whitfield, M.B.; Chinn, M.S.; Veal, M.W. Processing of materials derived from sweet sorghum for biobased products. *Ind. Crops Prod.* **2012**, *37*, 362–375. [CrossRef]
25. Wirawan, F.; Cheng, C.; Kao, W.; Lee, D.; Chang, J. Cellulosic ethanol production performance with SSF and SHF processes using immobilized Zymomonas mobilis. *Appl. Energy* **2012**, *100*, 19–26. [CrossRef]
26. Sluiter, A.; Ruiz, R.; Scarlata, C.; Sluiter, J.; Templeton, D. *Determination of Extractives in Biomass, Laboratory Analytical Procedure (LAP)*; Technical Report; Midwest Research Institute: Kansas City, MO, USA, 2008.
27. Xu, Y.; Li, J.; Moore, C.; Xin, Z.; Wang, D. Physico-chemical characterization of pedigreed sorghum mutant stalks for biofuel production. *Ind. Crops Prod.* **2019**, *124*, 806–811. [CrossRef]
28. Bals, B.; Dale, B.; Balan, V. Enzymatic hydrolysis of distiller's dry grain and solubles (DDGS) using ammonia fiber expansion pretreatment. *Energy Fuels* **2006**, *20*, 2732–2736. [CrossRef]
29. Dahadha, S.; Amin, Z.; Bazyar Lakeh, A.A.; Elbeshbishy, E. Evaluation of different pretreatment processes of lignocellulosic biomass for enhanced biomethane production. *Energy Fuels* **2017**, *31*, 10335–10347. [CrossRef]
30. Xu, Y.; Wang, D. Integrating starchy substrate into cellulosic ethanol production to boost ethanol titers and yields. *Appl. Energy* **2017**, *195*, 196–203. [CrossRef]
31. Yan, Z.; Li, J.; Chang, S.; Cui, T.; Jiang, Y.; Yu, M. Lignin relocation contributed to the alkaline pretreatment efficiency of sweet sorghum bagasse. *Fuel* **2015**, *158*, 152–158. [CrossRef]
32. Yan, Z.; Li, J.; Li, S.; Chang, S.; Cui, T.; Jiang, Y. Impact of lignin removal on the enzymatic hydrolysis of fermented sweet sorghum bagasse. *Appl. Energy* **2015**, *160*, 641–647. [CrossRef]
33. Yu, J.; Zhang, T.; Zhong, J.; Zhang, X.; Tan, T. Biorefinery of sweet sorghum stem. *Biotechnol. Adv.* **2012**, *30*, 811–816. [CrossRef]
34. Yu, M.; Li, J.; Li, S.; Du, R.; Jiang, Y.; Fan, G. A cost-effective integrated process to convert solid-state fermented sweet sorghum bagasse into cellulosic ethanol. *Appl. Energy* **2014**, *115*, 331–336. [CrossRef]
35. Zhang, J.; Ma, X.; Yu, J.; Zhang, X.; Tan, T. The effects of four different pretreatments on enzymatic hydrolysis of sweet sorghum bagasse. *Bioresour. Technol.* **2011**, *102*, 4585–4589. [CrossRef]

© 2020 by the authors. Licensee MDPI, Basel, Switzerland. This article is an open access article distributed under the terms and conditions of the Creative Commons Attribution (CC BY) license (http://creativecommons.org/licenses/by/4.0/).

applied sciences

MDPI

Article

Improvement of Anaerobic Digestion of Hydrolysed Corncob Waste by Organosolv Pretreatment for Biogas Production

Belkis Sulbarán-Rangel [1], Jaime Santiago Alarcón Aguirre [1], Luz Breton-Deval [2], Jorge del Real-Olvera [3] and Kelly Joel Gurubel Tun [1,*]

1 Department of Water and Energy, University of Guadalajara Campus Tonalá, Tonalá 45425, Mexico; Belkis.sulbaran@academicos.udg.mx
2 Catedras Conacyt-Instituto de Biotecnología, Universidad Nacional Autónoma de México, Cuernavaca, Morelos 62210, Mexico
3 Environmental Technology, Center of Research and Assistance in Technology and Design of the State of Jalisco, Normalistas 800, Guadalajara, Jalisco 44680, Mexico
* Correspondence: joel.gurubel@academicos.udg.mx; Tel.: +52-3320002300

Received: 19 March 2020; Accepted: 9 April 2020; Published: 17 April 2020

Abstract: This paper describes an organosolv pretreatment of corncob waste to improve its anaerobic digestion for biogas production. Through a thermochemical process based on the use of ethanol and acetic acid, it was possible to separate the fractions of lignin, considered to be a natural inhibitor of anaerobic digestion processes. In addition, with this organosolv pretreatment, the available sugars in the carbohydrates present as monosaccharides, or simple sugars, were depolymerised, facilitating the digestion process. The obtained results include the chemical characterisation of the corncob, the hydrolysate, and the mixture with cow manure, finding that these substrates have potential to be used in anaerobic digestion. The total reducing sugars consumed were 96.8%, and total sugars were 85.75%. It was clearly observed that with the use of pretreatment with organosolv, the production of biogas was superior, because 484 NmL/gVS was obtained compared to the other reported treatments. It was also observed that adding the hydrolysate organosolv increased the production because the values of the control without hydrolysate were 120 NmL/gVS in the bottle experiment. When the experiment was scaled to the 5L reactor, the total volumes of biogas that were accumulated in 15 days of production were 5050 NmL/gVS and 1212 NmL/gVS with and without hydrolysate, respectively. This indicates that the organosolv pretreatment of corncob waste is effective in improving biogas production.

Keywords: anaerobic digestion; methane; corn residue; organosolv pretreatment

1. Introduction

Energy demand, as well as the use of water resources, is growing proportionally as the population increases. As a sustainable alternative, the use of renewable such as biogas has been proposed. The process of biogas production involves anaerobic digestion, which is characterised as a process that transforms organic matter into biogas (methane, 60%–70%; carbon dioxide, 30%–40%) [1,2]. Anaerobic digestion is a very complex process in terms of both the number of biochemical reactions that take place and the microorganisms involved in them. In fact, many of these reactions occur simultaneously [3]. During the process, several groups of microorganisms participate to carry out a coordinated metabolism in four stages. The first is hydrolysis, where complex polymers present in the organic matter are broken down into simple monomers. Fermentation occurs in the second step with acidogenesis, which produces organic acids, and the third step is acetogenesis, which is the conversion of organic acids to acetate. Propionate and butyrate are products of acidogenesis/fermentation.

The fourth stage is methanogenesis, where microorganisms produce methane through the acetoclastic and hydrogenotrophic pathways [4].

This fermentation process is usually hindered by various factors related to the nature of the substrate, such as nutrient balance, pH, or the presence of toxins and inhibitors [5]. The organic matter that can be used in the anaerobic digestion process is diverse [6]. Studies have reported various solid wastes being used, such as waste sludge from water treatment plants, agricultural livestock waste, and lignocellulosic waste [7]. With the aim of achieving an adequate final disposal of organic waste and/or a reduction thereof in an environmentally sustainable way, it is relevant to conduct research on the feasibility of a process of digestion of widely distributed organic waste and to stabilise it for the production of gaseous biofuels [2].

As a lignocellulosic organic residue, corncob was chosen due to its high availability and limited use otherwise. In Mexico, it is estimated that the annual quantity of residues fluctuates around 45 million tonnes of dry material among the ten principal crops (corn, sorghum, wheat, beans, rice, barley, soy, cotton, safflower, and sesame). Stubble and corncob represent 25,500,000 tonnes that are usually used to feed livestock at times when traditional food is scarce [8]. Although corncob could be used as a substrate for biogas production, its lignocellulosic nature presents difficulties for its direct use [9].

The specific pretreatment needed to improve anaerobic digestion depends on the type of organic matter. In the case of lignocellulosic material, a large part of its composition is carbohydrates that are not readily available because they are embedded in a lignin matrix, so pretreatment is necessary before the digestion process [2,4,10]. Organosolv pretreatment is an economic and efficient method to improve the anaerobic digestion process [11]. This is a method that utilises organic solvents such as methanol, ethanol, acetone, or ethylene glycol mixed with or without an inorganic catalyst at high temperatures to remove the lignin from the lignocellulosic material. In this way, the material is fractionated into a liquid hydrolysate containing dissolved lignin and hemicellulose and a solid fraction known as holocellulose, which comprises hemicellulose with cellulose [12]. The main advantage of organosolv over other chemical pretreatments is that relatively pure and low-molecular-weight lignin is recovered as a by-product, and the solvents that are used can be recovered and recycled to reduce the operation costs [2,13].

The main objective of this study was to improve the performance of the anaerobic digestion of hydrolysed corncob by applying organosolv pretreatment for biogas production. The effects of the pretreatment parameters, that is, temperature and duration, on the methane yield were determined by liquid-state batch anaerobic digestion assays. The biogas production concentration and total volume were investigated for the pretreated substrate.

2. Materials and Methods

Corncob was obtained from the area of Nextipac, in the municipality of Zapopan Jalisco, Mexico. The inoculum used was obtained from cow manure collected from a farm in the area of Tonalá, Mexico. For the organosolv method, the following reagents were used: ethanol (CH_3CH_2OH un-denaturalised 96%), acetic acid (CH_3COOH 99.7%), and 3.5 dinitrosalicylic acid (DNS), all of them supplied by Sigma-Aldrich (Toluca, México).

2.1. Corncob Characterisation

The chemical characterisation of the corncob was determined using the standards of the Technical Association of the Pulp and Paper Industry (TAPPI). The cellulose content was determined using the T 203 cm-99 standard. The acid soluble lignin content was determined using the T 222 om-02 standard. The extractable compounds in acetone were determined according to T 280 pm-99, the extractable compounds in water using T 207 cm-99, and the ash content with T 211 om-02 [14]. An elemental analysis (C, H, N, and S) of the corncob was also performed. The elemental analysis was based on the volatilisation of a sample by total combustion under an atmosphere of pure oxygen, releasing the elements to be measured in the form of CO_2, H_2O, NOx, and SOx, respectively.

2.2. Organosolv Hydrolysate from Corncob

An ethanol to acetic acid ratio of 1:10 was used as an organic solvent. The process included high-temperature cooking inside a Jaime type reactor. This consisted of using 400 g (dry weight) of corncob with 50% solution of ethanol (2 L of distilled water and 2 L of ethanol) and 12 mL of acetic acid at a temperature of 175 °C and boiling for 2.5 h [15]. The liquid fraction, containing dissolved lignin and hemicelluloses, was separated. The hydrolysate was then broken up to evaporate to remove the lignin and to recover the used solvents.

To compare the results of the organosolv hydrolysate, a hydrolysate with soda was prepared, which is a widely used pretreatment. First, 400 g (dry weight) of corncob and 70 g of NaOH in a ratio of 1:10 with water as a solution were added in a Jaime type reactor at 195 °C for 1 h. As with the organosolv treatment, the liquid fraction was taken after the cooking time.

2.3. Characterisation of Corncob Hydrolysate

The reducing sugars (RS) present in the hydrolysate were determined by the DNS colorimetry method using a UV–visible spectrophotometer, for which a calibration curve was created from a standard glucose solution (2.0 g. L^{-1}) [16]. Total sugars (TS) were also determined by the colorimetric phenol–sulphuric technique [17]. This technique is based on the formation of furfural (aldose) and 5-hydroxymethylfurfural (ketose) in a strongly acidic and dehydrating medium.

The determination of monosaccharides or individual sugars was determined by TAPPI Standard T 249 cm-09 [14]. The carbohydrate composition of extractive-free wood and wood pulp was determined by gas-liquid chromatography using Agilent 1200 equipment for high-performance liquid chromatography (HPLC). The bromatological analysis was performed following the techniques established by the Association of Official Analytical Chemists [18] and was performed by the Laboratory of Chemical and Bromatological Analysis of CIAJ, Guadalajara. The percentages of moisture, ethereal extract, crude protein, ash, and raw fibre were determined. During hydrolysis, the chemical oxygen demand (COD) was also measured to determine the amount of hydrolysate to use to feed the bacteria in the reactor. The COD was determined by using the colorimetric method according to standard methods [19]. Volatile fatty acids (VFAs) were measured before the fermentation stage and at day 17 of the fermentation stage by HPLC [20]. The inhibitory compounds, such as furfural, were also measured at day 17 by HPLC [21].

2.4. Preparation of Inoculum from Cow Manure

A mixture was made in three equal parts with soil, cow manure, and water and left for stirring for 24 h at 37 °C for the microorganisms to achieve stabilisation in the conditions of development within a mesophilic environment (between 34 and 36 °C). Organic composition parameters, such as total solids (TS), volatile suspended solids (VSS), and pH [20], were evaluated.

2.5. Experimental Design

Batch anaerobic digestion experiments were carried out in three bottles with a liquid volume of 300 mL and then escalated to a Labfors batch-type biological reactor with a capacity of 5 L. To both reactors was added a mixture of water, methanogenic inoculum of cow manure, and hydrolysed corncob and non-hydrolysed corncob at a concentration of 10 g COD/L. Additionally, a blank test without the addition of corncob hydrolysate was carried out. The process was performed in batches, and the operating conditions were the following: pH 7, agitation 300 rpm, time 20 days, and temperature 36 °C. All tests were performed in triplicate, and the variables were controlled and monitored by an automated system with sensors.

Biogas production was recorded with a gas meter, where the amount of gas generated for 20 days was quantified. For gas production in the period of time when the accumulation of produced biogas could be measured, a measurement unit of mL per day (mL/d) was used or quantified per unit of raw material in normalised biogas in mL per gram of volatile solid (NmL/g VS) to determine the effectiveness of the anaerobic digestion process [22]. To analyse the composition of the biogas obtained in the digestion process, biogas samples were taken and analysed with an Agilent Technologies 6820 GC gas chromatograph with a DB-35 capillary column to determine the type and percentage of biogas generated.

3. Results and Discussion

The organosolv process is a delignification process that consists of breaking the internal lignin and hemicellulose bonds of the lignocellulose material; in this case, a corncob. Table 1 shows the chemical composition of corncob before performing the organosolv pretreatment: 45% cellulose, 34% hemicellulose, 16% lignin, and 6% extractives. These values are comparable to the data published by several authors, who indicated that the cellulose, hemicellulose and lignin ratios were 45%, 35% and 15%, respectively [23,24]. The high percentage of hemicelluloses in the corncob makes it a suitable material for thermochemical pretreatments such as organosolv. The pretreated corncob had 22% cellulose, 72% hemicellulose, 5% lignin, and 2% extractives, from which it could be verified that the organosolv treatment solubilised and de-polymerised lignin and hemicelluloses, dividing them into separate liquid streams (hydrolysate) and leaving behind solid fractions rich in cellulose for other applications [25,26].

Table 1. Corncob characterisation.

Parameters		Values% ± SD
Cellulose		45 ± 2.25
Hemicellulose		34 ± 1.7
Lignin		16 ± 0.8
Extractives		6 ± 0.29
Elemental analysis	C	46.6 ± 2.33
	H	6.2 ± 0.31
	O	46.8 ± 2.34
	N	0.4 ± 0.02
Ash		1.74 ± 0.08

The percentages (%; w/w) obtained for C: 46.6 and N: 0.4, as shown in Table 1, are in agreement with data reported in the literature (C: 41.7% and N: 0.1%) [22], suggesting that corncob is a material that can be used in an anaerobic process. However, the C:N ratio of corncob is 61:0.5, which is considerably higher than the optimal value reported in the literature for other biomasses used in anaerobic digestion [3]. The decomposition of materials with high carbon content, greater than 35:1, occurs more slowly, because the multiplication and development of bacteria are low, due to the lack of nitrogen, but the biogas production period is longer. On the other hand, with a C:N ratio less than 8:1, bacterial activity is inhibited due to the formation of an excessive ammonium content, which in large quantities is toxic and inhibits the process [4].

3.1. Characterisation of Corncob Hydrolysate

To confirm the solubility of the simple sugars in corncob hydrolysate, the hydrolysate obtained after the organosolv pretreatment was analysed. Table 2 represents the carbohydrates of the hydrolysate corresponding to the available sugars for use by the inoculum. It is important to know that the parameters as proteins constitute a very important substrate in the anaerobic digestion process because, in addition to being a source of carbon and energy, the amino acids derived from their hydrolysis have high nutritional value.

Table 2. Bromatological analysis of corncob hydrolysate.

Parameters	Values ± SD
Proteins (%)	0.20 ± 2.81
Fats (%)	0.77 ± 0.54
Ash (%)	0.47 ± 0.96
Carbohydrates (%)	5.95 ± 0.29

Table 3 shows the results of the sugars present in the organosolv hydrolysate and soda hydrolysate, showing that, with the organosolv pretreatment, the total reducing sugars (TRS) were higher (65.84 g/L) than the TRS with the soda pretreatment (4.64 g/L). This indicates that the hydrolysate is suitable for use in the anaerobic reactor. As with the TRS, the concentration of total sugar in the organosolv pretreatment was 157.7 g/L, which is higher than after the soda pretreatment (20.80 g/L), meaning that a greater amount of total sugars was extracted. Organosolv pretreatment improved the formation of sugars or the ability to subsequently form sugars by enzymatic hydrolysis, thus preventing the loss or degradation of carbohydrates and avoiding the formation of inhibitors, the by-products of subsequent hydrolysis and fermentation processes [27]. The non-reducing sugars obtained concentrations that exceeded the totals of TRS, demonstrating an incomplete depolymerisation from the organosolv pretreatment.

Table 3. Characterisation of corncob hydrolysate.

Parameters		Organosolv Hydrolysate	Soda Hydrolysate
Total Reducing Sugars (g/L)		65.84 ± 3.29	4.64 ± 0.23
Total sugars (g/L)		157.7 ± 3.86	20.80 ± 1.04
Non-reducing sugars (g/L)		91.86 ± 4.56	16.16 ± 0.80
Sugars (Monomers)	Glucose (g/L)	6.23 ± 0.34	4.12 ± 0.20
	Xylose (g/L)	4.12 ± 0.34	3.54 ± 0.17
	Galactose (g/L)	2.42 ± 0.17	1.9 ± 0.09
	Mannose (g/L)	1.66 ± 0.41	0.5 ± 0.02
Chemical oxygen demand (g/L)		690 ± 5.67	195 ± 6.56

The chemical oxygen demand (COD) of the corncob hydrolysate was 690 g/L. This result is considered high when compared with soda hydrolysate (195 g/L) and lower results of residues commonly present in the paper industry and its effluents without prior treatment, not exceeding a value of 52 g/L, taking into account that thermochemical processes are also carried out in this industry. The COD of the organosolv hydrolysate exceeded the results characteristic of residues from other types of pretreatment, where values ranging from 179 to 193 g/L for soluble COD and total COD, respectively, have been reported [28]. The COD increased with the organosolv process because the organic solvents used for the removal of lignin in the corncob leave the hydrolysate rich in depolymerised hemicelluloses (glucose, xylose, galactose, and mannose), thus increasing the organic matter available for the digestion process.

3.2. Anaerobic Digestion Process

The percentage of total solids contained in the mixture with which the digester is loaded is an important factor to consider to ensure that the process is carried out satisfactorily. The accessibility of methanogenic bacteria within the substrate is increasingly limited as the solids content increases, and the gas efficiency and production can therefore be affected. Table 4 shows the results of the characterisation of the substrate of cow manure and corncob hydrolysate obtained in this project. Experimentally, it has been shown that the load in semicontinuous digesters should not have more than

8% to 12% total solids to ensure the proper functioning of the process, unlike discontinuous digesters, which can handle between 40% and 60% total solids. In this case, the total solids value was 14.82%, which is attributed to the cow manure [29].

Table 4. Characterisation of substrate (cow manure + corncob hydrolysate).

Parameters	Values ± SD (%; w/w)
Solids totals	14.82 ± 2.81
Volatile solids	12.25 ± 0.54
Fixed solids	2.57 ± 0.96

Figure 1 shows the maximum value of biogas production in the bottles reactor obtained with the hydrolysate corncob at 15 days compared with a control corncob without the hydrolysate organosolv pretreatment and with different pretreatments in the bottles reactor reported by other authors [9,30,31]. Hua (2016) reported the effect of microbial pretreatment on the biogas production of non-sterile, rotted silage maize straw, finding 305 NmL/gVS [30], while Perez-Rodríguez (2017) reported extrusion and enzymatic hydrolysis, finding 279 and 290 NmL/gVS, respectively [9]. Shah (2018) reported 360 NmL/gVS using lime ($Ca(OH)_2$) soaking as a pretreatment [31]. It was clearly observed that, with the use of pretreatment with organosolv, the production of biogas was superior, because 484 NmL/gVS was obtained compared to the other reported treatments. It was also observed that adding the hydrolysate organosolv increased the production because the values of the control without hydrolysate were 120 NmL/gVS.

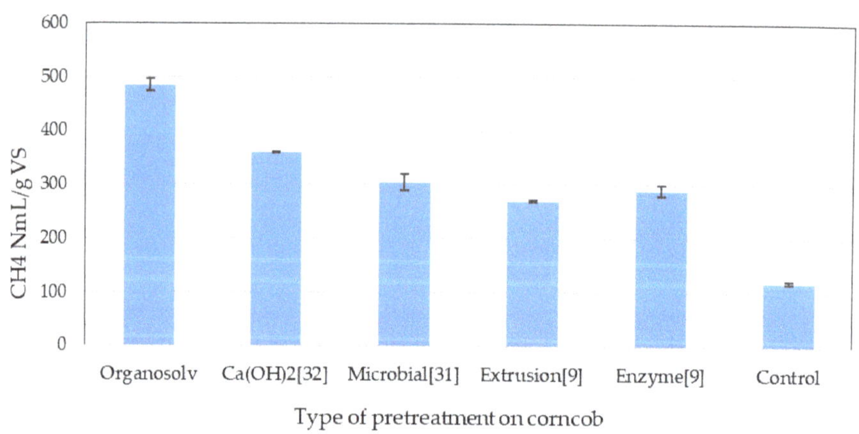

Figure 1. Comparison of different pretreatments in the production of biogas from corncob.

Figure 2 shows a representative evolution in cumulative biogas production in the Labfors batch-type biological reactor. In this section, the values found with and without corncob hydrolysate are compared. The biogas production was observed to be very low for the first nine days in both cases, but began from day 10. In the corncob hydrolysate test, the biogas production was higher than the non-hydrolysate test. Biogas production was monitored until day 20, but there was no production from days 15 to 20, so there was a constant line, and all the added organic substrate was consumed by day 15. The total biogas accumulated in 15 days of production was 5050 ± 253 NmL/gVS and 1212 ± 120 NmL/g VS with and without hydrolysate, respectively. The increase in biogas production was notable when the organosolv hydrolysate of corncob was added.

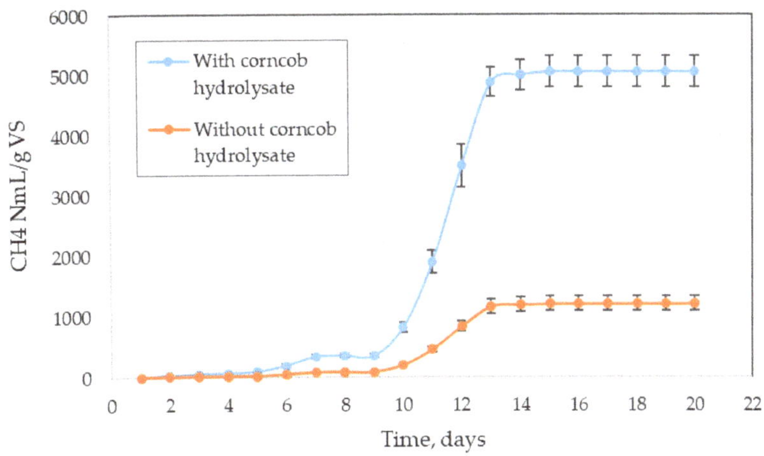

Figure 2. Cumulative biogas production in daily and total accumulated intervals.

Table 5 makes clear that the concentration of total reducing sugars is one of the factors that can affect methanogenic production in the anaerobic digestion process, directly and correlationally [32]. The consumption of TRS and total sugars during the anaerobic digestion was 96.8% and 85.75%, respectively. The COD was reduced to 79% with a COD reduction of 7.9 g, which is proportional to 639.24 NmLbiogas/g COD. Clearly confirming the influence of sugars in biogas production, the sugar content had the maximum reduction on the final day, demonstrating the potential of corncob residues, rich in sugars, as viable substrates in biogas production.

Table 5. Results of concentrations of total reducing sugars (TRS), Total sugars (TS) and chemical oxygen demand (COD) of the digestate with respect to the substrate.

Stage	TRS		TS		COD	
	(g/L)	%	(g/L)	%	(g/L)	%
Initial hydrolysate	65.84 ± 3.92	100	157.70 ± 7.88	100	10 ± 0.01	100
Final digestate	2.11± 0.10	3.20	22.48 ± 1.2	14.25	2.10 ± 0.10	21
Total degradation	63.73	96.80	135.22	85.75	7.9	79

Table 6 shows the results obtained from the daily biogas production and composition. The biogas composition was CH_4 with a concentration of 35.7% to 72.80% (v / v), CO_2 of 63.8% to 26.18%, and 0.5% to 1% (v/v) for other biogas (H_2, O_2, etc.) during anaerobic digestion (15 days). The CH_4 concentration is in line with values reported by previous studies of the contents of biogas, ranging from 50% to 65% (v/v) for CH_4 and from 35% to 50% (v/v) for CO_2 during sludge anaerobic digestion [33]. The production of CH_4 indicated the presence of methanogens, and the anaerobic digestion was effective.

Other important parameters during anaerobic digestion are the concentrations of volatile fatty acids (VFA), which increased with respect to the values obtained in the recovery of the substrate material. This is a clear indicator of the microbiological activity of the system during the time of the experimentation with a single feeding and periodic evaluation of the volume of biogas. During anaerobic degradation, complex organic matter is hydrolysed and fermented in low-molecular-weight compounds, including short-chain fatty acids (C_2 C_6). As shown in Table 7, the VFAs found in both the substrate (initial) and digestate (day 17) were acetic acid in the highest proportion, followed by butyric and propionic acid and lesser quantities of furfural.

Table 6. Composition of biogas produced.

Days	Biogas Production (NmL/gVS)	Biogas Composition (%)		
		CH_4	CO_2	Other
2	30	35.70	63.80	0.50
10	640	51.40	47.88	0.72
11	1259	57.30	41.90	0.80
12	2240	70.60	28.41	0.99
14	2360	72.80	26.18	1.02

Table 7. Characterisation of volatile fatty acids in the substrate and digestate.

Volatile Fatty Acids (VFA)	Substrate + Inoculum	Digestate
Acetic (mg/L)	1280 ± 370	2480 ± 30
Propionic (mg/L)	290 ± 60	510 ± 50
Butyric (mg/L)	410 ± 60	137 ± 40
Furfural (mg/L)	0.00	20 ± 10

During the hydrolytic–acidogenic stage, different organic acids are generated and even biogas rich in H_2, depending on the operating conditions. Table 7 shows the concentration of the main metabolic intermediaries detected at this stage. These precursors are metabolised to CH_4 during the methanogenesis stage. It should be noted that during the assays, about 70% of the volume of biogas corresponded to CH_4, which implies benefits for its subsequent exploitation. Finally, no accumulation of organic acid was detected throughout the methanogenesis operating period, indicating that the anaerobic digestion process was carried out efficiently with an overall COD balance of 79%.

4. Conclusions

Lignocellulosic residues from agroindustrial sources, such as corncob, are a raw material of the second generation with a permanent supply for anaerobic digestion. Thermochemical pretreatment with organosolv based on organic solvents is effective for the delignification and fractionation of sugars to monomers present in corncob. The corncob hydrolysate obtained in the organosolv pretreatment is a highly digestible carbohydrate substrate that increases the efficiency of biogas production. The results show that organosolv pretreatment is an efficient process for the improvement of anaerobic digestion from corncob lignocellulosic materials, removing lignin as an important inhibitor; however, its effectiveness greatly depends on the type of lignocellulose material.

Author Contributions: Methodology and experimental work: J.S.A.A. and L.B.-D.; writing—review and editing: K.J.G.T. and J.d.R.-O.; writing—original draft preparation: J.S.A.A. and B.S.-R.; conceptualisation and supervision: K.J.G.T. and B.S.-R. Funding acquisition: K.J.G.T. All authors have read and agreed to the published version of the manuscript.

Funding: This research was funded by the National Council of Science and Technology México (CONACyT) project number 258960 Basic Science.

Acknowledgments: The authors are grateful for the financial support of student grants from CONACyT.

Conflicts of Interest: The authors declare no conflicts of interest.

References

1. Gurubel, K.; Sanchez, E.N.; Coronado-Mendoza, A.; Zuniga-Grajeda, V.; Sulbaran-Rangel, B.; Breton-Deval, L. Inverse optimal neural control via passivity approach for nonlinear anaerobic bioprocesses with biofuels production. *Optim. Control Appl. Methods* **2019**, *40*, 848–858. [CrossRef]

2. Millati, R.; Wikandari, R.; Ariyanto, T.; Putri, R.U.; Taherzadeh, M.J. Pretreatment technologies for anaerobic digestion of lignocelluloses and toxic feedstocks. *Bioresour. Technol.* **2020**, *304*, 122998. [CrossRef] [PubMed]
3. Ward, A.J.; Hobbs, P.J.; Holliman, P.J.; Jones, D.L. Optimisation of the anaerobic digestion of agricultural resources. *Bioresour. Technol.* **2008**, *99*, 7928–7940. [CrossRef] [PubMed]
4. Kainthola, J.; Kalamdhad, A.S.; Goud, V.V. A review on enhanced biogas production from anaerobic digestion of lignocellulosic biomass by different enhancement techniques. *Process Biochem.* **2019**, *84*, 81–90. [CrossRef]
5. Bah, H.; Zhang, W.; Wu, S.; Qi, D.; Kizito, S.; Dong, R. Evaluation of batch anaerobic co-digestion of palm pressed fiber and cattle manure under mesophilic conditions. *Waste Manag.* **2014**, *34*, 1984–1991. [CrossRef] [PubMed]
6. Paudel, S.R.; Banjara, S.P.; Choi, O.K.; Park, K.Y.; Kim, Y.M.; Lee, J.W. Pretreatment of agricultural biomass for anaerobic digestion: Current state and challenges. *Bioresour. Technol.* **2017**, *245*, 1194–1205. [CrossRef]
7. Shilpi, S.; Lamb, D.; Bolan, N.; Seshadri, B.; Choppala, G.; Naidu, R. Waste to watt: Anaerobic digestion of wastewater irrigated biomass for energy and fertiliser production. *J. Environ. Manag.* **2019**, *239*, 73–83. [CrossRef]
8. SAGARPA. Plan de manejo de residuos generados en actividades agrícolas primera etapa: Diagnóstico nacional. In Secretaría de Agricultura, Ganadería, Desarrollo Rural, Pesca y Alimentación (SAGARPA) y Universidad Autónoma de Zacatecas "Francisco Garcia Salinas" México. 2015. Available online: https://www.gob.mx/cms/uploads/attachment/file/346978/Manejo_de_residuos_Detallado.pdf (accessed on 19 March 2020).
9. Perez-Rodríguez, N.; García-Bernet, D.; Domínguez, J.M. Extrusion and enzymatic hydrolysis as pretreatments on corn cob for biogas production. *Renew. Energy* **2017**, *107*, 597–603. [CrossRef]
10. Triolo, J.M.; Sommer, S.G.; Møller, H.B.; Weisbjerg, M.R.; Jiang, X.Y. A new algorithm to characterize biodegradability of biomass during anaerobic digestion: Influence of lignin concentration on methane production potential. *Bioresour. Technol.* **2011**, *102*, 9395–9402. [CrossRef]
11. Ferreira, J.A.; Taherzadeh, M.J. Improving the economy of lignocellulose-based biorefineries with organosolv pretreatment. *Bioresour. Technol.* **2020**, *299*, 122695. [CrossRef]
12. Kabir, M.M.; Rajendran, K.; Taherzadeh, M.J.; Horváth, I.S. Experimental and economical evaluation of bioconversion of forest residues to biogas using organosolv pretreatment. *Bioresour. Technol.* **2015**, *178*, 201–208. [CrossRef] [PubMed]
13. Zheng, Y.; Zhao, J. Pretreatment of lignocellulosic biomass for enhanced biogas production. *Prog. Energy Combust. Sci.* **2014**, *42*, 35–53. [CrossRef]
14. TAPPI. *Technical Association of the Pulp & Paper Industry TAPPI Test Methods*; TAPPI test methods; TAPPI: Atlanta, GA, USA, 1988.
15. Hernández, J.A.; Romero, V.H.; Escalante, A.; Toriz, G.; Rojas, O.J.; Sulbarán, B.C. Agave tequilana bagasse as source of cellulose nanocrystals via organosolv treatment. *BioResources* **2018**, *13*, 3603–3614. [CrossRef]
16. Miller, G.L. Use of dinitrosalicylic acid reagent for determination of reducing sugar. *Anal. Chem.* **1959**, *31*, 426–428. [CrossRef]
17. Dubois, M.; Gilles, K.A.; Hamilton, J.K.; Rebers, P.A.; Smith, F. A colorimetric method for the determination of sugars. *Nature* **1951**, *168*, 167. [CrossRef]
18. Baur, F.J.; Ensminger, L.G. The Association of Official Analytical Chemists (AOAC). *J. Am. Oil Chem. Soc.* **1977**, *54*, 171–172. [CrossRef]
19. Baird, R.; Bridgewater, L. *Standard Methods for the Examination of Water and Wastewater*; American Public Health Association, Ed.; American Public Health Association: Washington, DC, USA, 2017.
20. Breton-Deval, L.; Méndez-Acosta, H.O.; González-Álvarez, V.; Snell-Castro, R.; Gutiérrez-Sánchez, D.; Arreola-Vargas, J. Agave tequilana bagasse for methane production in batch and sequencing batch reactors: Acid catalyst effect, batch optimization and stability of the semi-continuous process. *J. Environ. Manag.* **2018**, *224*, 156–163. [CrossRef]
21. Lins, P. Reduction of accumulated volatile fatty acids by an acetate-degrading enrichment culture. *Fems Microbiol. Ecol.* **2010**, *71*, 469–478. [CrossRef]
22. Surra, E.; Bernardo, M.; Lapa, N.; Esteves, I.; Fonseca, I.; Mota, J.P. Maize cob waste pre-treatments to enhance biogas production through co-anaerobic digestion with OFMSW. *Waste Manag.* **2018**, *72*, 193–205. [CrossRef]
23. Prasad, S.; Singh, A.; Joshi, H.C. Ethanol as an alternative fuel from agricultural, industrial and urban residues. *Resour. Conserv. Recycl.* **2007**, *50*, 1–39. [CrossRef]

24. Zabed, H.; Sahu, J.N.; Suely, A.; Boyce, A.N.; Faruq, G. Bioethanol production from renewable sources: Current perspectives and technological progress. *Renew. Sustain. Energy Rev.* **2017**, *71*, 475–501. [CrossRef]
25. Haghighi Mood, S.; Golfeshan, A.H.; Tabatabaei, M.; Jouzani, G.S.; Najafi, G.H.; Gholami, M.; Ardjmand, M. Lignocellulosic biomass to bioethanol, a comprehensive review with a focus on pretreatment. *Renew. Sustain. Energy Rev.* **2013**, *27*, 77–93. [CrossRef]
26. Kumari, D.; Singh, R. Pretreatment of lignocellulosic wastes for biofuel production: A critical review. *Renew. Sustain. Energy Rev.* **2018**, *90*, 877–891. [CrossRef]
27. Sun, Y.; Cheng, J. Hydrolysis of lignocellulosic materials for ethanol production: A review. *Bioresour. Technol.* **2002**, *83*, 1–11. [CrossRef]
28. Houbron, E.; Sandoval Rojas, M.E.; Hernández Muñoz, A.F. Tratamiento de vinazas en un reactor de lecho fluidizado inverso anaerobio. *Rev. Int. Contam. Ambient.* **2016**, *32*, 255–266. [CrossRef]
29. Zhang, C.; Xiao, G.; Peng, L.; Su, H.; Tan, T. The anaerobic co-digestion of food waste and cattle manure. *Bioresour. Technol.* **2013**, *129*, 170–176. [CrossRef]
30. Hua, B.; Dai, J.; Liu, B.; Zhang, H.; Yuan, X.; Wang, X.; Cui, Z. Pretreatment of non-sterile, rotted silage maize straw by the microbial community MC1 increases biogas production. *Bioresour. Technol.* **2016**, *216*, 699–705. [CrossRef]
31. Shah, T.A.; Tabassum, R. Enhancing biogas production from lime soaked corn cob residue. *Int. J. Renew. Energy Res.* **2018**, *18*, 761–766.
32. Zou, S.; Kang, D. Relationship between anaerobic digestion characteristics and biogas production under composting pretreatment. *Renew. Energy* **2018**, *125*, 485–494. [CrossRef]
33. Yang, G.; Zhang, P.; Zhang, G.; Wang, Y.; Yang, A. Degradation properties of protein and carbohydrate during sludge anaerobic digestion. *Bioresour. Technol.* **2015**, *192*, 126–130. [CrossRef]

© 2020 by the authors. Licensee MDPI, Basel, Switzerland. This article is an open access article distributed under the terms and conditions of the Creative Commons Attribution (CC BY) license (http://creativecommons.org/licenses/by/4.0/).

applied sciences

MDPI

Article

Effect of Particulate Disintegration on Biomethane Potential of Particle-Rich Substrates in Batch Anaerobic Reactor

Fasil Ayelegn Tassew *, Wenche Hennie Bergland, Carlos Dinamarca and Rune Bakke

Department of Process, Energy and Environmental Technology, University of South-Eastern Norway, Kjølnes Ring 56, NO 3918 Porsgrunn, Norway

* Correspondence: fasil.a.tassew@usn.no

Received: 24 June 2019; Accepted: 15 July 2019; Published: 18 July 2019

Featured Application: The findings in this article contribute to understanding solid particle disintegration and hydrolysis kinetics and how the presence of solid particulates in the form of lignocellulosic substances affect biomethane production rate and yield. It has a potential application in anaerobic digestion of particle-rich feeds in high-rate reactors.

Abstract: An investigation of particle disintegration was carried out using batch anaerobic reactors and a particle-rich substrate from pig manure supernatant. Two types of samples were applied, one high in suspended particles (raw feed) and another low in suspended particle content (centrifuged feed). Both feeds were digested with and without cellulase enzyme addition to obtain a better understanding of particle degradation mechanisms. An automatic methane potential test system (AMPTS) was used to carry out batch reactions at 35 °C. The raw feed with high-suspended solids had higher biomethane potential than the centrifuged feed but the conversion rate and methane yield was lower. The addition of cellulase increased biomethane production rates in both high- and low-particle content samples enhancing yield by 54% and 40%, respectively and converting 69% and 87% of feed chemical oxygen demand (COD), respectively. This implies that the feed particles have high contents of cellulose. This is also the case for the smaller particles remaining after centrifugation. Comparisons of anaerobic digestion model no. 1 (ADM1) simulations with experimental data reveal that classifying substrate particles into a fast and a slow degrading fraction with separate disintegration kinetics fit the experimental data better than lumping all particles into one parameter.

Keywords: anaerobic digestion; particle-rich substrate; suspended solids disintegration; disintegration kinetics; cellulase

---

1. Introduction

Biomethane potential (BMP) test is an anaerobic digestion carried out, normally, in batch reactors for a prolonged time in order to estimate the ultimate biomethane or biogas potential of a specific substrate. There is no defined volume for batch reactors but volumes 0.5–1 L are often used. The substrate and inoculum used during anaerobic digestion are characterized in terms of total and soluble chemical oxygen demand (COD), total and volatile solids (TS and VS) as well as various other parameters (Table 1). The theoretical biomethane potential is calculated using various chemical equation relationships and compared with the estimate from the BMP tests in terms of yield, such as L CH_4/g VS or g CH_4 COD/g feed COD. BMP tests are widely used due to their low cost, simplicity and repeatability. Even though BMP tests take a relatively long time, usually longer than 30 days [1], they are crucial in assessing design parameters for full-scale anaerobic reactors. Full-scale reactors, especially those that are high-rate, often face difficulty in achieving the full biomethane potential of particle-rich substrates due to slow degradation of solid particles. Significant parts of the organic substances contained in the substrate remain undigested, limiting the efficiency of the reactors. It is important to unlock the biomethane potential of such substrates. Particle-rich substrates such as the organic fraction of municipal solid waste (OFMSW) and manure are abundantly available resources that are prime candidates for anaerobic digestion and biomethane production. If the problem of slow solid disintegration were solved, the efficiency of high-rate digestion of particle-rich substrates such as manure would be greatly improved. Estimating the BMP of particle-rich substrates is one of the steps towards achieving that goal. In this article, we aim to clarify the effect of solid particle content on disintegration and hydrolysis of substrates by comparing batch test results from high-particle and low-particle substrates. The tests were carried out with and without the addition of enzyme to obtain a better understanding of the limiting factors in disintegration of particulates. Finally, we aim to establish a simple but adequate kinetic model that uses classification of complex particulates into fast and slow degrading fractions to accurately represent the disintegration of particle-rich substrates.

Table 1. Feed sample characteristics.

Property	Raw Feed (RF)	Centrifuged Feed (RF)
TS (g/L)	21.5	12.2
VS (g/L)	13.8	5.9
TSS (g/L)	14.2	2.5
VSS (g/L)	12.0	2.3
TDS (g/L)	7.3	9.7
VDS (g/L)	1.7	3.6
COD_{total} (g/L)	33.2	19.7
$COD_{soluble}$ (g/L)	16.6	11.4
NH_4^+ (g/L)	1.8	1.3
pH	7.0	7.0

TS: Total Solids; VS: Volatile Solids; TSS: Total Suspended Solids; VSS: Volatile Suspended Solids; TDS: Total Dissolved Solids; VDS: Volatile Dissolved Solids; COD: Chemical Oxygen Demand.

1.1. Lignocellulosic Substances

Presence of lignocellulosic substances in substrates is one of the main reasons for the low conversion efficiency of particle-rich substrates. Lignocellulosic substances consist of three biopolymers called cellulose, hemicellulose and lignin that are present in the cell walls of plant matter. The relative composition of the polymers differs from plant to plant. Hardwoods and softwoods contain a relatively high amount of cellulose whereas straws and grass contain higher hemicellulose content [2]. Glucose molecules are linked through beta-(1,4) glycosidic bonds to form a disaccharide that is polymerized into cellulose (Figure 1). Cellulose is homogenous because it is formed from a single monosaccharide. Hemicellulose, on the other hand, is formed from several monosaccharides including xylose and glucose. This results in a heterogeneous polymer that is more amorphous and has a more hydrolysable structure than that of cellulose. Lignin is made up of phenol-based monomers that are cross-linked to form a large and complex chemical structure that is chemically and biologically resistant to degradation.

Particle-rich substrates such as manure slurry contain a significant amount of lignocellulosic substances [3]. The source of such lignocellulosic substances is plant matter that is fed to the animals and used as bedding material for the animals. Readily biodegradable material in the animal feed is absorbed in the intestine and the leftover manure is composed of a substantial amount of lignocellulosic matter that is difficult to biodegrade. Up to 40–50% of the total solids in manure are lignocellulosic substances [3]. Lignocellulosic substances are difficult to biodegrade because their composite structure limits the accessibility of substrates by hydrolyzing enzymes [4]. A total of 20–300 monomers of cellulose are bound together by hydrogen and Van der Waals forces to make packed cellulosic microfibrils. The microfibrils are mostly in crystalline form and their outer layer is covered with hemicellulose chains. Lignin polymer binds the cellulosic microfibrils and hemicelluloses together and acts like a "glue" to form a rigid macromolecular structure that is inaccessible for enzymatic attack. Due to this reinforced concrete-like structure, disintegration and hydrolysis are difficult. By some estimates, up to 80% of lignocellulosic substances remain undegraded in biogas reactors [5]. The composition of the lignocellulosic content of manure differs from animal to animal as well as the age of the animal. For swine manure, typical cellulose, hemicellulose and lignin contents are 30–50%, 20–30% and 10–20%, respectively (Figure 2). Despite favorable qualities such as abundance, easy availability and being renewable, lignocellulosic substances have not been efficiently used for biogas production due to their strong resistance to biodegradation. Various physical, chemical and biological methods were tried to unlock the biogas potential of lignocellulosic substances with various degrees of success. One of these methods involves the addition of enzymes to facilitate the breakdown of lignocellulose components into their monosaccharides. Bacteria naturally secrete enzymes such as cellulase and hemicellulase that facilitate hydrolysis of lignocellulosic substances. Both cellulase and hemicellulase are groups of several enzymes that can carry out cellulolysis and hemicellulolysis. Other microorganisms such as fungi are also known to produce enzymes that hydrolyze lignocellulose substances. Identifying and isolating enzymes for lignocellulose hydrolysis is a growing field of research due to the advantages associated with enzymatic hydrolysis of lignocellulose such as increased biogas yield and low energy demand [6,7]. Several authors reported an increased biogas yield due to the addition of hydrolytic enzymes [8–10]. The increase in biogas yield due to the addition of enzymes depends on the type and concentration of enzymes added, temperature, pH and other parameters. There are commercially available cocktails of hydrolytic enzymes that are extracted from various microorganisms including fungi.

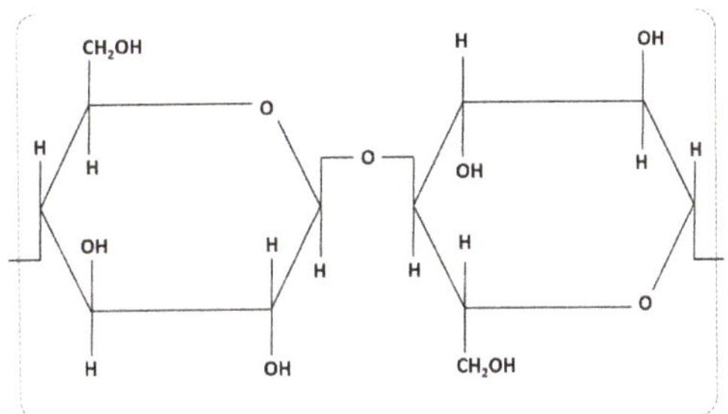

Figure 1. Building block of cellulose polymer.

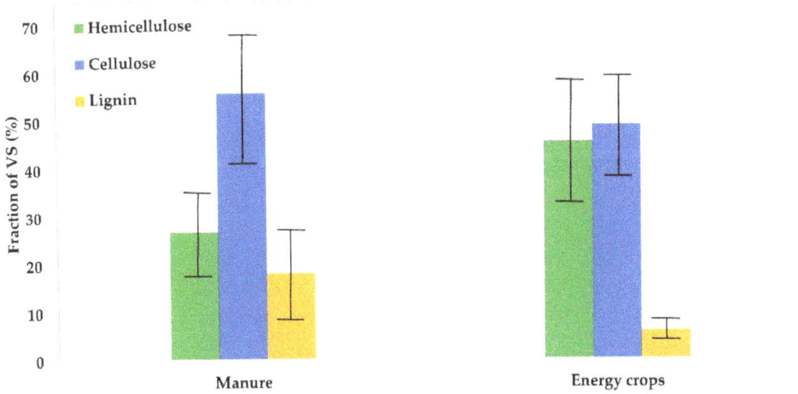

Figure 2. Composition of hemicellulose, cellulose and lignin contents as a percentage of volatile solids (VS) in manure and energy crops (data from Triolo et al. [11]).

1.2. Hydrolysis Kinetics

Hydrolysis is often assumed to be a first-order reaction [12,13] and its rate can be determined using batch reactor tests. Biomethane potential (BMP) and hydrolysis rate constant (K_h) are obtained by performing data fitting from batch reactor data. First-order kinetics is the simplest and most common hydrolysis rate expression.

$$\frac{dX}{dt} = K_h\, X \tag{1a}$$

$$\frac{dX_{ch}}{dt} = K_{h,ch}\, X_{ch} \tag{1b}$$

$$\frac{dX_{pr}}{dt} = K_{h,pr}\, X_{pr} \tag{1c}$$

$$\frac{dX_{li}}{dt} = K_{h,li}\, X_{li} \tag{1d}$$

where dX/dt is hydrolysis rate in kg $CODm^{-3}d^{-1}$, K_h is hydrolysis rate constant in d^{-1}, X is the particulate component in kg $CODm^{-3}$ and subscripts ch, pr, and li denote carbohydrate, protein and lipid, respectively. Angelidaki et al. [14] proposed a protocol for the determination of K_h from batch tests using an integrated form of the generalized hydrolysis rate expression (Equation (1a)).

$$\ln\frac{X_\infty - X}{X_\infty} = -K_h\, t \tag{2}$$

$$X = X_\infty\left(1 - e^{-K_h\, t}\right) \tag{3}$$

where, X_∞ is the value of ultimate methane production and X is the amount of methane produced at a given time, t. After batch test data are collected, a graph is plotted where K_h is determined as a slope of $\ln\frac{X_\infty - X}{X_\infty}$ and t. The last day of the batch test should be when the difference between biogas productions at day n and day n + 1 is less than or equal to 1% of the cumulative biogas production. This is in accordance with the German Guideline VDI 4630 for BMP estimation [15]. The value of K_h is important because it is a unique characteristic of a substrate and it can be used to assess the suitability of a given substrate for anaerobic digestion. It tells us how much time it takes to reach a certain percentage of the ultimate methane production [16]. There are also other methods to determine K_h experimentally such as the one suggested by Eastman and Ferguson [17].

2. Materials and Methods

Two swine manure slurry samples, with high- and low-suspended particle content, were applied. Both samples were digested with and without cellulase enzyme addition. Automatic methane potential test system (AMPTS) was used to carry out batch reactions at 35 °C. The digestion, without cellulase enzyme addition, was simulated in anaerobic digestion model no. 1 (ADM1) using two disintegration constants to describe fast and slow digestible particles.

2.1. Sample Preparation

Swine manure slurry was collected at a swine production farm in Porsgrunn, Norway. Samples were collected at various depths in intermediate indoor storage and mixed. In order to avoid the thick solid mass found at the bottom of the storage, sampling was made only in the top half of the storage. The samples still contained a substantial amount of suspended solids. This mixed sample was labelled "Raw feed" (RF). One additional sample called "Centrifuged feed" (CF) was prepared by centrifuging the raw feed sample and discarding most of the solids, thereby reducing the total and suspended solid contents. A high-speed centrifuge was used to carry out centrifugation (Beckman J-25, with JA-10 rotor). All samples were characterized immediately after preparation and kept in a refrigerator at 4 °C until they were transferred to the reactor bottles. Two separate sample groups were also prepared by adding 1.2 g cellulase in 75 mL of both RF and CF samples to form RF-cellulase (RF-CEL) and CF-cellulase (CF-CEL), respectively (Table 2). Two types of blanks were prepared, one that contained only distilled water and inoculum (BLANK) and another one that contained distilled water, inoculum and 1.2 g cellulase (BLANK-CEL). Preliminary tests, as well as reviews of works by other authors, indicated that low enzyme concentration might lead to an insignificant increase in biogas yield. As a result, we decided to use a relatively high concentration of enzyme so that the enzymatic effects are sufficiently noticeable (used 1.2 g enzyme/1.03 g VS for RF and 1.2 g Enzyme/0.44 g VS for CF).

Table 2. Sample preparation for batch test initial conditions.

Sample Name	Sample Description	Feed (mL)	Granule (mL)	Total (mL)	Headspace (mL)
RF1	Raw feed parallel 1	75 (sample)	200	275	200
RF2	Raw feed parallel 2	75 (sample)	200	275	200
RF3	Raw feed parallel 3	75 (sample)	200	275	200
RF-CEL1	Raw feed and cellulase parallel 1	1.2 g cellulase + 75 (sample)	200	275	200
RF-CEL2	Raw feed and cellulase parallel 2	1.2 g cellulase + 75 (sample)	200	275	200
RF-CEL3	Raw feed and cellulase parallel 3	1.2 g cellulase + 75 (sample)	200	275	200
CF1	Centrifuged feed parallel 1	75 (sample)	200	275	200
CF2	Centrifuged feed parallel 2	75 (sample)	200	275	200
CF3	Centrifuged feed parallel 3	75 (sample)	200	275	200
CF-CEL1	Centrifuged feed and cellulase parallel 1	1.2 g cellulase + 75 (sample)	200	275	200
CF-CEL2	Centrifuged feed and cellulase parallel 2	1.2 g cellulase + 75 (sample)	200	275	200
CF-CEL3	Centrifuged feed and cellulase parallel 3	1.2 g cellulase + 75 (sample)	200	275	200
BLANK1	Blank parallel 1	75 (distilled water)	200	275	200
BLANK2	Blank parallel 2	75 (distilled water)	200	275	200
BLANK-CEL	Blank and cellulase parallel 1	1.2 g cellulase + 75 (distilled water)	200	275	200

RF: Raw Feed; CF: Centrifuged Feed; CEL: Cellulase; BLANK: Distilled water and inoculum.

2.2. *Sample Analysis*

Total solids (TS), volatile solids (VS), total suspended solids (TSS) and volatile suspended solids (VSS) of samples were measured according to the American public health association standard method 2540 (APHA 1999) [18]. Total and soluble COD of feed samples were also measured according to the APHA standard (method 5220 D). Sample pH was measured using a Beckman 300 pH meter equipped with Sentix-82 pH electrode. Ammonium–nitrogen content (NH_4^+-N) was measured according to APHA 4500-NH_3. Both COD and NH_4^+-N concentrations were measured using commercially available test kits and Spectroquant Pharo 300 spectrophotometer (Darmstadt, Germany). Total and individual VFA (volatile fatty acid) content of samples were measured using an Agilent gas chromatography flame ionization detector (GC-FID). Sample characterization results are provided in Table 1 and VFA concentrations are provided in Table 3. Total and volatile solids contents of granular sludge were also measured as a mass percentage (% *w*/*w*) according to AMPTS II manual (Bioprocess control 2016) [19].

Table 3. Volatile fatty acid (VFA) concentration in feed samples.

VFA	Concentration (g/L)
Acetic acid	3.9
Propionic acid	0.2
Isobutyric acid	0.0
Butyric acid	0.6
Isovaleric acid	0.2
Valeric acid	0.1
Isocaproic acid	0.0
Caproic acid	0.0
Heptanoic acid	0.0
Total	4.9

2.3. Reactor and Experimental Procedure

Automatic methane potential test system (AMPTS II) from Bioprocess control, Sweden was used to carry out batch anaerobic digestion experiments [19]. The instrument includes a water bath, a CO_2 removal set-up using NaOH, adjustable motor stirrers and an apparatus for the measurement of methane flow (Figure 3). In addition, the instrument provides software to control/monitor reactor settings and plot gas measurement. The batch test was carried out in three parallels for RF, RF-CEL, CF and CF-CEL samples. The maximum number of batch reactors in the AMPTS II set up was 15 allowing two parallels for BLANK and only one parallel for BLANK-CEL. The experiment was carried out at 35 °C. Reactor contents were stirred every hour to allow proper mixing and to facilitate gas removal from the reactors.

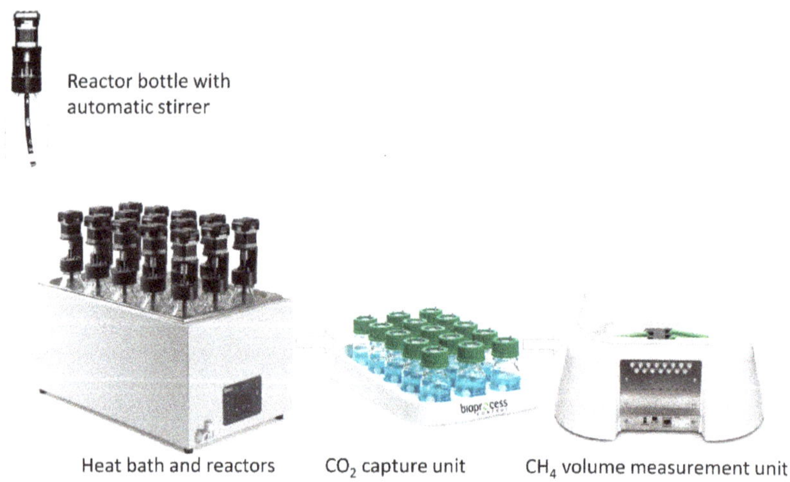

Figure 3. Automatic methane potential test system (AMPTS) II batch reactor experimental setup (pictures from Bioprocess control's homepage).

2.4. Granular Sludge Degassing

A mixture of granules from various sources was used as inoculum for the batch experiments (mainly granules that have been used to treat wastewater from pulp and paper industry mixed with granules obtained from econvert Water & Energy, Heerenveen, the Netherlands). The granules were degassed before samples were added [14]. Degassing was performed by placing granule containers in a water bath at the reaction temperature (35 °C) for 10 days. The total solid and volatile solid contents of the granule were measured after degassing.

2.5. CO_2 Removal

The CO_2 removal set-up requires solution preparation and the following solutions were prepared [19]. A solution of 1.2 L NaOH (3 M) was prepared by mixing 144 g of NaOH in 1.2 L distilled water. pH indicator thymolphthalein solution (0.4%) was prepared by mixing 40 mg of thymolphthalein in 9 mL of ethanol (99.5%) and 1 mL distilled water. A total of 6 mL of the thymolphthalein solution was added into 1.2 L NaOH solution. The resulting mixture was transferred to CO_2 removal bottles, one for each batch test and each with an 80 mL mixture. The thymolphthalein-NaOH mixture has a bright blue color and when enough CO_2 is absorbed, the blue color fades and becomes colorless. Thymolphthalein is bright blue in basic solutions but it turns colorless in acidic or neutral solutions. At this point, a new mixture has to be used. According to the CO_2 removal manual, this method absorbs more than 98% of CO_2 produced during the biogas production process. The removal is based on the following reaction [20]:

$$CO_{2\,(g)} \rightarrow CO_{2\,(aq)} \tag{4}$$

$$CO_{2\,(aq)} + OH^-_{(aq)} \rightarrow HCO_3{}^-_{(aq)} \tag{5}$$

$$HCO_3{}^-_{(aq)} + OH^-_{(aq)} \rightarrow CO^{2-}_{3\,(aq)} + H_2O_{(l)} \tag{6}$$

2.6. Theoretical Methane Yield

Theoretical methane yield was calculated based on the total COD of RF and CF samples. Theoretical calculation was performed as follows [21].

a. Determine COD equivalent of methane:

One mole of methane requires two moles of oxygen, meaning the chemical oxygen demand of methane is:

$$CH_4 + 2O_2 \rightarrow CO_2 + 2H_2O \tag{7}$$

$$\text{COD/mole } CH_4 = 2 \times 32 \text{ g } O_2\text{/mole} = 64 \text{ g } O_2\text{/mole}$$

b. Determine the theoretical volume of methane based on g COD

The volume of a mole of methane gas at standard conditions of 0 °C and one atm (atmospheric pressure) is 22.4 L. The theoretical volume of methane that can be obtained from a gram of COD is calculated as:

$$22.4 \text{ L } CH_4/64 \text{ g COD} = 0.35 \text{ L } CH_4/\text{g COD}$$

To calculate the theoretical methane yield at 35 °C, we used the ideal gas law:

$$\frac{P_1V_1}{T_1} = \frac{P_2V_2}{T_2} = \text{Constant} \tag{8}$$

$$\text{Yield at 35 °C} = \frac{(1\text{ atm})(0.35\text{ LCH}_4/\text{g COD})(308.15\text{ K})}{(1\text{ atm})(298.15\text{ K})} = 0.36\text{ LCH}_4/\text{g COD}$$

c. Calculate theoretical methane yield of sample

Theoretical methane yields of all samples were calculated from total COD, sample volume and the value for yield at 35 °C.

$$CH_4\text{ yield of sample (L)} = (COD_{total})(V_{sample})(0.36\text{ LCH}_4/\text{g COD}) \tag{9}$$

d. Compare theoretical and experimental methane yield

Experimental methane yield was corrected by subtracting the average volume of methane produced by blank parallels (V_{B1}, V_{B2}, V_{B3}) from that of samples (V_{P1}, V_{P2}, V_{P3}).

$$\text{Experimental CH}_4\text{ yield (L)} = \frac{\sum (V_{P1} + V_{P2} + V_{P3})}{N_{sample}} - \frac{\sum (V_{B1} + V_{B2} + V_{B3})}{N_{blank}} \quad (10)$$

2.7. Anaerobic Digestion Model No. 1 (ADM1) Simulation

Aquasim software was used to implement the ADM1 model to simulate the batch reactors (Table 4). Two modes of disintegration kinetics were used. In the first mode, first-order disintegration kinetics was used, and all complex particulates were assumed to be equally degradable (single disintegration constant, K_{dis} used). In the second mode, the complex particulates were classified into fast degrading and slow degrading fractions, where two separate disintegration constants, K_{dis1} for fast degrading and K_{dis2} for slow degrading fractions, were used (Figure 4).

Table 4. Selected simulation parameters and values used to implement ADM1 in Aquasim.

Parameter	RF	RF	CF	CF
	K_{dis}	K_{dis1} and K_{dis2}	K_{dis}	K_{dis1} and K_{dis2}
Disintegration constant (d^{-1})	0.17	0.17, 0.075	0.17	0.17, 0.075
Amino acid degrading organisms (kg COD/m^3)	1.70	1.70	1.70	1.70
Acetate degrading organisms (kg COD/m^3)	2.23	2.23	2.23	2.23
Butyrate/valerate degrading organisms (kg COD/m^3)	0.69	0.69	0.69	0.69
Fatty acid degrading organisms (kg COD/m^3)	1.85	1.85	1.85	1.85
Hydrogen degrading organisms (kg COD/m^3)	1.05	1.05	1.05	1.05
Propionate degrading organisms (kg COD/m^3)	0.29	0.29	0.29	0.29
Sugar degrading organisms (kg COD/m^3)	1.68	1.68	1.68	1.68
Soluble amino acids (kg COD/m^3)	0.2	0.2	0.2	0.2
Soluble fatty acids (kg COD/m^3)	0.25	0.25	0.25	0.25
Soluble acetates (kg COD/m^3)	0.66	0.66	0.66	0.66
Soluble butyrates (kg COD/m^3)	0.15	0.15	0.15	0.15
Soluble propionates (kg COD/m^3)	0.05	0.05	0.05	0.05
Soluble valerates (kg COD/m^3)	0.08	0.08	0.08	0.08

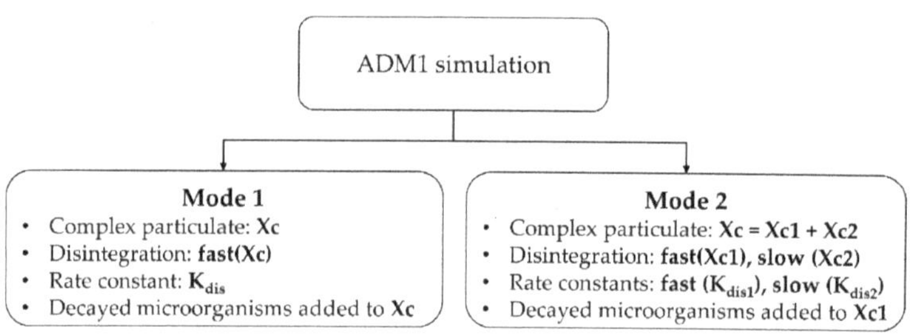

Figure 4. ADM1 simulation scheme for mode 1 and mode 2 simulations.

Classification of particulates into fast degrading and slow degrading was carried out as follows:

a. Calculate particulate COD (Xc)

$$\text{Particulate COD} = \text{COD}_{\text{total}} - \text{COD}_{\text{soluble}} \quad (11)$$

$$\text{COD}_{\text{total}} = \text{Total g COD in reactor}/V_{\text{reactor}} = \text{Sample COD}_{\text{total}} \times V_{\text{sample}}/V_{\text{reactor}} \quad (12)$$

For simplification of the simulation, soluble COD was assumed to equal to COD of volatile fatty acids (COD_{VFA}) and some minor constituents as seen in Table 4, according to Equation (13).

$$\text{COD}_{\text{soluble}} = \text{COD}_{\text{VFA}} + (\text{COD}_{\text{soluble amino acids}} + \text{COD}_{\text{soluble fatty acids}} + \text{COD}_{\text{soluble inerts}} + \text{COD}_{\text{soluble sugars}}) \quad (13)$$

b. Classify particulate COD into fast (Xc1) and slow (Xc2) degrading fractions:

Classification of Xc into Xc1 and Xc2 was carried out separately for RF and CF. The ratio of total dissolved solids to total solids was used as a basis to estimate the fast degrading fraction (Xc1) from which Xc2 was estimated (Xc2 = 1 − Xc1). Since RF contains a relatively large fraction of solid particulates, it was estimated that 85% of the COD comes from slowly degrading fragments and the rest from fast degrading fragments. In the case of CF, most of the solid particulates are removed due to centrifugation making COD from solid particles constitute a small part of the total COD. We estimated that 15% of COD comes from slow degrading and 85% comes from fast degrading particulates.

For RF:

$$\text{Particulate COD (Xc)} = (33.24 \text{ g COD/L sample} \times 0.075 \text{ L sample} / 0.275 \text{ L}) - (1.35 \text{ g/L}) = 7.72 \text{ g COD/L}$$

$$\text{Xc1} = 0.85 \times 7.72 \text{ g COD/L} = 6.56 \text{ g COD/L}$$

$$\text{Xc2} = 0.15 \times 7.72 \text{ g COD/L} = 1.16 \text{ g COD/L}$$

For CF:

$$\text{Particulate COD (Xc)} = (19.74 \text{ g COD/L sample} \times 0.075 \text{ L sample} / 0.275 \text{ L}) - (1.35 \text{ g/L}) = 4.03 \text{ g COD/L}$$

$$\text{Xc1} = 0.15 \times 4.03 \text{ g COD/L} = 0.605 \text{ g COD/L}$$

$$\text{Xc2} = 0.85 \times 4.03 \text{ g COD/L} = 3.43 \text{ g COD/L}$$

Based on suggestions from preliminary experimental data and literature survey [22], we used K_{dis} value of 0.17 d^{-1} for swine manure samples. For fast degrading fraction, K_{dis1} stays at 0.17 d^{-1} and for K_{dis2} we used 0.075 d^{-1} (~45% of K_{dis1}, estimated from biogas production data for straws, fibers and other solids). We used hydrolysis constants (10 d^{-1}) as suggested by Batstone et al. [23].

In the ADM1 model, decayed microorganisms are added into complex particulates (Xc). In the first mode of simulation, there is no change; all decayed microorganisms are added back to Xc, however, in the second mode, the decayed microorganisms are recycled back to the fast degrading (Xc1) fraction only.

3. Result and Discussion

The raw feed with high-suspended solids had higher biomethane potential per liter of substrate than the centrifuged feed but the conversion rate and methane yield (g COD_{CH4}/g $\text{COD}_{\text{total}}$) was lower. Addition of cellulase increased biomethane production rates and yields in both high- and low-particle content samples.

3.1. Yields

Measured average methane production for the four cases investigated and two blank cases are presented in Figure 5. Blank adjusted total methane productions after 40 d were 403 ± 73 mL for RF, 621 ± 54 mL for RF-CEL, 331 ± 61 mL for CF and 462 ± 57 mL for CF-CEL. The highest volumes of methane were produced by cellulase containing samples RF-CEL and CF-CEL. Cellulase enhanced the COD conversions from 45% to 69% and 62% to 87% for RF and CF samples, respectively (Table 5). As expected, centrifuged samples resulted in lower ultimate methane production but higher specific methane yield than their non-centrifuged counterparts. From RF to RF-CEL specific yield increased from 390 to 600 L CH_4/kg VS and from CF to CF-CEL it increased from 742 to 1037 L CH_4/kg VS (Table 6).

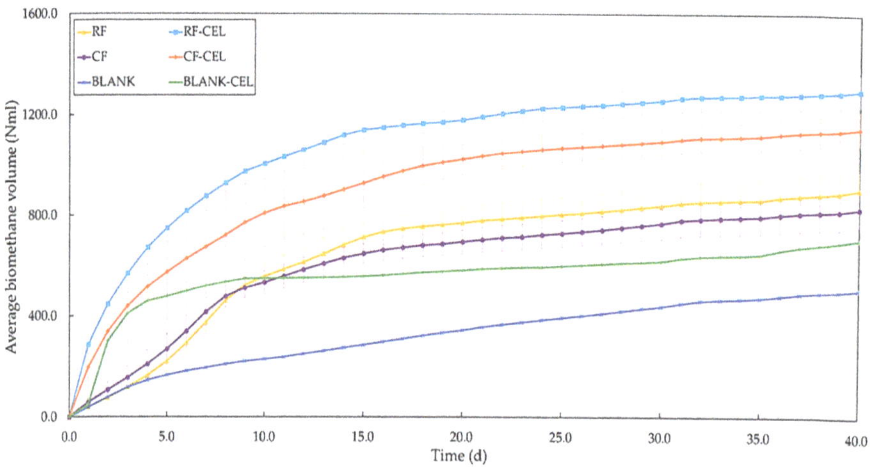

Figure 5. Average biomethane production for raw feed (RF) and centrifuged feed (CF), with and without cellulase, and two blank cases.

Table 5. Comparison of theoretical (assuming complete feed COD conversion) and experimental methane productions.

Sample	Experimental (mL)	Theoretical (mL)	Efficiency (%)
RF	403 ± 73	898	45
RF-CEL	621 ± 54	898	69
CF	331 ± 61	533	62
CF-CEL	462 ± 57	533	87

Table 6. Specific methane yield of samples.

Specific Methane Yield	RF	RF-CEL	CF	CF-CEL
L CH_4/g TS	0.25	0.38	0.36	0.51
L CH_4/g VS	0.4	0.6	0.7	1.0
L CH_4/g COD_{total}	0.16	0.25	0.22	0.31
L CH_4/g $COD_{soluble}$	0.32	0.5	0.39	0.54
g COD_{CH4}/g COD_{total}	0.44	0.69	0.61	0.86

3.2. Production Rates

Biomethane production rates peaked faster with much higher maximum production rates in samples with cellulase addition (RF-CEL and CF-CEL) than those without addition (Figure 6). Cellulase-added samples showed maximum biomethane production rate in the first 30 h of the experiment while the cases without enzyme addition had much lower maximum production and it was distributed over a longer time span, peaking after ~150 h. A brief peak during startup in all cases is assumed irrelevant (methane release from methane saturated inoculum due to temperature increase). RF-CEL reached a maximum of 34 mL/h at 17 h and CF-CEL reached 19 mL/h at 19 h followed by a decrease to ~3 mL/h at 250 h. The BLANK-CEL sample with cellulase but without added substrates reached a methane production rate of 48 mL/h at 28 h, showing that the cellulase itself has a significant BMP and can be degraded quickly. Fortunately, the cellulase was degraded after the main degradation peaks for the feeds, implying that it can carry out the intended enzymatic attack on the feed particulates before it is itself degraded and converted to methane. Cellulase can, therefore, be added to anaerobic digesters to enhance biomethane production from cellulose containing feeds but it is not analyzed here whether this is a sustainable solution.

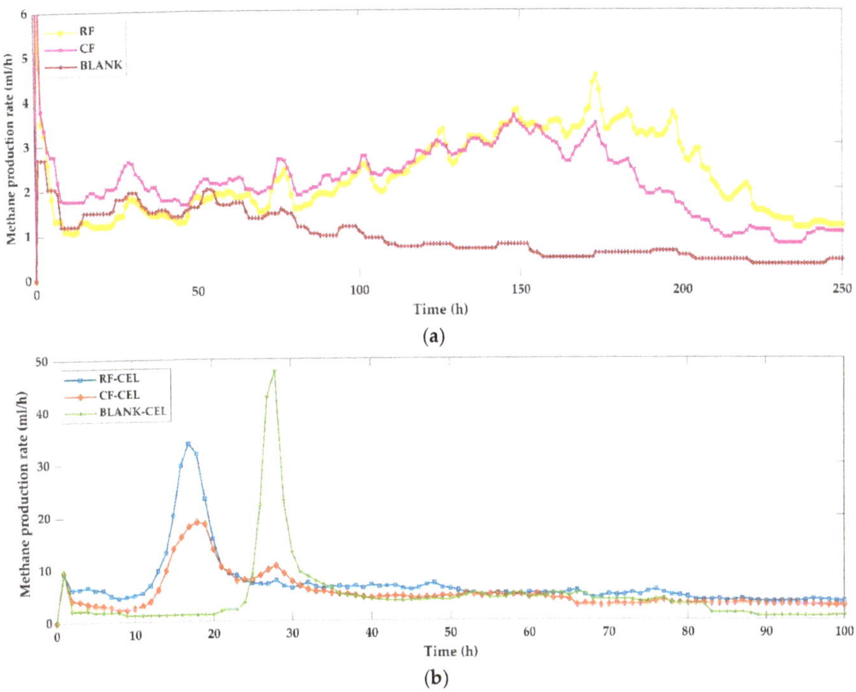

Figure 6. Measured average biomethane production rate: without cellulase (**a**) for the first 250 h and with cellulase (**b**) for the first 100 h.

3.3. *Effect of Cellulase Addition and Centrifugation*

Enzyme additions were expected to have a stronger effect on samples that contain more particles, as observed. However, methane yield increase from CF to CF-CEL was quite high (40%) considering that centrifugation removed more than 80% of feed VSS (Table 1). The observation that addition of cellulase enzyme had a positive and comparable impact on methane yield both in raw and centrifuged samples (specific methane yield increased by 54% and 40% for RF and CF, respectively (Table 5)) suggests that there are similar fractions of cellulose in large and small particles in such animal manure slurries. The small particles evidently needed to undergo a similar disintegration process as those removed by centrifugation, with maximum rates at approximately the same time both with (Figure 6b) and without (Figure 6a) enzyme addition. Hydrolysis rate constants, K_h, determined from the slope of the plot ln $\frac{X_\infty - X}{X_\infty}$ against t show how these observations can be included in process modeling. Only the first few days of the plot, where the curve was at its steepest was used to determine K_h in accordance with suggestions by Angelidaki et al. [14]. Enzyme addition led to much higher K_h values while centrifugation caused marginally larger K_h (Table 7).

Table 7. Hydrolysis rate constant values estimated using Equation (2).

Sample	K_h (d^{-1})
RF	0.088
RF-CEL	0.154
CF	0.094
CF-CEL	0.120
BLANK	0.062
BLANK-CEL	0.148

3.4. Simulation Results

Using a single disintegration constant assumes all particulates disintegrate equally. K_{dis} values in this mode of simulation are usually estimated based on the fast degrading fractions and this leads to an overestimation of methane production. In mode 1 simulation of the RF sample, methane production peaked very early in the digestion process (4–5 d) and continued to decline for the rest of the digestion, which differs significantly from the pattern of methane production observed in the experimental results. This phenomenon is visible in Figure 7a. Classifying particulates into fast degrading and slow degrading fractions seemed to rectify the overestimation of methane production (mode 2). Accounting for slow degrading particulates led to similar patterns in the timing of peak methane production. When the two modes of simulations are compared, it is apparent that the contribution of slow degrading particulates to the methane production became more and more significant at the later stages of digestion. Application of both modes of simulations on CF samples did not lead to significantly different results. Unlike RF, CF samples contain relatively small quantities of solid particulates, which are the main causes of reduced disintegration rate. Even if K_{dis}, K_{dis1} and K_{dis2} values for CF and RF are the same, the relative proportion of fast and slow degrading fractions are different. RF contains far more slow disintegrating particulates than CF. As a result, it is expected that simulations of CF in mode 1 (where all particulates are assumed to be degraded at K_{dis} of 0.17 d^{-1}) and mode 2 (where 85% of particulates are assumed to be degraded at K_{dis1} of 0.17 d^{-1}) lead to similar patterns of methane production. Both modes of simulations suggested peak methane productions slightly earlier than observed. Comparing both modes of simulations for CF samples (Figure 7b) it is noticeable that mode 1 simulation seemed a better fit for CF samples than mode 2, suggesting that classification of particulates into fast and slow degrading fractions may be better suited for particle-rich substrates than particle "free" substrates.

The patterns of degradation of particulates were compared using data from mode 1 and mode 2 simulation results (Figure 8). In both RF and CF mode 1 simulations, Xc followed a "logarithmic" decline throughout the course of the digestion process. In RF mode 2 simulation, slow degrading particulates decline in a similar fashion as the one observed by Xc in mode 1 simulations, however fast degrading particulates increased first (until 8–10 days) followed by a gradual decline. The increase in the fast degrading particulates is partly attributed to decaying microorganisms being added into Xc1. Particle size and presence of recalcitrant substances contribute heavily to the slow degradation of solid particulates. After disintegration, the rest of the anaerobic digestion process continues the same way whether the disintegrated particulate originated from slow or fast disintegrating fractions. As a result of this, the slowly disintegrated particulates are continuously being added into a rapidly disintegrated fraction that contributes to the increase in Xc1 at the beginning of the digestion. Mode 2 simulation of the CF sample did not show increasing Xc1 mainly due to the absence of enough slowly disintegrating particulates continuously added to it. In addition, the contribution from microbial decay was minimal.

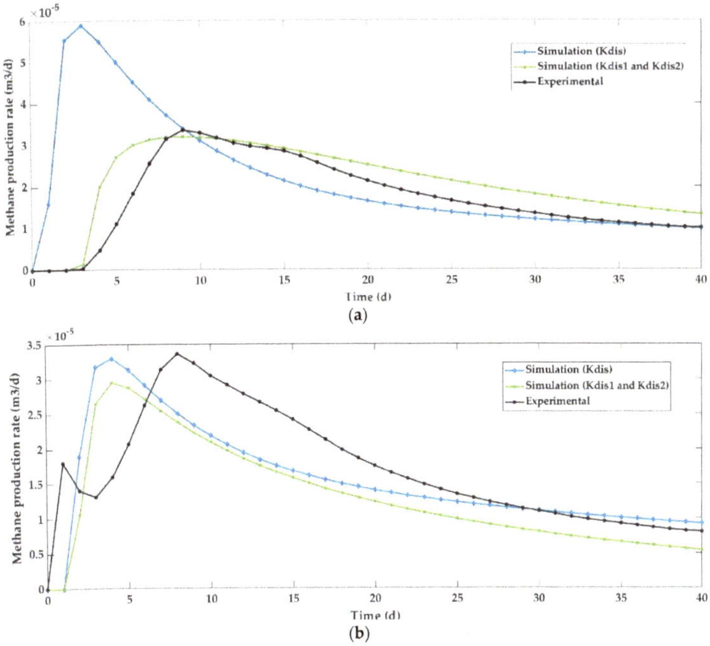

Figure 7. Methane production rate of simulation and experimental results: RF (**a**) and CF (**b**).

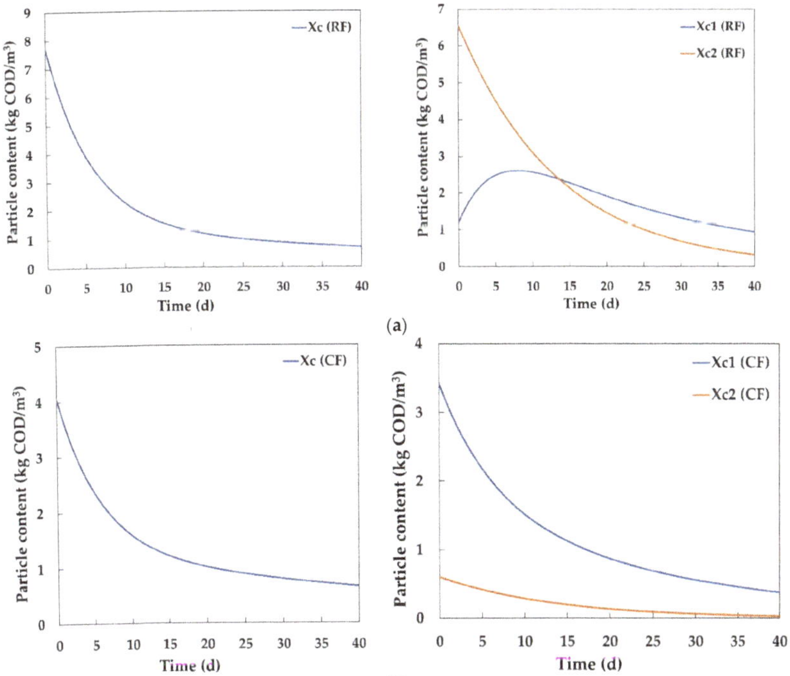

Figure 8. Comparison of particulate degradation in samples: RF (**a**) and CF (**b**).

Sensitivity analysis was carried out on K_{dis1} and K_{dis2} parameters in mode 2 simulation. Sensitivity analysis, which combines identifiability and uncertainty analysis, is used to check if K_{dis} parameters can be uniquely determined from available data [24]. The sensitivity function of methane production with respect to K_{dis1} and K_{dis2} are shown in Figures 9 and 10. In both RF and CF, the methane production was much more sensitive to $K_{dis}1$ than K_{dis2}. Sensitivity to K_{dis2} was more apparent in RF than CF. The sensitivity of various other variables to K_{dis1} and K_{dis2} is also given in the form of sensitivity functions (SensAR) in Appendix A. The absolute-relative sensitivity function (Equation (A1)), was used to measure the absolute change in y (methane production) for 100% change in p (K_{dis1}/K_{dis2}).

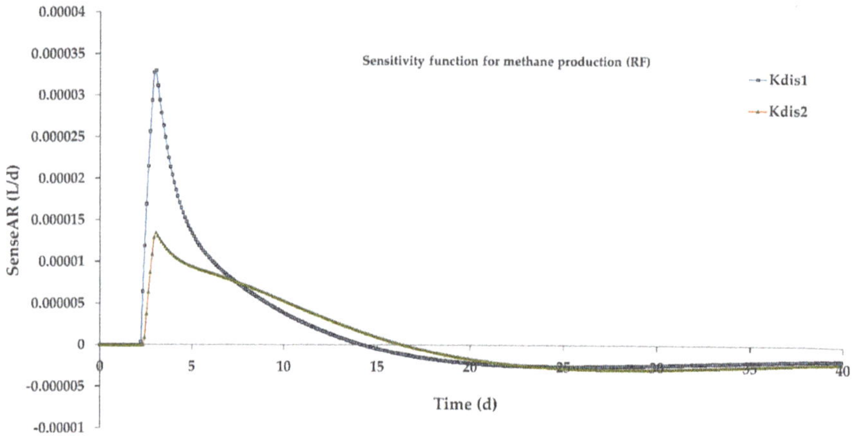

Figure 9. Sensitivity function of methane production with respect to K_{dis1} and K_{dis2} (RF).

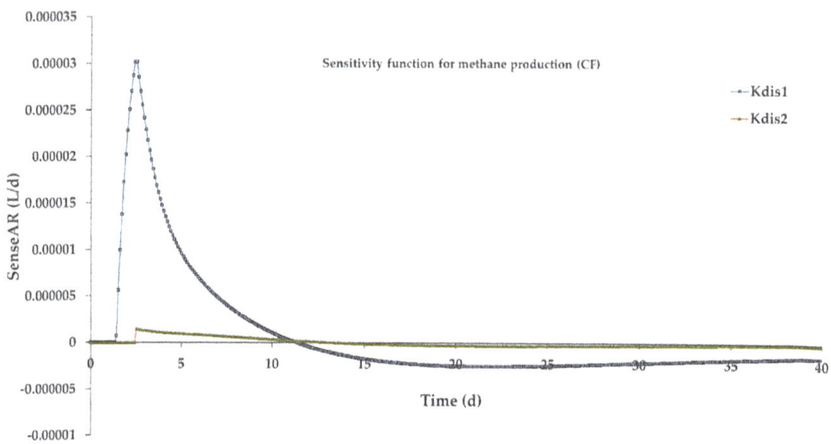

Figure 10. Sensitivity function of methane production with respect to K_{dis1} and K_{dis2} (CF).

Comparison of rates for RF and CF samples showed that the rate of biomethane production is faster in particles with lower solid particles, but it also showed that solid particulates breakdown slowly resulting in a steady biomethane production over a long period. In continuous reactors, solid accumulation may occur when substrates are added continuously without efficient solid disintegration and removal. Ideally, there should be a balance between rates of solid substrate addition and solid disintegration for a stable digestion process. In granular sludge bed reactors, solid particulates are often trapped in the sludge bed for long periods, meaning solid retention times much longer than the hydraulic retention time can be achieved, but appreciable disintegration of trapped solid may be hindered due to various reasons among which are large particle size, inefficient mixing, and mass transfer limitation. A carefully adjusted balance between influent solids and solid disintegration kinetics has to be established and considered during reactor design in order to use the granular sludge bed for particle-rich substrates. Reactor conditions such as volume temperature, HRT (Hydraulic Retention Time) and SRT (Solids Retention Time) have to consider possible solid accumulations. In addition, a combination periodic removal of excess solids, pretreatment of solid substrates before and during reaction and continuous monitoring of reactor conditions have to be maintained.

4. Conclusions

After conducting batch reactor tests and analyzing results from substrates with high- and low-suspended particle contents, we have made the following conclusions:

- High biomethane production was observed in samples with higher particle content however, specific biomethane yield was low compared to samples with low particle contents.
- Centrifugation of samples decreased the volume of methane produced but increased the rate of methane production regardless of the addition of cellulase.
- Cellulase addition improved overall and specific methane productions both in raw and centrifuged samples but the improvement was higher in samples that contained higher suspended solids.
- Simulation results revealed that classifying complex particulates into fast and slow disintegrating fractions led to a more accurate modeling of particle-rich substrates.

Suggestions for Future Work

This article is based on experimental results from batch anaerobic reactors. As a result, its applicability may be limited. In order to increase the validity of the findings in this work, the authors recommend future investigations on how the classification of complex particulates into fast and slow degrading fractions could be implemented in continuous anaerobic reactors. The effect of temperature on the disintegration of particulates would also be an interesting investigation to carry out.

Author Contributions: Conceptualization, F.A.T., W.H.B., C.D. and R.B.; methodology, F.A.T.; software, F.A.T.; validation, F.A.T.; formal analysis, All authors; investigation, All authors; resources, F.A.T.; data curation, F.A.T.; writing—original draft preparation, F.A.T.; writing—review and editing, All authors; visualization, F.A.T.; supervision, R.B. and W.H.B.; project administration, R.B.; funding acquisition, R.B.

Funding: This research was part of a PhD project funded by the European Regional Development Fund; Interreg BioGas2020.

Acknowledgments: The authors would like to thank Dag Øvrebø for his help during sample collection.

Conflicts of Interest: The authors declare no conflict of interest.

Appendix A

The absolute-relative sensitivity function is used in Aquasim software to measure the absolute change in y for 100% change in P. In this case, y is methane production and P is parameter K_{dis1} or K_{dis2}.

$$\text{Absolute} - \text{Relative sensitivity function} = P\frac{\partial y}{\partial P} \quad \text{(A1)}$$

In the tables below, the sensitivity function (SensAR) is expressed in: root mean square (r(av(SensAR^2))) and mean absolute (av(|SensAR|)) and for error contributions as: (av(|ErrCont|)). S_CH4, S_CO2 and S_H2 are concentrations of CH_4, CO_2 and H_2, respectively.

Table A1. Variables ranked based on sensitivity to K_{dis1} and K_{dis2} in the headspace (RF).

Variable	r(av(SensAR^2))		av(|SensAR|)		av(|ErrCont|)	
	K_{dis1}	K_{dis2}	K_{dis1}	K_{dis2}	K_{dis1}	K_{dis2}
S_CH4	0.03	0.02	0.01	0	0.05	0.03
S_CO2	0	0	0	0	0	0
S_H2	0	0	0	0	0	0

Table A2. Variables ranked based on sensitivity to K_{dis1} and K_{dis2} in the bulk reactor (RF).

Variable	r(av(SensAR^2))		av(|SensAR|)		av(|ErrCont|)	
	K_{dis1}	K_{dis2}	K_{dis1}	K_{dis2}	K_{dis1}	K_{dis2}
S_CH4	0.001	0.001	0	0	0.002	0.001
S_CO2	0	0	0	0	0.002	0.002
S_H2	0	0	0	0	0	0
Other parameters						
Xc1	0.32	0.10	0.12	0.02	0.73	0.32
Xc2	0	0.25	0	0.06	0	0.83

Table A3. Variables ranked based on sensitivity to K_{dis1} and K_{dis2} in the headspace (CF).

Variable	r(av(SensAR^2))		av(|SensAR|)		av(|ErrCont|)	
	K_{dis1}	K_{dis2}	K_{dis1}	K_{dis2}	K_{dis1}	K_{dis2}
S_CH4	0.09	0	0.04	0	0.24	0.04
S_CO2	0	0	0	0	0.01	0
S_H2	0	0	0	0	0	0

Table A4. Variables ranked based on sensitivity to K_{dis1} and K_{dis2} in the bulk reactor (CF).

Variable	r(av(SensAR^2))		av(|SensAR|)		av(|ErrCont|)	
	K_{dis1}	K_{dis2}	K_{dis1}	K_{dis2}	K_{dis1}	K_{dis2}
S_CH4	0.003	0	0.001	0	0.008	0.001
S_CO2	0.001	0	0.001	0	0.003	0.001
S_H2	0	0	0	0	0	0
Other parameters						
Xc1	0.63	0.03	0.53	0.02	3.09	0.28
Xc2	0	0.07	0.00	0.06	0	0.75

References

1. Esposito, G. Bio-Methane Potential Tests to Measure The Biogas Production From The Digestion and Co-Digestion of Complex Organic Substrates. *Open Environ. Eng. J.* **2012**, *5*, 1–8. [CrossRef]
2. Bajpai, P. *Pretreatment of Lignocellulosic Biomass for Biofuel Production*; Springer Science and Business Media LLC: Berlin/Heidelberg, Germany, 2016.
3. Bruni, E.; Jensen, A.P.; Angelidaki, I. Steam treatment of digested biofibers for increasing biogas production. *Bioresour. Technol.* **2010**, *101*, 7668–7671. [CrossRef] [PubMed]
4. Sun, Y.; Cheng, J. Hydrolysis of lignocellulosic materials for ethanol production: A review. *Bioresour. Technol.* **2002**, *83*, 1–11. [CrossRef]
5. Pandey, A. *Handbook of Plant-Based Biofuels*; CRC Press: Boca Raton, FL, USA, 2008.
6. Maitan-Alfenas, G.P.; Visser, E.M.; Guimarães, V.M. Enzymatic hydrolysis of lignocellulosic biomass: Converting food waste in valuable products. *Curr. Opin. Food Sci.* **2015**, *1*, 44–49. [CrossRef]
7. Christy, P.M.; Gopinath, L.; Divya, D. A review on anaerobic decomposition and enhancement of biogas production through enzymes and microorganisms. *Renew. Sustain. Energy Rev.* **2014**, *34*, 167–173. [CrossRef]

8. Romano, R.T.; Zhang, R.; Teter, S.; McGarvey, J.A. The effect of enzyme addition on anaerobic digestion of Jose Tall Wheat Grass. *Bioresour. Technol.* **2009**, *100*, 4564–4571. [CrossRef] [PubMed]
9. Malayil, S.; Chanakya, H. Fungal Enzyme Cocktail Treatment of Biomass for Higher Biogas Production from Leaf Litter. *Procedia Environ. Sci.* **2016**, *35*, 826–832. [CrossRef]
10. Quiñones, T.S.; Plöchl, M.; Budde, J.; Heiermann, M. Results of Batch Anaerobic Digestion Test–Effect of Enzyme Addition. *Agric. Eng. Int. CIGR J.* **2012**, *14*, 38–50.
11. Triolo, J.M.; Sommer, S.G.; Møller, H.B.; Weisbjerg, M.R.; Jiang, X.Y. A new algorithm to characterize biodegradability of biomass during anaerobic digestion: Influence of lignin concentration on methane production potential. *Bioresour. Technol.* **2011**, *102*, 9395–9402. [CrossRef] [PubMed]
12. Koch, K.; Drewes, J.E. Alternative approach to estimate the hydrolysis rate constant of particulate material from batch data. *Appl. Energy* **2014**, *120*, 11–15. [CrossRef]
13. Vavilin, V.; Rytov, S.; Lokshina, L. A description of hydrolysis kinetics in anaerobic degradation of particulate organic matter. *Bioresour. Technol.* **1996**, *56*, 229–237. [CrossRef]
14. Angelidaki, I.; Alves, M.; Bolzonella, D.; Borzacconi, L.; Campos, J.L.; Guwy, A.J.; Kalyuzhnyi, S.; Jenicek, P.; Van Lier, J.B. Defining the biomethane potential (BMP) of solid organic wastes and energy crops: A proposed protocol for batch assays. *Water Sci. Technol.* **2009**, *59*, 927–934. [CrossRef] [PubMed]
15. VDI—The Association of German Engineers. *Vergärung organischer Stoffe. Substratcharakterisierung, Probenahme, Stoffdatenerhebung, Gärversuche*; Beuth Verlag: Berlin, Germany, 2006.
16. Macé, S.; Bolzonella, D.; Cecchi, F.; Mata-Álvarez, J. Comparison of the biodegradability of the grey fraction of municipal solid waste of Barcelona in mesophilic and thermophilic conditions. *Water Sci. Technol.* **2003**, *48*, 21–28. [CrossRef] [PubMed]
17. Eastman, J.A.; Ferguson, J.F. Solubilization of Particulate Organic Carbon during the Acid Phase of Anaerobic Digestion. *J. (Water Pollut. Control Fed.)* **1981**, *53*, 352–366.
18. APHA; Water Environment Federation; American Water Works Association. *Standard Methods for the Examination of Water and Wastewater*; American Public Health Association: Washington, DC, USA, 1998.
19. Bioprocess Control. *AMPTS II Light Automatic Methane Potential Test System Operation and Maintenance Manual*; Bioprocess Control: Lund, Sweden, 2016.
20. Darmana, D.; Henket, R.; Deen, N.G.; Kuipers, J. Detailed modelling of hydrodynamics, mass transfer and chemical reactions in a bubble column using a discrete bubble model: Chemisorption of CO_2 into NaOH solution, numerical and experimental study. *Chem. Eng. Sci.* **2007**, *62*, 2556–2575. [CrossRef]
21. Tchobanoglous, G.; Burton, F.L.; Stensel, H.D. *Wastewater Engineering Treatment and Reuse*; McGraw-Hill Education: Boston, MA, USA, 2003.
22. Gali, A.; Benabdallah, T.; Astals, S.; Mata-Alvarez, J. Modified version of ADM1 model for agro-waste application. *Bioresour. Technol.* **2009**, *100*, 2783–2790. [CrossRef] [PubMed]
23. Batstone, D.; Keller, J.; Angelidaki, I.; Kalyuzhnyi, S.; Pavlostathis, S.; Rozzi, A.; Sanders, W.; Siegrist, H.; Vavilin, V. The IWA Anaerobic Digestion Model No 1 (ADM1). *Water Sci. Technol.* **2002**, *45*, 65–73. [CrossRef] [PubMed]
24. Reichert, P. *Aquasim 2.0-User Manual*; Swiss Federal Institute for Environmental Science and Technology: Dubendorf, Switzerland, 1998.

© 2019 by the authors. Licensee MDPI, Basel, Switzerland. This article is an open access article distributed under the terms and conditions of the Creative Commons Attribution (CC BY) license (http://creativecommons.org/licenses/by/4.0/).

applied sciences

MDPI

Review

Improvement of Anaerobic Digestion of Lignocellulosic Biomass by Hydrothermal Pretreatment

Banafsha Ahmed [1], Kaoutar Aboudi [2], Vinay Kumar Tyagi [1], Carlos José Álvarez-Gallego [2], Luis Alberto Fernández-Güelfo [3], Luis Isidoro Romero-García [2,*] and A. A. Kazmi [1]

1 Department of Civil Engineering, Indian Institute of Technology Roorkee, Roorkee 247667, India

2 Department of Chemical Engineering and Food Technology, Faculty of Sciences. University of Cádiz, P.O. Box n 40, Puerto Real, 11510 Cádiz, Spain

3 Department of Environmental Technologies, Faculty of Marine Sciences. University of Cádiz, P.O. Box n 40, Puerto Real, 11510 Cádiz, Spain

* Correspondence: luisisidoro.romero@uca.es

Received: 23 July 2019; Accepted: 29 August 2019; Published: 13 September 2019

Abstract: Lignocellulosic biomass, comprising of cellulose, hemicellulose, and lignin, is a difficult-to-degrade substrate when subjected to anaerobic digestion. Hydrothermal pretreatment of lignocellulosic biomass could enhance the process performance by increasing the generation of methane, hydrogen, and bioethanol. The recalcitrants (furfurals, and 5-HMF) could be formed at high temperatures during hydrothermal pretreatment of lignocellulosic biomass, which may hinder the process performance. However, the detoxification process involving the use of genetically engineered microbes may be a promising option to reduce the toxic effects of inhibitors. The key challenge lies in the scaleup of the hydrothermal process, mainly due to necessity of upholding high temperature in sizeable reactors, which may demand high capital and operational costs. Thus, more efforts should be towards the techno-economic feasibility of hydrothermal pre-treatment at full scale.

Keywords: lignocellulosic biomass; anaerobic digestion; hydro-thermal pretreatment; biofuels

1. Lignocellulosic Biomass

Lignocellulosic biomass (LB), namely agri-wastes and energy crops, have been gaining much attention as candidate feedstocks for producing bioenergy and biobased products [1]. LB offers a promising alternative to satisfy future energy demand, since it is a widely abundant and potentially carbon-neutral source for bioenergy production [2,3]. Moreover, the agricultural activities generate large amounts of wastes, which are considered as the most important feedstock source of LB for energy production.

There are several sorts of LB. A broad classification into woody and non-woody biomass considers only the chemical composition and physical properties of biomass [4,5]. However, based on the context of the use of LB for bioenergy production, a classification according to LB origin/source seems to be more suitable [6]. Biomass is mainly generated in rural (agriculture, forestry, and livestock), urban (sewage sludge and municipal solid wastes) and industrial (cellulose and agri-food industries) areas [7,8]. Each of these biomass generation areas are comprised of different types of biomass [9]. Figure 1 gives an overview about the different LB sources used for bioenergy production [7,10,11].

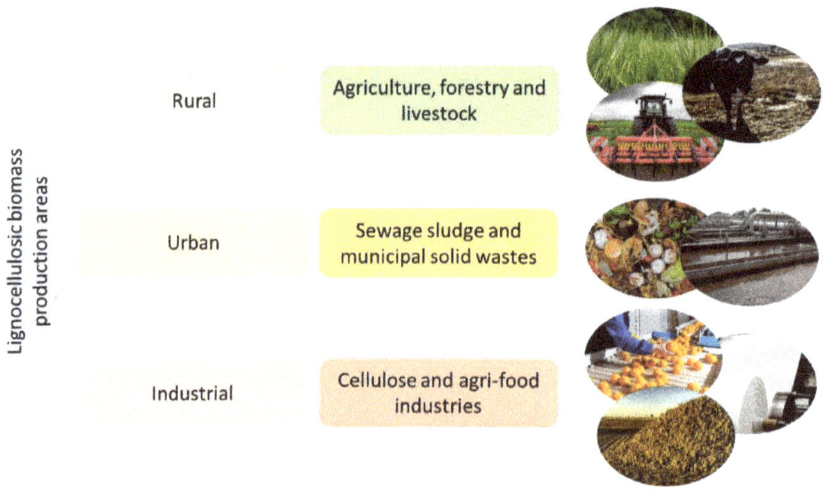

Figure 1. Lignocellulosic biomass production areas and wastes included in each category.

Notwithstanding the LB is a promising source for bioenergy generation, not all LB types are suitable for biogas production through the Anaerobic Digestion (AD) process. Therefore, some forestry and woody residues are not suitable for biogas production due to their very high lignin composition and the poor lignin degradation, which prevent an adequate hydrolysis of macromolecules to reach high process efficiency and methane generation [12–14]. This is similarly applied to most of the hardwood trees, which cannot be degraded through the AD process. Poplar, willow and populous are well known short rotation tree species used as solid biomass for direct combustion.

2. Composition of Lignocellulosic Biomass

The chemical composition of biomass depends strongly on its source [6,15]. The main components of LB are polysaccharides, which are cellulose (40–50%), hemicellulose (25–35%), and lignin (15–20%) [16,17]. The ratios between these fractions vary with the plant age, stage of growth and other conditions [3,16,18,19]. These polymers are organized in complex non-uniform three-dimensional structure with different degrees and varying relative composition [20]. Table 1 resumes the average content of these three components in some LB types.

Cellulose is the major mass constituent of most natural biomass and it is found in the walls of the cells of plants. It is a structural polysaccharide in plants and it is a part of supporting tissues.

The wall of a young plant cell contains approximately 40–45% cellulose; wood 50%, while the purest example of cellulose is cotton with a percentage greater than 90% [21,22]. It is made up of interlinked glucose units which are β-1,4-*O*-glucosidic bonds. These bonds result in significant hydrogen bonding, both intramolecular and intermolecular cellulose molecules, which makes cellulosic material relatively hard to hydrolyze [23]. Thus, cellulose has crystalline and amorphous parts [19].

Hemicelluloses are the second most abundant polymers on the world after cellulose in lignocellulosic materials [4,24]. Hemicelluloses are heteropolysaccharides (i.e., containing more than one type of sugar unit) with branches attached to the main backbone. Nevertheless, hemicellulose is formed by a single type of monosaccharides linked by β-bonds (1–4), mostly containing pentoses (D-xylose and L-arabinose), hexoses (D-glucose, D-mannose and D-galactose) and smaller amounts of L-rhamnose, in addition to uronic acids such as glucuronic acid, 4-*O*-methyl-D-glucuronic acid, and galacturonic acid. The proportions of these substituents may vary according to biomass kinds [3]. Hemicelluloses connect cellulose and lignin fibers to give consistency and flexibility to the structure of the cell wall [3,19].

Lignin is a structurally important polymer in biomass and is one of the most abundant organic substances on earth, next to cellulose [25]. Lignins are major structural components of higher plants and the primary cell wall provides structural support, impermeability confers to biomass its resistance to hydrolysis and microbial degradation [26,27]. Structurally, lignin is a complex three-dimensional polymer of phenylpropane units (approx. 25). The phenylpropane units are derivatives of carbohydrates, coming from the dehydration and cyclisation of sugars. They are mostly either 4-hydroxycinnamyl alcohol (para-coumaryl alcohol, H) or its 3- and/or 3,5-methoxylated derivatives—coniferyl (guaiacyl, G) alcohol, and sinapyl alcohol (or syringal, S), respectively. The ratio of these units varies according to the plant and biomass type. According to previous studies, lignin content may vary both between species and among different tissues of an individual plant [3]. It has been found that a high lignin content is correlated with the recalcitrance of polysaccharides to enzymatic hydrolysis, which makes pretreatments necessary previously to any biological process aiming to produce bioenergy [13,15]. The lignin content and its partial degradation is related with some problems in the AD process.

Table 1. Average composition of the main lignocellulosic biomass (LB) categories (expressed as percentage in dry basis).

Lignocellulosic Biomass (per Group)	Cellulose (%)	Hemicellulose (%)	Lignin (%)
Cellulose and paper wastes	76–99	0–13	0–11
Energy crops and agriculture wastes	16–48	19–36	8–25
Animal wastes	5–28	12–21	2–14
Wood	41–54	11–36	16–28

Source of data from [19,23,24,26,28].

The elemental analysis of biomass provides the mass concentrations of the major elements (carbon, oxygen, hydrogen, nitrogen). These concentrations depend on the chemical composition of the LB and allow the estimation of the theoretical biomethane yield and the subsequent bioenergy yield using the Buswell equation [29]. The average composition of the main wastes used in the AD process is shown in Table 2.

Table 2. Average elemental analysis of some organic wastes classified as LB (expressed as percentage in dry basis).

Lignocellulosic Biomass	C	H	O	N	S
Sewage sludge	26.8–32.2	3.8–5.1	18.2–20.1	3.7–5.4	1.6–2.0
Municipal solid wastes	35.4–41.7	4.7–5.9	15.8–20.9	1.4–2.2	0.2–0.8
Animal manure	28.7–44.2	3.3–6.1	2.4–43.2	0.8–3.8	1.9–3.1
Agricultural wastes	38.3–54.1	5.4–6.7	1.0–45.1	0.2–5.5	0.6–2.9
Crop residues	33.8–49.1	4.6–6.3	32.8–51.4	0.2–0.8	0.1–0.3

Source of data from [30–34].

3. Anaerobic Digestion of Lignocellulosic Biomass

The AD process consists in the transformation of the organic matter contained in the waste into a gaseous effluent (biogas) together with a semisolid stabilized effluent named 'digestate'. The process is developed by a very complex microbial population operating in absence of molecular oxygen in the medium. Different stages can be distinguished in the overall AD process: hydrolysis, acidogenesis, acetogenesis, and methanogenesis. The microorganisms responsible of those stages should synchronize their metabolic rates and work synergistically to break down the complex structures (i.e., lignocellulose) in the organic matter and hence, to obtain a stable process with a high generation of biogas [35].

Regarding the availability of feedstock for the AD process, it should be taken into account that the amount of lignocellulosic agro-industrial wastes and byproducts produced is very high and, consequently, the exploitation of LB to obtain bioenergy or high-value bioproducts constitutes a priority. However, the complexity of the LB structure greatly hinders its AD due to the high level of crystallinity of cellulose, as well as the cross-linking of carbohydrates and lignin [1,36]. Thus, lignocellulosic biomass is constituted mainly by cellulose, hemicellulose, and lignin and the above-mentioned interactions among these fractions lead to a very stable and recalcitrant structure. This complex structure of lignocellulosic material makes it resistant to enzymatic attack [37].

Solid material and macromolecules constituting organic matter should be hydrolyzed and solubilized in the medium in order to be used by the microorganisms. Therefore, this is a key step for the appropriate development of the overall AD process. Moreover, for lignocellulosic solid wastes, hydrolysis is normally the rate-limiting step of the global process [38].

This stage is performed by extracellular enzymes excreted by the acidogenic microorganisms. However, it should be taken into account that lignin is the most difficult biodegradable component in lignocellulose materials and its cross-linking with the carbohydrates reduces the surface area available for enzyme attack [39]. In addition, Yu et al. [40] reported that hemicellulose can be preferably attacked and, hence, broken down before than cellulose or lignin by the anaerobic microbiota. The structure of LB is weakened by the hemicellulose removal and, thus, the enzymatic solubilization of the cellulosic fraction is favored [41].

Therefore, considering the extraordinary relevance of this stage on the performance of the global process, multiple pretreatments of the feedstock (mechanical, physical, chemical and biological) have been tested, despite the increase in costs that may result by applying them [2,3,42,43]. The aim of these pretreatments is to promote and enhance the organic matter solubilization and the subsequent transformation into bioenergy or biobased products [2–4,42,43]. Consequently, the application of these pretreatments becomes relevant to enhance the methane production from LB and their effectiveness has been proved in several studies [44,45].

Among the pretreatment technologies studied, the hydrothermal pretreatment is considered an environmentally-friendly process due to not using any chemicals. It also decreases the formation of fermentation inhibitors, which are formed mainly through sugar degradation at high temperature [37].

4. Hydrothermal Pre-Treatment

The biodegradability of lignocellulosic biomass is affected by the cellulose crystallinity, exposure of the surface to enzymes and structure of lignin. The main component of cellulose is β(1→4 linked D-glucose units), which solubilises at temperature >200 °C into sugars, aldehydes, phenols, ketones, and acid groups. Similarly, hemicellulose is comprised of xyloglucans, xylans, mannans, glucomannans, and β(1→3, 1→4)-glucans, solubilising at temperature >150 °C into sugars, aldehydes, phenols, ketones, and acid groups. Lignin is made of p-coumaryl, coniferyl, and sinaphyl alchohol, solubilising at temperature at 180 °C into phenolics and oils [46].

4.1. Hydrothermal Pre-Treatment: (Principle and Mechanism)

The hydro-thermal pre-treatment of the LB can be an interesting option to achieve a high organic matter solubilization, increase in acidogenic and methanogenic biodegradability, and subsequent improvement in CH_4 production. Hydrothermal pre-treatment of organic feedstocks at elevated temperatures/pressures (150–300 °C, initial pressure of 0–60 bar, 2–40 min) has garnered consideration for the production of biofuels from lignocellulosic substrate as it eliminates chemical addition and corrosion-resistant material requirements for hydrolysis reactors [47]. The recalcitrant structure of the lignocellulosic fraction gets easily broken, hemicellulose and lignin is degraded and the cellulose is hydrolysed effectively in the hydrothermal pre-treatment, thus rendering them as soluble fraction in the anaerobic digestion process (Figure 2). Thus, hydrothermal pre-treatment has been successfully applied in the anaerobic digestion of lignocellulosic biomass for the production of biogas, bioethanol, and other value-added products (VAPs) like hydrogens, hydrochar, Polyhydroxyalkanoates (PHAs) and volatile acids.

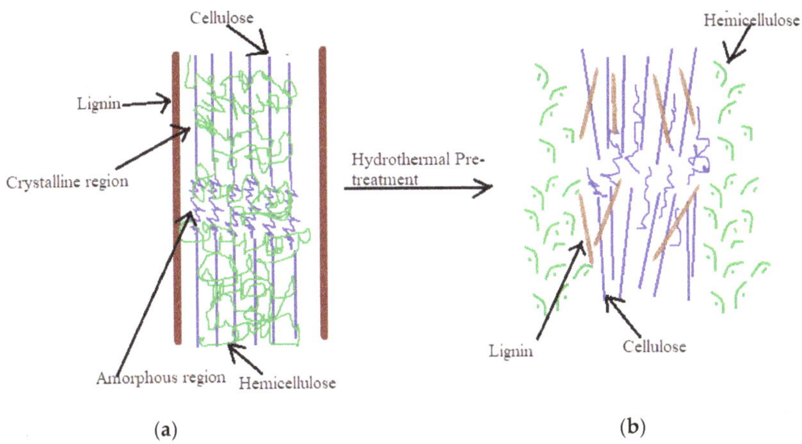

Figure 2. Effect of hydrothermal pre-treatment on the lignocellulosic biomass: (**a**) biomass without pre-treatment; (**b**) hydrothermal pre-treatment of lignocellulose (adapted with permission from Kumar et al. [48]. Copyright (2009) American Chemical Society).

4.2. Effect of Pre-Treatment on Cellulose

Cellulose constitutes the 11–53% (dry basis) of the total lignocellulosic biomass [46]. The cellulose chains contain a number of hydroxylic groups, which leads to the formation of hydrogen bonds in the same chains or in the chains in the vicinity. Interlinking of cellulose chains by hydrogen bonds and van der Waals forces, results in high tensile strength microfibrils [49]. Cellulose molecules have different levels of crystallinity due to its different orientation throughout the structure-low or amorphous crystallinity and high or crystalline crystallinity [50]. If the crystallinity index of cellulose crystals is high, then it is difficult to degrade it and vice versa. The cellulose microfibrils are also attached to each other by pectin and hemicellulose, and they are covered by lignin. This complicated structure of cellulose makes it difficult to be degraded by chemical and biological attacks [51]. The bonds are so strong that they do not break even at high temperature boiling water [52]. The disruption of the inter and intra-hydrogen bonds by the hydrothermal pre-treatment can change the crystallinity structure of cellulose [53]. However, complete disintegration is not possible [54]. The cellulose crystal structure, amount of bound water and the degree of intermolecular regularity determines the intensity of the hydrogen bonds [52]. Sakaki et al. [55], reported that cellulose begins to decompose at temperature of 230 °C, and the reaction was completed at 295 °C. Jin et al. [56] stated that cellulose got hydrolysed to glucose in 120 s at temperature of 300 °C, and pressure of 8.9 MPa. Gao et al. [57] studied the hydrothermal pre-treatment of cellulose at temperature of 200 °C and 400 °C and reaction time of 5 min to 2 h, whereby, the solubilised products contained aldehyde, sugars, phenols, ketones, and acid groups. Girolamo et al. [58], studied the batch thermophilic anaerobic digestion of Giant reed, hydro-thermo-chemically pre-treated at temperature of 150 °C and 180 °C, time of 10 min and 20 min, and with/without sulphuric acid (2% *w*/*w*) as catalyst. The catalyst addition was done in two ways: one immediately before the hydrothermal treatment and another as 24 h prior to the steam cooking. Cellulose, owing to its crystalline and thermal-resistant structure, dissolved just modestly in the pre-soaking thermo-chemical pre-treatment. Whereas hemicellulose dissolved to a great extent at the same temperatures (150 and 180 °C).

4.3. Effect of Pre-Treatment on Hemicellulose

The second most important component of lignocellulosic biomass is hemicellulose, which are polysaccharides and contains xylans, mannans, xyloglucans, glucomannans and (1–3, 1–4)-glucans [46]. The hemicellulose contains hexoses and pentoses, which can be degraded into HMF, levulinic acid and formic acid [59]. On the contrary, pentose, and more specifically D-xylose and L-arabinose, is responsible for the formation of furfural. The molecular weight of hemicellulose is low and the lateral chains are short, and so they are easily hydrolysable. The hemicellulose and lignin are solubilised at temperatures greater than 150 °C and 180 °C, respectively. Under high pressure, water acts as an acid, which breaks up the biomass molecules, enhances the rate of hydrolysis of cellulose and thus leading to solubilisation of hemicellulose and lignin [46]. Costa et al. [60] studied the hydrothermal pre-treatment of sugarcane baggase at 150–200 °C for 10–30 min, and observed that soluble sugars were formed as a result from hydrolysis of hemicellulose. As per findings, when lignocellulosic biomass was pre-treated at 100 °C, the hemicellulose remained in the solid fraction, whereas at temperatures above 150 °C, hemicellulose was solubilised into the liquid part. Hydrothermal pre-treatment of hemicellulose resulting in the formation of acetic acid, which acts as a catalyst in the hydrolysis thus further degrade the biomass and leads to an increase in the sugar yield [46].

4.4. Effect of Pre-Treatment on Lignin

Lignin is an amorphous and water-insoluble heteropolymer. It is composed of phenylpropane units (coniferyl, p-coumaryl and sinapyl alcohol) held together by different linkages [61]. It binds cellulose and hemicellulose together and helps in supporting of plant structure and prevents them against microbial attack. The fermentation of lignocellulosic biomass is difficult due to the high recalcitrant lignin and inadequate accessibility of enzymes [46]. Lignin derivatives having aldehyde groups are inhibitory to methanogenesis stage [62]. The precursors in lignin, i.e., p-coumaryl, coniferyl, sinapyl alcohol, decide the solubility of lignin in acidic, neutral or alkaline environment [63]. When the hydrothermal pre-treatment of lignocellulosic biomass is performed, the dissolved lignin may inhibit the activity of cellulase, xylanase and glucosidase [64]. Recent researches conclude that irrespective of its solubilisation, the lignin content change is related to the solidification and re-deposition due to cooling after severe pre-treatment. Therefore, only re-allocation of lignin takes place, instead of lignin removal during pre-treatment at high temperature and pressure [65,66]. In the hydrothermal pre-treatment of switchgrass and paper tube residuals, such a re-allocation of lignin was observed [67,68]. Besides this, the lignin present in raw corncob has a negative yield of sugars with furans produced from the same. However, if the corncob would have undergone hydrothermal pre-treatment, the delignification can give high sugar yields, which indicates that lignin dominates in biomass conversion giving sugar yields [69].

5. Hydrothermal Pre-Treatment: Energy and Value-Added Products Recovery

5.1. Methane and Hydrogen

In order to generate a sustainable energy system, it is predicted that hydrogen will soon be in global usage. Hydrogen is environment-friendly, safe to produce, quite useful for heating purpose in residences, and fuel for non-polluting transport systems and aircraft. The energy content of hydrogen (122 kJ/g) is 2.7 times higher than the gasoline. Similarly, methane gas, produced from anaerobic reactors and digesters is also used as a fuel to generate heat and light. It is a clean fuel thus used in industries, transportation, appliances and power generation [70]. Chandra et al. [71], studied the mesophilic anaerobic digestion of rice straw for the production of methane, which involved thermo-chemical pre-treatment. Hydrothermal treatment was given for 10 min at 200 °C to the ground rice straw. For the untreated sample, the methane production was 59.8 L/kg VS. Hydrothermal pre-treatment with 5% NaOH addition (for maintaining suitable pH for biogas production) enhanced the hydrolysis step of the anaerobic digestion thereby giving a methane yield of 132.7 L/kg VS, i.e., 2.22 times higher than the untreated substrate. Girolamo et al. [58], studied the hydrothermal pre-treatment of Giant reed biomass at temperature combinations of 150 °C and 180 °C, for a time of 10 and 20 min, in the presence and absence of catalyst. The catalyst used was 2% *w*/*w* sulphuric acid, immediately before the hydrothermal pre-treatment and 24 h soaking in the catalyst. The untreated assay showed a methane yield of 273 mL/g VS, whereas the assay with no catalyst showed an increase of 10%, 7%, 23% and 4% in the 150 °C/10 min, 150 °C/20 min, 180 °C/10 min and 180 °C/20 min combination, respectively. The pre-treatments in the presence of the catalyst showed an inhibition in the methanogenic activity due to high sulphate concentration, which, led to the growth of the sulphate reducing bacteria. Li et al. [72], studied the hydrothermal pre-treatment on the methane production potential of antibiotic mycelial residue (AMR). Varying the treatment time from 0 to 60 min and temperature from 80 °C to 180 °C. The best digestion results were observed at a temperature of 120 °C for 60 min with a methane yield of 290 mL/g VS, which was almost 3 times the methane yield of raw AMR (100 mL/g VS).

Eskicioglu et al. [47] studied the hydrothermal pre-treatment followed by anaerobic digestion of five kinds of lignocellulosic biomass, namely, wheat straw, rice straw, and Douglas fir bark. Carbon dioxide was used as the catalyst in the experiment and the methane and hydrogen yield was also observed. Temperature and pressure variation of pre-treatment was from 26–175 °C and from 25–102 bars, respectively. 23–42% of hemicellulose destruction and 0–12% of delignification was observed. A 20–30% enhancement of hydrolysis rate was observed except for Douglas fir bark, which showed a 172% enhancement in digestion rate. For hydrothermally pre-treated biomass, a methane yield of 269 mL CH_4/g VS_{added} for wheat straw, 319 mL CH_4/g VS_{added} for rice straw, and 136 mL CH_4/g VS_{added} for Douglas fir bark was observed. Methane yield for the pre-treated substrate was higher by 5.1% in wheat straw, 2.6% in rice straw, and 39% in Douglas fir bark in comparison with non-pre-treated biomass. Phuttaro et al. [73] studied the effect of hydrothermal pre-treatment on Napier grass and the anaerobic digestion. The highest methane yield in batch studies was observed to be 248.2 NmL/VS added at a pre-treatment temperature of 175 °C, which was 35% greater yield in comparison with untreated biomass. He et al. (2014) [74] studied the hydrothermal pretreatment of rice straw (20% TS) at 150 °C and 210 °C for 0 to 30 min operation time. The soluble carbohydrates was observed to be 80 mg/ g VS at a temperature of 210 °C and 0 min holding time. 28mL H_2/VS was produced, which was 93 times higher than the control. Kongjan et al. [75], studied the hydrothermal pre-treatment of wheat straw at a temperature of 180 °C and observed a hydrogen yield of 1.59 mol/mol hexose. Jung et al. [76], also studied the hydrothermal treatment of marine algae (*Laminaria japonica*) at temperature 170 °C and observed a hydrogen yield of 110 L/kg COD, i.e., about 60% greater than non-pre-treated substrate.

Therefore, on average, a methane yield of 250–350 mL/g VS and hydrogen yield of 100–150 mL/g VS is produced from the hydrothermal pre-treatment of lignocellulosic biomass. It can be concluded that a methane yield has been reported to increase by 2% to 300% as compared to the non-pre-treated lignocellulosic biomass and hydrogen yield increased from 0.6 times to 93 times as compared to the non-pre-treated substrate. These numbers are an indicator that regulating the temperature and time in hydrothermal treatment, substantial methane and hydrogen yield could be achieved.

5.2. Hydrochar

Hydrochar is a high value-added carbonaceous matter obtained from hydrothermal pre-treatment of biomass at temperature ranges between 150–350 °C [77]. It is used for soil amendments, as a catalyst and in energy storage application. In the field of energy storage, hydrochar is processed to be used as electrodes in Li-ion batteries [78]. Kambo et al. [79], studied the hydrothermal carbonisation (HTC) of Miscanthus for the production of hydrochar. The treatment temperature of 190, 225, and 260 °C, reaction time of 5, 15, and 30 min and substrate to water ratio of 1:6 and 1:12 was varied. It was observed that the reaction temperature was the most important factor controlling the biomass properties. The hydrochar produced at a hydrothermal carbonisation of 260 °C gave the maximum energy density of 26–30 MJ/kg. Kim et al. [80] studied the HTC of cellulose, and observed that the pre-treatment contributed in high carbon content and high calorific value. HTC at 220 °C led to an increase in fixed carbon content of cellulose increased from 6.1% to 35.0%, which implies that decomposition of cellulose began at this temperature. Along with the fixed carbon, the calorific value of cellulose also was enhanced at temperatures of 180, 200, 220, 280 °C showing an increase from 16.5 to 18.9, 23.1, 26.5 and 27.7 MJ/kg, respectively. The effect of HTC on the quality of biochar obtained from the lignocellulosic biomass can be calculated by increasing the calorific value and this effect can be due to the pyrolysis of cellulose during HTC reactions leading to hydrolysis, chemical dehydration, and decarboxylation reactions. Thus, HTC which is a combination of high pressure and temperature uses biomass as an energy resource for the production of hydrochar, having energy density in the range of 25–30 MJ/kg. Through HTC, biomass with a low calorific value and high water content can be upgraded to a valuable carbon-rich-solid, lignite-like fuel [80].

5.3. Bioethanol

Bioethanol is a non-reactive and environment-friendly reactant to produce biodiesel. Bioethanol production through biomass fermentation makes it a sustainable energy resource, which can be a substitute to fossil fuels [81]. Use of bioethanol as a fuel in the transportation sector contributes to reduce GHG emissions thus helps in climate change mitigation. Being a liquid biofuel produced from the fermentation of corn, wheat, cane, beet and wood, bioethanol is mixed with gasoline with a gasoline: bioethanol ratio ranging from 90:10 to 15:8 [82]. Nitsos et al. [83] optimised the hydrothermal pre-treatment conditions for bioethanol production at various temperature (130–220 °C), reaction times (15–180 min) and water-to-solid ratio of 15. Hemicellulose was totally solubilised and the lignin was only partly delignified. The solubilisation of the ligno-cellulosic biomass led to an increase in surface area and pore volume to nearly 2.5 times. Mabee et al. [84] studied the hydrothermal pre-treatment, where heating of biomass under pressurised steam of 20–50 bar and temperature of 160–270 °C for few minutes followed by pressure release to the atmosphere was carried out, which leads to the desegregation of lignocellulosic biomass. The results were: hydrolysis of hemicellulose, transformation of lignin owing to high temperatures, and increase in crystallinity of cellulose. Reczey and Zacchi [85] studied the anaerobic digestion of hydrothermally pre-treated corn stover at 200 °C for 5 min with 2% sulphuric acid, which led to the four times enhanced enzymatic conversion of cellulose to glucose, and 90% higher yield of ethanol as compared to untreated raw materials. In another study, hydrothermal pre-treatment of wheat straw at 180 °C for 10 min with 0.9% sulphuric acid, gave a sugar yield of 85% of the total sugar present in the raw material.

Petersen et al. [86] studied the hydrothermal pre-treatment of wheat straw for production of bioethanol and concluded that the optimum temperature and time for pre-treatment was 195 °C and 6–12 min. Under optimized conditions, 70% of hemicellulose and 93% cellulose were solubilized, where 89% of cellulose was converted to bioethanol. Kumar et al. [68] conducted the experiments on hydrothermal pre-treatment of switchgrass and corn-stover for the production of ethanol and carbon microspores. Upon pre-treatment at 190 °C for 20 min, more than 80% of glucan was digested. Adding 0.4–0.9% of potassium carbonate could allow the digestion to occur at lower temperature whereby pre-treatment efficiency was also enhanced. On increasing the temperature from 150 °C to 190 °C, the hydrolysis of hemicellulose to water-soluble products showed an increase from 30% to 77%. 40% lignin was solubilized, however, it remained constant at temperature ranges from 150 to 190 °C. Hence, bioethanol generated from the digestion of glucan, hydrolysis of hemicellulose and delignification of lignin, fulfils the aim of hydrothermal treatment of the lignocellulosic biomass.

6. Intermediators-Inhibitory By-Products

6.1. Furfural and 5-HydroxyMethyl Furfural Production

Furfurals are produced during thermo-chemical processing of lignocellulosic biomass. 5-HydroxyMethyl Furfural (HMF) and furfurals are the most important members of furfurals. HMF is formed from hexoses and their polymers, including cellulose. Furfurals are produced from pentoses, which are obtained from hemicelluloses [87]. Acids inhibit cell growth, specifically weak non-dissociated acids cross the cell wall and energy is required to be exported out of the cell [88]. Furfural decreases specific growth rate, and HMF has a mechanism similar to furfural but produces a longer delay phase during growth. Phenolics interact with the cell membrane causing a loss of the integrity of the membrane and decreasing its permeability. However, methanogenic bacteria are able to adapt to such compounds in a certain period of time, up to a certain concentration [89].

In cellulose, the D-glucose is dehydrated to 5-Hydroxymethylfurfural (HMF) by hydrothermal pre-treatment, and then decomposed to formic acid and levulinic acid. In supercritical water, the 99% conversion of glucose to HMF and then to formic and levulinic acids occurred at 0.01 s without the use of catalyst [52]. According to Lopez-Gonzalez et al. [90], when switchgrass was pre-treated at 200 °C for 10 min, production of furfural was observed at a rate of 0.72 g/100 g switchgrass. It was observed in several studies that when temperature was varied between 200–220 °C for 5–15 min, the furfural concentration was between 0.2 and 3.1 g/100 g switchgrass [83,91–93]. The hydrothermal pre-treatment of sugar cane press mud for the production of methane was studied. The methane concentration decreased at temperature >200 °C, accounting to inhibition to methanogenesis due to recalcitrant compounds (furfural) formation at high temperatures. The furfural production was observed at the rate of 0.73 g/100 g of press mud. Bougrier et al. [94], reported that Maillard reaction was responsible for the formation of recalcitrant compounds at high temperature. This was due to the polymerisation between the carbohydrates and amino acids. These recalcitrant compounds inhibited the methanogenesis and reduced the overall process efficiency.

Some soluble sugars like xylose, glucose, cell-oligomers, and xylo-oligomers, weak acids like formic, levulinic and acetic acid, furan derivatives like furfurals and HMF and phenol compounds like ferulic acid and vanillin are formed as a result of hydrothermal pre-treatment of lignocellulosic biomass. These compounds are inhibitory for the metabolic activities of substrate degrading microbes and enzymes [95].

The concentration of inhibitory compounds vary as per the type of pre-treatment applied on the lignocellulosic biomass, and the solid loading of lignocellulose on the substrate. of pre-treatment [96–98]. Kim et al. [99] studied that hydrothermal pre-treatment of maple wood. Pre-treatment of maple wood at 230 g/L with hot water at 200 °C for 20 min led to the formation of xylooligomers and xylose with a concentration greater than 11.2 and 9.2 g/L, respectively. The concentration of HMF and furfurals were 4.1 g/L, while as the phenol compounds were present in 1.3 g/L concentration. Garcia-Aparicio et al. [100], stated that hydrothermal pre-treatment of barley star at a temperature of 210 °C for 5 min, generated inhibitors like furfural (0.7 g/L), acetic acid (2.1 g/L), phenolic compounds (0.2 g/L) and HMF (0.2 g/L), which affected the enzymatic hydrolysis of the substrate.

Therefore, hydrothermal pre-treatment of ligno-cellulosic biomass is sometimes accompanied by the formation of inhibitors like furfurals and 5-HMF, which act as recalcitrant leading to inhibition of the biological process.

6.2. *Inhibitors to Enzymatic Hydrolysis and Fermentation Derived from Hydrothermal Pre-Treatment*

6.2.1. Lignin

Lignin inhibited the hydrolysis by forming physical barriers and non-productive adsorption of cellulase enzymes [36]. Thus, lignin restricts the enzymes from reaching to cellulose and thereby the active enzymes for cellulose hydrolysis also reduces [101]. Ko et al. [102,103] observed that 80–90% of lignin was recovered from solid fraction of hardwood, upon hydrothermal pre-treatment at 180–220 °C. Therefore, as the severity of hydrothermal pre-treatment increases, the lignin content in the pre-treated solids also increases due to the simultaneous de- and re-polymerization reactions of lignin. Ko et al. [102] stated that hydrothermal pre-treatment changes the structure of lignin to a more heterogeneous and condensed form, and at the same time lignin is more inhibitory to cellulase than it is before pre-treatment. At low enzyme loading, the non-productive adsorption of enzymes to lignin act as an obstacle to the efficient enzymatic hydrolysis of cellulose. Ko et al. [102,103], investigated that the hydrolysis of Avicel (a microcrystalline cellulose from wood pulp) in the presence of lignin isolated from hardwood, showed a decrease in glucose yield from 62% to 51%. Nakagame et al. [104] also observed that hydrolysis of Avicel decreased to 40% in the presence of lignin isolated from softwood. In a study, where incubation of 10 mg cellulase with lignin isolated from hydro-thermally pre-treated hardwood, a loss of 50–60% total cellulase was seen due to adsoption of lignin. Apart from this, the commercial cellulase Cellic CTec2 having the enzymatic components, >90% of β-glucosidase activity was lost due to lignin adsorption [103].

6.2.2. Lignin-Derived Phenolics

A considerable inhibitory effect on enzymes during cellulose conversion was observed due to the presence of phenolic compounds derived from lignocellulosic biomass. Ximenes et al. [105] reported that soluble phenolics may be formed at a high solid (up to 20%) hydro-thermal pre-treatment of lignocellulosic slurry, which inhibits the enzymatic hydrolysis. The phenolics, which can be extracted from pre-treated slurries can also cause inhibition to enzymatic hydrolysis. The phenolics are considered as the strongest inhibitors among all the toxic compounds derived from the hydrothermal pre-treatment. Kim et al. [99] studied that when the enzyme loading is 1 and 25 mg/g glucan, then the phenolic compounds at 1.3 g/L reduced the rate and strength of cellulose loading by half.

6.2.3. Furan Aldehydes and Weak Acids

During hydrothermal pre-treatment, degradation of sugars to furan aldehydes has been observed. The furan aldehydes include furfural and 5 HMF formed from pentose and hexose, respectively [106,107]. Furan aldehydes degradation generates weak acids like levulinic acid and formic acids. [107,108]. Although the furan aldehyde and acetic acid are inhibitors of fermenting microorganisms, they have little effect on the cellulase activity. Kim et al. [99] observed that acetic acid (13 g/L) and furfural (4 g/L) had no effect on enzymatic hydrolysis. Ask et al. [109] stated that furfurals and 5-HMF inhibit the cell growth and ethanol production of xylose-utilising *Sachharomyces cerevisiae* (yeast). Inhibition of yeast fermentation is caused by furan aldehydes i.e., by reduction of enzymatic and biological activities. Larsson et al. [110] demonstrated that HMF and furfurals decreased the volume of ethanol production by yeast, during the dilute acid hydrolysis of softwood, whereby furfural inhibition was greater on the cell growth than on ethanol production.

7. Challenges and Opportunities

The hydrothermal pre-treatment of lignocellulosic biomass comes with some challenges, for example: production of inhibitory compounds, toxification of the biomass, and production of 5-HMF and furfurals. Various studies show that xylo-oligomers and oligomer sugars can be removed by hydrolysis and fermentation of pentose [111], but other inhibitory compounds like weak acids, furan aldehydes and phenolic compounds are still inhibitory to enzymatic process and fermentation [112]. Over liming, i.e., pH adjustment by using an alkali, vacuum evaporation, sulphite addition, and adsorbent treatment are some of the methods stated by previous studies in order to remove volatile inhibitors [111]. Removal of phenolic compounds and giving higher yield of ethanol can be achieved by various chemical detoxification processes like addition of activated charcoal adsorbents and polymeric resin [113]. At the same time, these detoxification processes may also lead to high manufacturing cost, generation of waste, and fermentable sugar wastage [114,115]. Bio-abatement methods, using laccase and peroxide as microbial enzymes are also used as detoxication against furan aldehydes, phenolic compounds, and weak acids. *C. ligniaria* NRRL30616, an ascomyecete, was used as a mitigation bio-abatement against the inhibitors, mainly phenolics, formed during hydrothermal pre-treatment of corn stover slurries. More than 95% acetic acid and >50% HMF, phenolics, and furfurals were removed from the slurries using this technique and on the contrary, 16% higher yield of cellulose conversion was observed [114].

One of the methods for detoxification is the evolutionary engineering of fermenting microorganisms to fight the inhibitors. This method is necessary as it requires no extra detoxification treatment technique, and is based on successive cultivation [116]. *S. cerevisiae*, an ethanol-producing microorganism, possesses an inborn tolerance to inhibitors like furans and phenolics, and it also convert them to less harmful compounds [117,118]. HMF is reduced to 2,5-bis HMF and the furfurals are reduced to furfural alcohols [119]. Several other studies show that adaptation of yeast to hydrolysates increases the ethanol production and microbial growth [119–121]. Also, in order to improve microbial performance in the presence of inhibitors, genetically or metabolically engineered of yeast strains can be used. For example, for the detoxification of phenolic compounds from hydrolysates, *S. cerevisiae* mutants have been developed [122]. Along with this, strains of *S. cerevisiae*, which shows resistance to furan aldehydes, have been constructed by many oxidoreductases, such as alcohol dehydrogenases [119].

The sources and component of enzyme determine the loss of cellulase activity due to non-productive adsorption of cellulases to lignin. Upon incubation of Cellic CTec2 with lignin, the major part of β-glucosidase was adsorbed to the lignin, while 50–60% of the initial cellobiohydrolase and endoglucanase remained in the enzyme supernatant [95]. The enzymes need to be engineered to become more resistant to adsorption to insoluble lignin. This is one of the solutions for overcoming enzymatic inhibition [64]. Ko et al. [95] stated that adsorption of β-glucosidase to lignin is dependent on the pH and the salt ions concentration of the medium, thus indicating the presence of electrostatic interactions. So, by changing the surface charge, the adsorption of enzyme to lignin can be reduced.

8. Conclusions

(1) Lignocellulosic biomass is a difficult-to-degrade substrate when subjected to anaerobic digestion.

(2) A considerable increase in methane, hydrogen and bioethanol production could be achieved, when opting for hydrothermal pretreatment of lignocellulosic biomass.

(3) The production of recalcitrant such as furfurals, and 5-HMF takes place during hydrothermal pretreatment of lignocellulosic biomass at very high temperature, which leads to process inhibition.

(4) In order to convert the inhibitors into less toxic compounds, evolutionary engineering is being applied using genetically engineered microbes (e.g., *S. cerevisiae*, a yeast) as a promising detoxification process.

(5) Hydrothermal pre-treatment comes out to be a beneficial option for production of bio-energy from lingo-cellulosic biomass, which otherwise is difficult to degrade in anaerobic digestion alone.

Author Contributions: B.A., K.A., V.K.T. and L.A.F.-G. reviewed the literature and wrote the different sections of the paper. C.J.Á.-G., L.I.R.-G. and A.A.K. reviewed and edited the paper to enhance the scientific standard.

Funding: This research was supported by the project CTM2016-79071-R funded by the Spanish "Agencia Estatal de Investigación—AEI" and by the European Regional Development Fund (ERDF).

Acknowledgments: Authors are thankful to Department of Biotechnology-GoI (Grant No. BT/RLF/Re-entry/12/2016) for financial support to this research.

Conflicts of Interest: The authors declare no conflict of interest.

Abbreviations

AD	anaerobic digestion
LB	lignocellulosic biomass
PHAs	polyhydroxyalkanoates
TS	total solids
VAPs	value added products
HMF	hydroxy methyl furfurals
AMR	antibiotic mycelial residue
VS	volatile solids
HTC	hydrothermal carbonisation

References

1. Sawatdeenarunat, C.; Surendra, K.C.; Takara, D.; Oechsner, H.; Khanal, S.K. Anaerobic digestion of lignocellulosic biomass: Challenges and opportunities. *Bioresour. Technol.* **2015**, *178*, 178–186. [CrossRef] [PubMed]
2. Sen, S.M.; Han, J.; Luterbacher, J.S.; Alonso, D.M.; Dumesic, J.A.; Maravelias, C.T. Synthesis of catalytic biomass-to-fuels strategies. *Comput. Aided Chem. Eng.* **2014**, *34*, 615–620.
3. De Jong, E.; Gosselink, R.J.A. Lignocellulose-Based Chemical Products. In *Bioenergy Research: Advances and Applications*; Elsevier: Amsterdam, The Netherlands, 2014; pp. 277–313.
4. Zhu, J.Y.; Pan, X.J. Woody biomass pretreatment for cellulosic ethanol production: Technology and energy consumption evaluation. *Bioresour. Technol.* **2010**, *101*, 4992–5002. [CrossRef] [PubMed]
5. Tye, Y.Y.; Lee, K.T.; Wan Abdullah, W.N.; Leh, C.P. The world availability of non-wood lignocellulosic biomass for the production of cellulosic ethanol and potential pretreatments for the enhancement of enzymatic saccharification. *Renew. Sustain. Energy Rev.* **2016**, *60*, 155–172. [CrossRef]
6. Popa, V.I. Biomass for Fuels and Biomaterials. In *Biomass as Renewable Raw Material to Obtain Bioproducts of High-Tech Value*; Elsevier: Amsterdam, The Netherlands, 2018; pp. 1–37.
7. International Renewable Energy Agency. *Global Bioenergy Supply and Demand Projections*; A Working Paper for REmap 2030; International Renewable Energy Agency: Abu Dhabi, UAE, 2014.
8. Saxena, R.C.; Adhikari, D.K.; Goyal, H.B. Biomass-based energy fuel through biochemical routes: A review. *Renew. Sustain. Energy Rev.* **2009**, *13*, 167–178. [CrossRef]
9. Rettenmaier, N.; Schorb, A.; Koppen, S. *Status of Biomass Resource Assessments*; Biomass Energy Europe, Institute for Energy and Environmental Research: Heidelberg, Germany, 2010.
10. *Key Findings: Statistical Report*; Bioenergy Europe: Brussels, Belgium, 2018.
11. Nanda, S.; Azargohar, R.; Dalai, A.K.; Kozinski, J.A. An assessment on the sustainability of lignocellulosic biomass for biorefining. *Renew. Sustain. Energy Rev.* **2015**, *50*, 925–941. [CrossRef]
12. Cai, J.; He, Y.; Yu, X.; Banks, S.W.; Yang, Y.; Zhang, X.; Yu, Y.; Liu, R.; Bridgwater, A.V. Review of physicochemical properties and analytical characterization of lignocellulosic biomass. *Renew. Sustain. Energy Rev.* **2017**, *76*, 309–322. [CrossRef]
13. Paul, S.; Dutta, A. Challenges and opportunities of lignocellulosic biomass for anaerobic digestion. *Resour. Conserv. Recycl.* **2018**, *130*, 164–174. [CrossRef]
14. Čater, M.; Zorec, M.; Marinšek Logar, R. Methods for Improving Anaerobic Lignocellulosic Substrates Degradation for Enhanced Biogas Production. *Springer Sci. Rev.* **2014**, *2*, 51–61. [CrossRef]
15. Bajpai, P. *Pretreatment of Lignocellulosic Biomass for Biofuel Production*; Springer: Berlin/Heidelberg, Germany, 2016; p. 8. ISBN 978-981-10-0686-9.
16. Bernal, M.P.; Sommer, S.G.; Chadwick, D.; Qing, C.; Guoxue, L.; Michel, F.C. Current Approaches and Future Trends in Compost Quality Criteria for Agronomic, Environmental, and Human Health Benefits. *Adv. Agron.* **2017**, *144*, 143–233.
17. Smith, A.D.; Landoll, M.; Falls, M.; Holtzapple, M.T. Chemical production from lignocellulosic biomass: Thermochemical, sugar and carboxylate platforms. In *Bioalcohol Production*; Woodhead Publishing: Swaston/Cambridge, UK, 2010; pp. 391–414.
18. Hassan, S.S.; Williams, G.A.; Jaiswal, A.K. Emerging technologies for the pretreatment of lignocellulosic biomass. *Bioresour. Technol.* **2018**, *262*, 310–318. [CrossRef] [PubMed]
19. Barakat, A.; Monlau, F.; Solhy, A.; Carrere, H. Mechanical dissociation and fragmentation of lignocellulosic biomass: Effect of initial moisture, biochemical and structural proprieties on energy requirement. *Appl. Energy* **2015**, *142*, 240–246. [CrossRef]
20. Isikgor, F.H.; Becer, C.R. Lignocellulosic biomass: A sustainable platform for the production of bio-based chemicals and polymers. *Polym. Chem.* **2015**, *6*, 4497–4559. [CrossRef]
21. Keshk, S.M. Bacterial Cellulose Production and its Industrial Applications. *J. Bioprocess. Biotech.* **2014**, *4*, 1–10. [CrossRef]
22. Cosgrove, D.J. Growth of the plant cell wall. *Nat. Rev. Mol. Cell Biol.* **2005**, *6*, 850–861. [CrossRef]
23. Saini, J.K.; Saini, R.; Tewari, L. Lignocellulosic agriculture wastes as biomass feedstocks for second-generation bioethanol production: Concepts and recent developments. *3 Biotech* **2015**, *5*, 337–353. [CrossRef] [PubMed]

24. Yang, S.T. Bioprocessing from Biotechnology to Biorefinery. In *Bioprocessing for Value-Added Products from Renewable Resources*; Elsevier: Amsterdam, The Netherlands, 2007; pp. 1–24.
25. Faruk, O.; Sain, M.; Patil, N.D.; Tanguy, N.R.; Yan, N. Lignin Interunit Linkages and Model Compounds. In *Lignin in Polymer Composites*; William Farino, Andrew Mitchell & Company: Boston, MA, USA, 2016; pp. 27–47.
26. Jung, S.-J.; Kim, S.-H.; Chung, I.-M. Comparison of lignin, cellulose, and hemicellulose contents for biofuels utilization among 4 types of lignocellulosic crops. *Biomass Bioenergy* **2015**, *83*, 322–327. [CrossRef]
27. Aboudi, K.; Álvarez-Gallego, C.J.; Romero-García, L.I. Improvement of Exhausted Sugar Beet Cossettes Anaerobic Digestion Process by Co-Digestion with Pig Manure. *Energy Fuels* **2015**, *29*, 754–762. [CrossRef]
28. Wen, Z.; Liao, W.; Chen, S. Hydrolysis of animal manure lignocellulosics for reducing sugar production. *Bioresour. Technol.* **2004**, *91*, 31–39. [CrossRef]
29. Tarvin, D.; Buswell, A.M. The Methane Fermentation of Organic Acids and Carbohydrates1,2. *J. Am. Chem. Soc.* **1934**, *56*, 1751–1755. [CrossRef]
30. Werle, S.; Dudziak, M. Analysis of Organic and Inorganic Contaminants in Dried Sewage Sludge and By-Products of Dried Sewage Sludge Gasification. *Energies* **2014**, *7*, 462–476. [CrossRef]
31. Shen, J.; Zhao, C.; Liu, Y.; Zhang, R.; Liu, G.; Chen, C. Biogas production from anaerobic co-digestion of durian shell with chicken, dairy, and pig manures. *Energy Convers. Manag.* **2018**. [CrossRef]
32. Li, K.; Liu, R.; Sun, C. Comparison of anaerobic digestion characteristics and kinetics of four livestock manures with different substrate concentrations. *Bioresour. Technol.* **2015**, *198*, 133–140. [CrossRef] [PubMed]
33. Yang, Y.; Heaven, S.; Venetsaneas, N.; Banks, C.J.; Bridgwater, A.V. Slow pyrolysis of organic fraction of municipal solid waste (OFMSW): Characterisation of products and screening of the aqueous liquid product for anaerobic digestion. *Appl. Energy* **2018**, *213*, 158–168. [CrossRef]
34. Angelo Basile, F.D. (Ed.) *Second and Third Generation of Feedstocks*; Elsevier: Amsterdam, The Netherlands, 2019; ISBN 978-0-12-815162-4.
35. Weiland, P. Biogas production: Current state and perspectives. *Appl. Microbiol. Biotechnol.* **2010**, *85*, 849–860. [CrossRef] [PubMed]
36. Chang, V.S.; Holtzapple, M.T. Fundamental factors affecting biomass enzymatic reactivity. In *Twenty-First Symposium on Biotechnology for Fuels and Chemicals*; Humana Press: Totowa, NJ, USA, 2000; Volume 84, pp. 5–37.
37. Taherzadeh, M.J.; Karimi, K. Pretreatment of lignocellulosic wastes to improve ethanol and biogas production: A review. *Int. J. Mol. Sci.* **2008**, *9*, 1621–1651. [CrossRef] [PubMed]
38. Khanal, S.K. *Anaerobic Biotechnology for Bioenergy Production: Principles and Applications*; John Wiley & Sons: Hoboken, NJ, USA, 2008; ISBN 9780813823461.
39. Agbor, V.B.; Cicek, N.; Sparling, R.; Berlin, A.; Levin, D.B. Biomass pretreatment: Fundamentals toward application. *Biotechnol. Adv.* **2011**, *29*, 675–685. [CrossRef] [PubMed]
40. Yue, Z.; Teater, C.; Liu, Y.; MacLellan, J.; Liao, W. A sustainable pathway of cellulosic ethanol production integrating anaerobic digestion with biorefining. *Biotechnol. Bioeng.* **2010**, *105*, 1031–1039. [CrossRef]
41. Yue, Z.; Teater, C.; MacLellan, J.; Liu, Y.; Liao, W. Development of a new bioethanol feedstock Anaerobically digested fiber from confined dairy operations using different digestion configurations. *Biomass Bioenergy* **2011**, *35*, 1946–1953. [CrossRef]
42. Europe Observer Barometer 18th Annual Overview Barometer. Available online: https://www.eurobserv-er.org/category/2018/ (accessed on 28 May 2019).
43. *Renewable Energy Policy Network for the 21st Century. Renewables 2016 Global Status Report*; REN21: Paris, France, 2016.
44. Holm-Nielsen, J.B. Introduction to biomass supply chains. In *Biomass Supply Chains for Bioenergy and Biorefining*; Woodhead Publishing: Cambridge, UK, 2016; pp. 3–13.
45. World Bioenergy Association. *Global Bioenergy Statistics 2018*; World Bioenergy Association: Stockholm, Sweden, 2018.
46. He, L.; Huang, H.; Zhang, Z.; Lei, Z. A Review of Hydrothermal Pretreatment of Lignocellulosic Biomass for Enhanced Biogas Production. *Curr. Org. Chem.* **2015**, *19*, 437–446. [CrossRef]
47. Eskicioglu, C.; Monlau, F.; Barakat, A.; Ferrer, I.; Kaparaju, P.; Trably, E.; Carrere, H. Assessment of hydrothermal pre-treatment of various lignocellulosic biomass with CO_2 catalyst for enhanced methane and hydrogen production. *Water Res.* **2017**, *120*, 32–42. [CrossRef] [PubMed]

48. Kumar, P.; Barrett, D.M.; Delwiche, M.J.; Stroeve, P. Methods for Pretreatment of Lignocellulosic Biomass for Efficient Hydrolysis and Biofuel Production. *Ind. Eng. Chem. Res.* **2009**, *48*, 3713–3729. [CrossRef]
49. Ha, M.A.; Apperley, D.C.; Evans, B.W.; Huxham, I.M.; Jardine, W.G.; Viëtor, R.J.; Reis, D.; Vian, B.; Jarvis, M.C. Fine structure in cellulose microfibrils: NMR evidence from onion and quince. *Plant J.* **1998**, *16*, 183–190. [CrossRef] [PubMed]
50. Atalla, R.H.; Vanderhart, D.L. Native cellulose. A composite of two distinct crystalline forms. *Science* **1984**, *223*, 283–285. [CrossRef] [PubMed]
51. Zheng, Y.; Zhao, J.; Xu, F.; Li, Y. Pretreatment of lignocellulosic biomass for enhanced biogas production. *Prog. Energy Combust. Sci.* **2014**, *42*, 35–53. [CrossRef]
52. Fan, S.; Zhang, P.; Li, F.; Jin, S.; Wang, S.; Zhou, S. A Review of Lignocellulose Change During Hydrothermal Pretreatment for Bioenergy Production. *Curr. Org. Chem.* **2016**, *20*, 2799–2809. [CrossRef]
53. Xiang, H.; Sun, R.C. Unraveling the structural characteristics of lignin in hydrothermal pretreated fibers and manufactured binderless boards from Eucalyptus grandis. *Sustain. Chem. Process.* **2014**, *9*, 1–12.
54. Kumar, R.; Wyman, C.E. Access of cellulase to cellulose and lignin for poplar solids produced by leading pretreatment technologies. *Biotechnol. Prog.* **2009**, *3*, 807–813. [CrossRef]
55. Sakaki, T.; Shibata, M.; Sumi, T.; Yasuda, S. Saccharification of cellulose using a hot-compressed water-flow reactor. *Ind. Eng. Chem. Res.* **2002**, *41*, 661–665. [CrossRef]
56. Jin, F.M.; Zhou, Z.Y.; Enomoto, H.; Moriya, T.; Higashijima, H. Conversion mechanism of cellulosic biomass to lactic acid in subcritical water and acidbase catalytic effect of subcritical water. *Chem. Lett.* **2004**, *33*, 126–127. [CrossRef]
57. Gao, Y.; Wang, X.H.; Yang, H.P.; Chen, H.P. Characterization of products from hydrothermal treatments of cellulose. *Energy* **2012**, *42*, 457–465. [CrossRef]
58. Di Girolamo, G.; Grigatti, M.; Barbanti, L.; Angelidaki, I. Effects of hydrothermal pre-treatments on Giant reed (Arundo donax) methane yield. *Bioresour. Technol.* **2013**, *147*, 152–159. [CrossRef] [PubMed]
59. Scheller, H.V.; Ulvskov, P. Hemicelluloses. *Annu. Rev. Plant Biol.* **2010**, *61*, 263–289. [CrossRef] [PubMed]
60. Costa, A.G.; Pinheiro, G.C.; Pinheiro, F.G.C.; Dos Santos, A.B.; Santaella, S.T.; Leitão, R.C. The use of thermochemical pretreatments to improve the anaerobic biodegradability and biochemical methane potential of the sugarcane bagasse. *Chem. Eng. J.* **2014**, *248*, 363–372. [CrossRef]
61. Hendriks, A.T.W.M.; Zeeman, G. Pretreatments to enhance the digestibility of lignocellulosic biomass. *Bioresour. Technol.* **2009**, *100*, 10–18. [CrossRef] [PubMed]
62. Den Camp, H.J.M.O.; Verhagen, F.J.M.; Kivaisi, A.K.; de Windt, F.E.; Lubberding, H.J.; Gijzen, H.J.; Vogels, G.D. Effects of lignin on theanaerobic degradation of (ligno) cellulosic wastes by rumen microorganisms. *Appl. Microbiol. Biotechnol.* **1988**, *29*, 408–412. [CrossRef]
63. Grabber, J.H. How do lignin composition, structure, and cross-linking affect degradability? A review of cell wall model studies. *Crop Sci.* **2005**, *45*, 820–831. [CrossRef]
64. Berlin, A.; Balakshin, M.; Gilkes, N.; Kadla, J.; Maximenko, V.; Kubo, S.; Saddler, J. Inhibition of cellulase, xylanase and beta-glucosidase activities by softwood lignin preparations. *J. Biotechnol.* **2006**, *125*, 198–209. [CrossRef]
65. Negro, M.J.; Manzanares, P.; Ballesteros, I.; Oliva, J.; Caba_as, A.; Ballesteros, M. Hydrothermal pretreatment conditions to enhance ethanol production from popular biomass. *Biotechnol. Appl. Biochem.* **2003**, *105*, 87–100. [CrossRef]
66. Kristensen, J.B.; Thygesen, L.G.; Felby, C.; Jørgensen, H.; Elder, T. Cellwall structural changes in wheat straw pretreated for bioethanol production. *Biotechnol. Biofuels* **2008**, *1*, 1–9. [CrossRef]
67. Teghammar, A.; Yngvesson, J.; Lundin, M.; Taherzadeh, M.J.; Horváth, I.S. Pretreatment of paper tube residuals for improved biogas production. *Bioresour. Technol.* **2010**, *101*, 1206–1212. [CrossRef] [PubMed]
68. Kumar, S.; Kothari, U.; Kong, L.; Lee, Y.Y.; Gupta, R.B. Hydrothermal pretreatment of switchgrass and corn stover for production of ethanol and carbon microspheres. *Biomass Bioenergy* **2011**, *35*, 956–968. [CrossRef]
69. Daorattanachai, P.; Viriya-empikul, N.; Laosiripojana, N.; Faungnawakij, K. Effects of kraft lignin on hydrolysis/dehydration of sugars, cellulosic and lignocellulosic biomass under hot compressed water. *Bioresour. Technol.* **2013**, *144*, 504–512. [CrossRef] [PubMed]
70. Chynoweth, D.P.; Owens, J.M.; Legrand, R. Renewable methane from anaerobic digestion of biomass. *Renew. Energy* **2000**, *22*, 1–8. [CrossRef]

71. Chandra, R.; Takeuchi, H.; Hasegawa, T. Hydrothermal pretreatment of rice straw biomass: A potential and promising method for enhanced methane production. *Appl. Energy* **2012**, *94*, 129–140. [CrossRef]
72. Li, C.; Zhang, G.; Zhang, Z.; Ma, D.; Wang, L.; Xu, G. Hydrothermal pretreatment for biogas production from anaerobic digestion of antibiotic mycelial residue. *Chem. Eng. J.* **2015**, *279*, 530–537. [CrossRef]
73. Phuttaro, C.; Sawatdeenarunat, C.; Surendra, K.C.; Boonsawang, P.; Chaiprapat, S.; Khanal, S.K. Anaerobic digestion of hydrothermally-pretreated lignocellulosic biomass: Influence of pretreatment temperatures, inhibitors and soluble organics on methane yield. *Bioresour. Technol.* **2019**, *284*, 128–138. [CrossRef]
74. He, L.; Huang, H.; Lei, Z.; Liu, C.; Zhang, Z. Enhanced hydrogen production from anaerobic fermentation of rice straw pretreated by hydrothermal technology. *Bioresour. Technol.* **2014**, *171*, 145–151. [CrossRef]
75. Kongjan, P.; Angelidaki, I. Extreme thermophilic biohydrogen production from wheat straw hydrolysate using mixed culture fermentation: Effect of reactor configuration. *Bioresour. Technol.* **2010**, *101*, 7789–7796. [CrossRef]
76. Jung, K.W.; Kim, D.H.; Shin, H.S. Fermentative hydrogen production from Laminaria japonica and optimization of thermal pretreatment conditions. *Bioresour. Technol.* **2011**, *102*, 2745–2750. [CrossRef]
77. Zhang, S.; Zhu, X.; Zhou, H.; Shang, S.; Luo, J.; Tsang, D.C. Hydrothermal Carbonization for Hydrochar Production and Its Applications. In *Biochar from Biomass and Waste*; Elsevier: Amsterdam, The Netherlands, 2019; Chapter 15; pp. 275–294.
78. Titirici, M.M.; Antonietti, M. Chemistry and materials options of sustainable carbon materials made by hydrothermal carbonization. *Chem. Soc. Rev.* **2010**, *39*, 103–116. [CrossRef] [PubMed]
79. Kambo, H.S.; Dutta, A. Comparative evaluation of torrefaction and hydrothermal carbonization of lignocellulosic biomass for the production of solid biofuel. *Energy Convers. Manag.* **2015**, *105*, 746–755. [CrossRef]
80. Kim, D.; Yoshikawa, K.; Park, K.Y. Characteristics of biochar obtained by hydrothermal carbonization of cellulose for renewable energy. *Energies* **2015**, *8*, 14040–14048. [CrossRef]
81. Balat, M.; Balat, H.; Öz, C. Progress in bioethanol processing. *Prog. Energy Combust. Sci.* **2008**, *34*, 551–573. [CrossRef]
82. López-Aparicio, S.; Hak, C. Evaluation of the use of bioethanol fuelled buses based on ambient air pollution screening and on-road measurements. *Sci. Total Environ.* **2013**, *452*, 40–49. [CrossRef] [PubMed]
83. Nitsos, C.K.; Matis, K.A.; Triantafyllidis, K.S. Optimization of Hydrothermal Pretreatment of Lignocellulosic Biomass in the Bioethanol Production Process. *ChemSusChem* **2013**, *6*, 110–122. [CrossRef] [PubMed]
84. Mabee, W.E.; Gregg, D.J.; Arato, C.; Berlin, A.; Bura, R.; Gilkes, N.; Mirochnik, O.; Pan, X.; Pye, E.K.; Saddler, J.N. Updates on softwood-to-ethanol process development. *Appl. Biochem. Biotechnol.* **2006**, *129–132*, 55–70. [CrossRef]
85. Reczey, V.K.; Zacchi, G. Optimization of steam pretreatment of corn stover to enhance enzymatic digestibility. *Biotechnol. Appl. Biochem.* **2004**, *113*, 509–523.
86. Petersen, M.Ø.; Larsen, J.; Thomsen, M.H. Optimization of hydrothermal pretreatment of wheat straw for production of bioethanol at low water consumption without addition of chemicals. *Biomass Bioenergy* **2009**, *33*, 834–840. [CrossRef]
87. Steinbach, D.; Kruse, A.; Sauer, J. Pretreatment technologies of lignocellulosic biomass in water in view of furfural and 5-hydroxymethylfurfural production-A review. *Biomass Convers. Biorefinery* **2017**, *7*, 247–274. [CrossRef]
88. Kim, T.H.; Lee, Y.Y. Pretreatment of corn stover by soaking in aqueous ammonia. *Biotechnol. Appl. Biochem.* **2005**, *124*, 1119–1131. [CrossRef]
89. Monlau, F.; Trably, E.; Barakat, A.; Quemeneur, M.; Steyer, J.P.; Carrère, H. Do by-products of thermochemical treatment of lignocellulosic materials inhibit anaerobic mixed cultures? Overview of recent findings. In Proceedings of the 13th World Congress on Anaerobic Digestion, Santiago de Compostela, Spain, 25–28 June 2013; p. 171.
90. López González, L.M.; Pereda Reyes, I.; Dewulf, J.; Budde, J.; Heiermann, M.; Vervaeren, H. Effect of liquid hot water pre-treatment on sugarcane press mud methane yield. *Bioresour. Technol.* **2014**, *169*, 284–290. [CrossRef] [PubMed]
91. Suryawati, L.; Wilkins, M.R.; Bellmer, D.D.; Huhnke, R.L.; Maness, N.O.; Banat, I.M. Effect of hydrothermolysis process conditions on pretreated switchgrass composition and ethanol yield by SSF with *Kluyveromyces marxianus* IMB4. *Process Biochem.* **2009**, *44*, 540–545. [CrossRef]

92. Yu, G.; Yano, S.; Inoue, H.; Inoue, S.; Endo, T.; Sawayama, S. Pretreatment of rice straw by a hot-compressed water process for enzymatic hydrolysis. *Appl. Biochem. Biotechnol.* **2010**, *160*, 539–551. [CrossRef] [PubMed]
93. Pérez, J.A.; González, A.; Oliva, J.M.; Ballesteros, I.; Manzanares, P. Effect of process variables on liquid hot water pretreatment of wheat straw for bioconversion to fuel-ethanol in a batch reactor. *J. Chem. Technol. Biotechnol.* **2007**, *82*, 929–938. [CrossRef]
94. Bougrier, C.; Delgenes, J.P.; Carrere, H. Effects of thermal treatments onfive different waste activated sludge samples solubilisation, physical propertiesand anaerobic digestion. *Chem. Eng. J.* **2008**, *139*, 236–244. [CrossRef]
95. Ko, J.K.; Um, Y.; Park, Y.C.; Seo, J.H.; Kim, K.H. Compounds inhibiting the bioconversion of hydrothermally pretreated lignocellulose. *Appl. Microbiol. Biotechnol.* **2015**, *99*, 4201–4212. [CrossRef]
96. Cantarella, M.; Cantarella, L.; Gallifuoco, A.; Spera, A.; Alfani, F. Effect of inhibitors released during steam-explosion treatment of poplar wood on subsequent enzymatic hydrolysis and SSF. *Biotechnol. Prog.* **2004**, *20*, 200–206. [CrossRef]
97. Dekker, R.F.H. Inhibitors of Trichoderma reesei β-glucosidase activity derived from auto-hydrolysis-exploded Eucalyptous regnans. *Appl. Microbiol. Biotechnol.* **1988**, *29*, 593–598.
98. Excoffier, G.; Toussaint, B.; Vignon, M.R. Saccharification of steam exploded poplar wood. *Biotechnol. Bioeng.* **1991**, *38*, 1308–1317. [CrossRef]
99. Kim, Y.; Ximenes, E.; Mosier, N.S.; Ladisch, M.R. Soluble inhibitors/deactivators of cellulase enzymes from lignocellulosic biomass. *Enzym. Microb. Technol.* **2011**, *48*, 408–415. [CrossRef] [PubMed]
100. García-Aparicio, M.P.; Ballesteros, I.; González, A.; Oliva, J.M.; Ballesteros, M.; Negro, M.J. Effect of inhibitors released during steamexplosion pretreatment of barley straw on enzymatic hydrolysis. *Appl. Biochem. Biotechnol.* **2006**, *129*, 278–288. [CrossRef]
101. Nakagame, S.; Chandra, R.P.; Saddler, J.N. The influence of lignin on the enzymatic hydrolysis of pretreated biomass substrates. In *Sustainable Production of Fuels, Chemicals, and Fibers from Forest Biomass*; Zhu, J.Y., Zhang, X., Pan, X.J., Eds.; American Chemical Society: Washington, DC, USA, 2011; pp. 145–167.
102. Ko, J.K.; Kim, Y.; Ximenes, E.; Ladisch, M.R. Effect of liquid hot water pretreatment severity on properties of hardwood lignin and enzymatic hydrolysis of cellulose. *Biotechnol. Bioeng.* **2015**, *112*, 252–262. [CrossRef] [PubMed]
103. Ko, J.K.; Ximenes, E.; Kim, Y.; Ladisch, M.R. Adsorption of enzyme onto lignins of liquid hot water pretreated hardwoods. *Biotechnol. Bioeng.* **2015**, *112*, 447–456. [CrossRef] [PubMed]
104. Nakagame, S.; Chandra, R.P.; Kadla, J.F.; Saddler, J.N. The isolation, characterization and effect of lignin isolated from steam pretreated Douglas-fir on the enzymatic hydrolysis of cellulose. *Bioresour. Technol.* **2011**, *102*, 4507–4517. [CrossRef] [PubMed]
105. Ximenes, E.; Kim, Y.; Mosier, N.; Dien, B.; Ladisch, M. Deactivation of cellulases by phenols. *Enzym. Microb. Technol.* **2011**, *48*, 54–60. [CrossRef] [PubMed]
106. Dunlop, A. Furfural formation and behavior. *Ind. Eng. Chem.* **1948**, *40*, 204–209. [CrossRef]
107. Ulbricht, R.J.; Sharon, J.; Thomas, J.A. A review of 5-hydroxymethylfurfural (HMF) in parental solutions. *Fundam. Appl. Toxicol.* **1984**, *4*, 843–853. [CrossRef]
108. Jönsson, L.J.; Alriksson, B.; Nilvebrant, N.O. Bioconversion of lignocellulose: Inhibitors and detoxification. *Biotechnol. Biofuels* **2013**, *6*, 16. [CrossRef]
109. Ask, M.; Bettiga, M.; Mapelli, V.; Olsson, L. The influence of HMF and furfural on redox-balance and energy-state of xylose-utilizing Saccharomyces cerevisiae. *Biotechnol. Biofuels* **2013**, *6*, 1–13. [CrossRef]
110. Larsson, S.; Palmqvist, E.; Hahn-Hägerdal, B.; Tengborg, C.; Stenberg, K.; Zacchi, G.; Nilvebrant, N.O. The generation of fermentation inhibitors during dilute acid hydrolysis of softwood. *Enzym. Microb. Technol.* **1999**, *24*, 151–159. [CrossRef]
111. Kim, Y.; Kreke, T.; Hendrickson, R.; Parenti, J.; Ladisch, M.R. Fractionation of cellulase and fermentation inhibitors from steam pretreated mixed hardwood. *Bioresour. Technol.* **2013**, *135*, 30–38. [CrossRef] [PubMed]
112. Gómez-Quiroga, X.; Aboudi, K.; Álvarez-Gallego, J.C.; Romero-García, I.L. Enhancement of Methane Production in Thermophilic Anaerobic Co-Digestion of Exhausted Sugar Beet Pulp and Pig Manure. *Appl. Sci.* **2019**, *9*, 1791. [CrossRef]
113. Karthikeyan, O.P.; Visvanathan, C. Bio-energy recovery from high-solid organic substrates by dry anaerobic bio-conversion processes: A review. *Rev. Environ. Sci. Bio/Technol.* **2013**, *12*, 257–284. [CrossRef]

114. Grant, T.M.; King, C.J. Mechanism of irreversible adsorption of phenolic compounds by activated carbons. *Ind. Eng. Chem. Res.* **1990**, *29*, 264–271. [CrossRef]
115. Cao, G.; Ximenes, E.; Nichols, N.N.; Zhang, L.; Ladisch, M. Biological abatement of cellulase inhibitors. *Bioresour. Technol.* **2013**, *146*, 604–610. [CrossRef] [PubMed]
116. Larsson, S.; Reimann, A.; Nilvebrant, N.O.; Jönsson, L.J. Comparison of different methods for the detoxification of lignocellulose hydrolyzates of spruce. *Appl. Biochem. Biotechnol.* **1999**, *77*, 91–103. [CrossRef]
117. Parawira, W.; Tekere, M. Biotechnological strategies to overcome inhibitors in lignocellulose hydrolysates for ethanol production: Review. *Crit. Rev. Biotechnol.* **2011**, *31*, 20–31. [CrossRef]
118. Almeida, J.R.; Modig, T.; Petersson, A.; Hähn-Hägerdal, B.; Lidén, G.; Gorwa-Grauslund, M.F. Increased tolerance and conversion of inhibitors in lignocellulosic hydrolysates by Saccharomyces cerevisiae. *J. Chem. Technol. Biotechnol.* **2007**, *82*, 340–349. [CrossRef]
119. Liu, Z.L.; Slininger, P.J.; Gorsich, S.W. Enhanced biotransformation of furfural and hydroxymethylfurfural by newly developed ethanologenic yeast strains. *Appl. Biochem. Biotechnol.* **2005**, *121*, 451–460. [CrossRef]
120. Liu, Z.L.; Moon, J.; Andersh, B.J.; Slininger, P.J.; Weber, S. Multiple gene-mediated NAD(P)H-dependent aldehyde reduction is a mechanism of in situ detoxification of furfural and 5-hydroxy methylfurfural by Saccharomyces cerevisiae. *Appl. Microbiol. Biotechnol.* **2008**, *81*, 743–753. [PubMed]
121. Keller, F.A.; Bates, D.; Ruiz, R.; Nguyen, Q. Yeast adaptation on softwood prehydrolysate. *Appl. Biochem. Biotechnol.* **1998**, *70*, 137–148. [CrossRef] [PubMed]
122. Larsson, S.; Cassland, P.; Jönsson, L.J. Development of a Saccharomyces cerevisiae strain with enhanced resistance to phenolic fermentation inhibitors in lignocellulose hydrolysates by heterologous expression of laccase. *Appl. Environ. Microbiol.* **2001**, *67*, 1163–1170. [CrossRef] [PubMed]

© 2019 by the authors. Licensee MDPI, Basel, Switzerland. This article is an open access article distributed under the terms and conditions of the Creative Commons Attribution (CC BY) license (http://creativecommons.org/licenses/by/4.0/).

applied sciences

MDPI

Article

Insights into Anaerobic Co-Digestion of Lignocellulosic Biomass (Sugar Beet By-Products) and Animal Manure in Long-Term Semi-Continuous Assays

Kaoutar Aboudi *, Xiomara Gómez-Quiroga, Carlos José Álvarez-Gallego and Luis Isidoro Romero-García *

Department of Chemical Engineering and Food Technology, Faculty of Sciences, Wine and Agrifood Research Institute (IVAGRO), The University of Cádiz, Republic Saharawi Avenue, Box nº 40, 11510 Puerto Real, Spain; Xiomara.gomez@uca.es (X.G.-Q.); carlosjose.alvarez@uca.es (C.J.Á.-G.)

* Correspondence: kaoutar.aboudi@uca.es (K.A.); luisisidoro.romero@uca.es (L.I.R.-G.)

Received: 30 June 2020; Accepted: 21 July 2020; Published: 26 July 2020

Abstract: Biogas production through anaerobic digestion has proven to be one of the most important pillars of the transition into the circular economy concept, a sustainable approach for biorefinery. This work aims to extend and improve knowledge in the anaerobic co-digestion of complementary substrates, given insights into wastes biodegradability and the influence of manure composition on the anaerobic process stability. Anaerobic co-digestion of sugar beet by-products with two kinds of animal manure (pig and cow) was investigated in semi-continuous assays, analyzing both common and non-classical parameters. Co-digestion with manure clearly mitigated the inhibitory effect of volatile fatty acids at high organic loading rates, leading to increases in methane production by 70% and 31% in comparison with individual digestion of sugar beet by-products, for co-digestion with pig and cow manure, respectively. Non-classical parameters could give more insight into the coupling/uncoupling of the anaerobic digestion phases and the involved microorganisms. Indirect parameters indicated that the process failure at the critical organic loading rates was mainly due to methanogenesis inhibition in the co-digestion with pig manure, while in co-digestion with cow manure or in individual digestion of sugar beet by-products, both hydrolysis–acidogenesis and methanogenesis phases were affected. Biomethanation degree refers to the maximum methane potential of organic wastes. Sugar beet by-products required a long digestion-time to reach high biodegradability. However, short digestion-times for co-digestion assays led to a high biomethanation degree.

Keywords: anaerobic co-digestion; sugar beet by-products; manure; semi-continuous feeding mode; methane improvement; non-classical parameters

1. Introduction

According to the European Economic and Social Committee, the success of the circular economy is based on three pillars. The first pillar is about energy consumption (i.e., the efficiency in energy sources). The second pillar is about reducing the cost of product processing (i.e., sustainable technological processes for waste-out systems designing), and the third one concerns employment.

Biogas production through anaerobic digestion (AD) has proven to provide direct benefits to the second pillar and indirect benefits to the rest of the pillars, achieving the transition into the circular economy, a sustainable approach for the biorefinery concept [1].

Currently, agriculture is one of the industrial sectors that generates the largest amount of wastes. Therefore, 1.3 billion tonnes of food are yearly discarded during production handling, storage processing, and distribution, with an estimated carbon footprint of 3.3 billion tonnes of CO_2 equivalent of greenhouse gases (GHG) released into the atmosphere per year [2]. Agroindustrial wastes such as sugarcane bagasse, wheat straw, wheat bran, and many others are cheapest and abundantly available. Their valorization by AD has long been investigated and implemented for bioenergy production [3,4]. By-products from sugar beet plant after the sugar extraction process may be bio-converted into bioenergy through AD [5].

Similarly, a large amount of manure produced in livestock facilities is a potential candidate for clean energy production such as biogas [6]. Moreover, livestock production in the European Union is high, being Spain and Germany the largest producers of pigs, while France held the largest number of bovines [7]. It has been reported that AD of animal manure can save up to 20% of the dairy processing facility's energy demand [8].

Despite their suitability for the anaerobic treatment, the use of manure as feedstock in AD could inhibit to methanogenic archaea due to the ammonia released during the process [9,10]. Manure has been traditionally used as a fertilizer for agricultural soils due to its high content in nitrogen and phosphorus. Nevertheless, this uncontrolled practice may have environmental problems, such as groundwater contamination due to infiltration, or greenhouse gas emissions [11]. Furthermore, lignocellulosic agroindustrial wastes and by-products lack of some macro and micronutrients [12,13].

The AD of both kinds of wastes separately may be improved and optimized by co-digestion as this biomass has different/complementary characteristics [14,15]. Sugar beet by-products (SBB) are rich in carbon while manure is a potential substrate to compensate nitrogen deficiency in the anaerobic process, together with minerals and essential nutrients. Moreover, manure provides a high buffer capacity, which could neutralize volatile fatty acids (VFAs) released by agroindustrial wastes in the AD process [16,17].

Besides, the manure type plays an important role in the AD process. Hence, some manure resulted in more effective in the co-digestion process with agroindustrial wastes than others [18]. This effect returns to several factors such as the composition of each manure according to its origin (animal digestive tract), the food given to animals in the course of animal husbandry, and the synergy between the manure and agroindustrial waste used in co-digestion [19–21]. Despite the numerous benefits of co-digestion, the mixture of substrates with different characteristics could create either synergistic interactions of microorganisms or antagonist interactions, which in this case, can worsen the anaerobic process [22].

To date, several previous studies have investigated anaerobic co-digestion of agroindustrial wastes with livestock manure to enhance biogas production and in most of these studies; synergistic effect between the two kinds of substrates has been suggested to be the reason of the process improvement. However, only a few studies have thoroughly investigated the effect of manure on the neutralization of acidity from agroindustrial wastes in semi-continuously operated reactors.

Cheng and Zhong [23] have demonstrated that the use of swine manure as a co-substrate helps to neutralize the high VFAs concentration released by cotton stalk, emphasizing advantages related to the high buffer capacity provided by swine manure. Other authors have used a different strategy to minimize the acidification by VFAs in a biological process, by adding some micronutrients [12,13].

The use of the semi-continuous feeding mode in anaerobic digesters operating for a long-term period is of great interest as it gives insight into the most viable options for the organic material treatment through the AD in a view of a real scale approach. Optimization of hydraulic retention times and organic loading rates, as well as information on feedstock combination, may help in future research and decisions on agricultural and manure management and treatment for efficient and long-lasting wastes treatment plants in rural and agricultural areas.

This work aims to extend and improve knowledge in the anaerobic co-digestion of complementary substrates, given insights into wastes biodegradability and the influence of manure composition on the AD stability. Anaerobic co-digestion of sugar beet by-products with two kinds of animal manure, pig and cow, (PM and CM) was investigated in semi-continuous assays, analyzing both common and non-classical parameters. This research addresses a deep analysis of the findings from long-term anaerobic digesters operating with agricultural and animal wastes.

2. Materials and Methods

2.1. Feedstock and Inoculum Characteristics

Sugar beet by-products were a mixture of exhausted pulp (85%) and molasses (15%). They were provided by the sugar processing company in the south of Spain (El Portal, Jerez de la Frontera, Andalusia). The used SBB have a total solids content of 80–90%, which was adjusted to 8% of TS to avoid the non-homogenization of the reactor content due to the rheological behavior of SBB [24]. For this purpose, the SBB as dried pellets were rehydrated with deionized water for 24 h previously to their use. Cow and pig manure were collected from two different livestock facilities in the same zone. The inoculum for the start-up of the single AD of SBB (Inoc1) was provided from a laboratory-scale reactor under semi-continuous operation that was fed with the organic fraction of municipal solid wastes (OFMSW), which was first adapted to the new substrate SBB for a long period before starting the herein studied assays [25]. The effluent from the reactor of single digestion of sugar beet by-products was used as inoculum for the start-up of co-digestion assays (Inoc2). In co-digestion reactors, the mixture between the two co-substrates has been established as 25% of SBB and 75% of pig manure or cow manure (v/v) [26,27].

The SBB, CM, PM, and inocula characteristics are presented in Table 1.

Table 1. Physicochemical properties of sugar beet by-products (Sugar beet by-products, cow manue, pig manure), and inocula.

Parameters (Units)	SBB	CM	PM	Inoc1	Inoc2
pH	5.7 ± 0.7	6.2 ± 1.3	6.5 ± 1.7	7.5 ± 0.7	7.6 ± 0.8
TS (g/kg)	880.3 ± 12.4	222.7 ± 11.5	225.3 ± 7.4	41.2 ± 1.3	36.1 ± 10.6
VS (%TS)	90.6 ± 1.8	77.7 ± 6.7	75.1 ± 5.2	16.3 ± 1.6	21.8 ± 3.3
CODS (gO_2/kg TS)	64.7 ± 2.5	77.7 ± 13.9	83.0 ± 7.9		246.5 ± 6.1
CODT (gO_2/kg TS)	165.8 ± 5.9	308.5 ± 6.3	220.6 ± 11.1		321.3 ± 6.9
DOC (gC/kg TS)	49.7 ± 5.1	33.7 ± 9.8	39.1 ± 6.6	150 ± 5.0	127.4 ± 19.4
TVFA (gHAc/kg TS)	2.5 ± 0.9	24.3 ± 4.9	20.4 ± 2.2	7.2 ± 2.4	47.1 ± 4.3
Alkalinity ($gCaCO_3$/kg TS)	3.3 ± 1.4	170.2 ± 0.9	169.9 ± 49.3		116.4 ± 69.3
N-NH4+ (gN/kg TS)	0.3 ± 0.1	15.7 ± 3.2	10.6 ± 2.6		30.5 ± 5.5
TN (gTN/kg TS)	14.5 ± 1.5	157.6 ± 6.7	145.2 ± 12.4		174.5 ± 3.6
Ratio C/N	38.9 ± 2.7	13.2 ± 1.0	13.5 ± 0.3	-	19.4 ± 2.4
Hemicellulose (%)	15.4 ± 0.3	13.9 ± 0.8	14.4 ± 0.1	-	-
Cellulose (%)	25.9 ± 0.3	23.2 ± 1.2	16.3 ± 1.2	-	-
Lignin (%)	1.6 ± 0.1	19.8 ± 1.1	16.8 ± 1.9	-	-
Mineral salts (%)	6.7 ± 0.2	26.2 ± 1.0	27.2 ± 3.1	-	-

The organic content of SBB was the highest since the total solids were almost all volatile (90%). The VS content of manure was also higher (75–77%) with a high fraction of nitrogenous material. The C/N ratio of SBB was much higher than for animal manure, which led us to expect a complementarity between the two kinds of substrates in terms of carbon and nitrogen contents. Concerning the lignocellulosic content of SBB and manure, all of them could be considered as lignocellulosic substrates with a high cellulose fraction in SBB and CM. Moreover, the mineral content of PM and CM is much higher than for SBB. Trace elements are also necessary for the AD process and are considered to stimulate the digestion of cellulosic material [26]. Therefore, manure are natural sources of trace elements and their mixture with SBB gives rise to an appropriate medium for the growth of microorganisms. Wintsche et al. [28] reported that supplementation of trace elements in anaerobic reactors leads to stable and more efficient methane production processes while deficits in these elements may cause process imbalances. Both manure have a high mineral content and important alkalinity indicating a high buffering capacity.

2.2. Semi-Continuous Digesters

Three semi-continuous stirred tank reactors (SSTR) were used for individual digestion of SBB, co-digestion of SBB with CM, and co-digestion of SBB with PM. Assays were carried out at mesophilic temperature (35 °C). Reactors are comprised of stainless steel with a useful volume of 10 L. The temperature was maintained by a recirculating bath (Ultraterm200-Selecta) through the reactor jacket. A motor installed at the top of the reactor (Heidolph-RZR-2102) with a stirring blade performed the mixing of the reactor content.

Table 2 depicts the hydraulic retention times (HRT) and organic loading rates (OLR) studied in each digester.

Table 2. The operation conditions (hydraulic retention times and organic loading rates) of the three semi-continuous stirred tank reactors (SSTR) digesters.

Reactors		Hydraulic Retention Times (Days)						
		20	18	15	12	8	6	5
SBB	Organic loading rates (gVS/Lreactor*d)	3.3	3.6					
SBB + PM		4.2	4.7	5.9	7.4	8.5	11.2	12.8
SBB + CM		3.7	4.2	4.9	6.2			

The first HRT tested was at 20 days according to literature in which the AD of lignocellulosic biomass and similar substrates to this study indicated the need for high HRT [29]. Demirer and Scherer applied HRTs with a range of 95–15 days to sugar beet silage and reported that the HRT around 25 days gave the highest biogas yield [29].

2.3. Analytical Methods

Total solids (TS), volatile solids (VS), and alkalinity were measured according to APHA Methods 2540B, 2540E, and 2320B, respectively [30]. The chemical oxygen demand (COD, total, and soluble) was determined by colorimetric techniques using a spectrophotometer instrument (HACH®, DR/4000U), according to the method 5220C [30]. Total chemical oxygen demand (TCOD) was measured directly in the sample while for soluble chemical oxygen demand (SCOD) and the rest of the soluble parameters; samples were lixiviated and filtered through a 0.47 μm filter. Total nitrogen (TN) and ammonium (N-NH4+) were analyzed by distillation (Selecta®, Pronitro II) according to the 4500-NH3E method [30]. Filtered samples for the volatile fatty acids (VFAs) analysis were filtered once more through a 0.22 μm Teflon® filter and analyzed with a gas chromatograph (Shimadzu®, GC-2010).

The dissolved organic carbon (DOC) was analyzed in an automatized carbon analyzer (Analytic-Jena®, multi N/C 3100), according to the combustion-infrared method 5310B [30]. The biogas was collected in a 10 L gas-bag (Tedlar®, SKC). The biogas volume was daily measured by a high precision drum-type gas meter (Ritter®, TG5). The composition of biogas was analyzed by a gas chromatograph (Shimadzu®, GC-2014) [15]. For the lignocellulosic content of substrates (cellulose, hemicellulose, lignin), the Van Soest method was applied by using an automatized analyzer (Foss®, FIBERTECTM 8000) [31].

2.4. Biomethanation Degree Calculations

The biomethanation degree (BD) of a substrate or different co-substrates refers to the potential of methane production from the selected organic wastes and by-products in specific conditions. To find out the biomethanation degree of the substrates in a semi-continuous reactor, the parameter defined in Equation (1) was used. Thus, BD was calculated as the percentage of methane yield achievable in a semi-continuous assay regarding the maximum methane yield attainable as the methane potential of organic wastes, which can be determined by a batch test in the same experimental conditions.

As previously mentioned, in an earlier study of Aboudi et al. [26], a series of batch assays were carried out to find out the best mixture ratio between the same substrates used in the present work.

Later on, the authors have studied the effect of the total solids content on the methane yield, also in a batch study with the same substrates at the best mixture ratios between them [27]. The ultimate methane potentials of SBB + CM and SBB + PM were 464.5 $LCH_4/kgVS_{added}$ and 451.4 $LCH_4/kgVS_{added}$, respectively. The maximum methane potential of SBB was 308.8 $LCH_4/kgVS_{added}$. These ultimate methane potential productions were in agreement with other works studying the co-digestion of manure and by-products from tuber plants [32,33].

$$Biomethanation\ degree\ (\%) = \frac{100 * SMP\_semicontinuous}{MBMP\_batch} \quad (1)$$

$SMP_{\_semicontinuous}$ is the specific methane production in SSTR, expressed as $LCH_4/kgVS_{added}$ and $MBMP_{\_batch}$ is the maximum biomethane potential obtained in batch assays in similar conditions and the same wastes, also expressed as $LCH_4/kgVS_{added}$.

2.5. *The Indirect Carbon-Related Parameter Calculations: Acidogenic Substrate as Carbon*

Non-classical parameters measured indirectly have shown to give substantial information about the AD process, giving interesting insights into the organic material degradation and bioconversion into methane [34]. In this sense, indirect parameters such as the acidogenic substrate as carbon (ASC), which refers to the organic carbon non-converted into VFAs, could be obtained according to Equation (2), from the subtraction of two directly measured parameters: dissolved organic carbon (DOC) and volatile fatty acidity (DAC).

$$ASC = DOC - DAC \quad (2)$$

DAC is accounting the carbon contained in the different VFAs from C2 to C7 and it is obtained according to Equation (3).

$$DAC = \sum_{i=2}^{i=7} \frac{AiH * ni * 12}{MWi} \quad (3)$$

where "AiH" is the concentration of each volatile fatty acid; "ni" and "MWi" are the numbers of carbon atoms and the molecular weight of each acid, respectively.

These parameters have been used as a useful tool to determine the critical stage of the AD process [15,34,35]. Thus, the accumulation of the ASC in the anaerobic digester usually indicates an imbalance between the hydrolysis and the acidogenic phases, while the accumulation of the DAC has been related to an imbalance between the acidogenesis and the methanogenesis phases.

3. Results and Discussion

3.1. *Methane Production Yields and Process Efficiency*

Daily methane productions (DMP) in the three SSTR are depicted in Figure 1A–C.

As shown in Figure 1A, individual digestion of SBB at 20-day HRT was stable for a long period of operation with values around 840 LCH_4/m^3reactor*d. The 18-day HRT showed similar productions for approximately one period of this HRT. However, production decreased by 44%, in comparison with the previous HRT. For the SBB + CM reactor Figure 1B, the decrease in HRT from 20 to 18 days and consecutively to 15 days, led to an improvement in methane production of 19% and 28%, respectively. However, operation at 12-day HRT induced a pronounced decline in methane generation by 61%. In the case of the SBB+PM reactor, daily methane productions increased by 24%, 35%, 46%, 55%, and 60% when HRT decreased to 18 days, 15 days, 12 days, 8 days, and 6 days, respectively, indicating a high system performance at lower HRTs and higher OLRs.

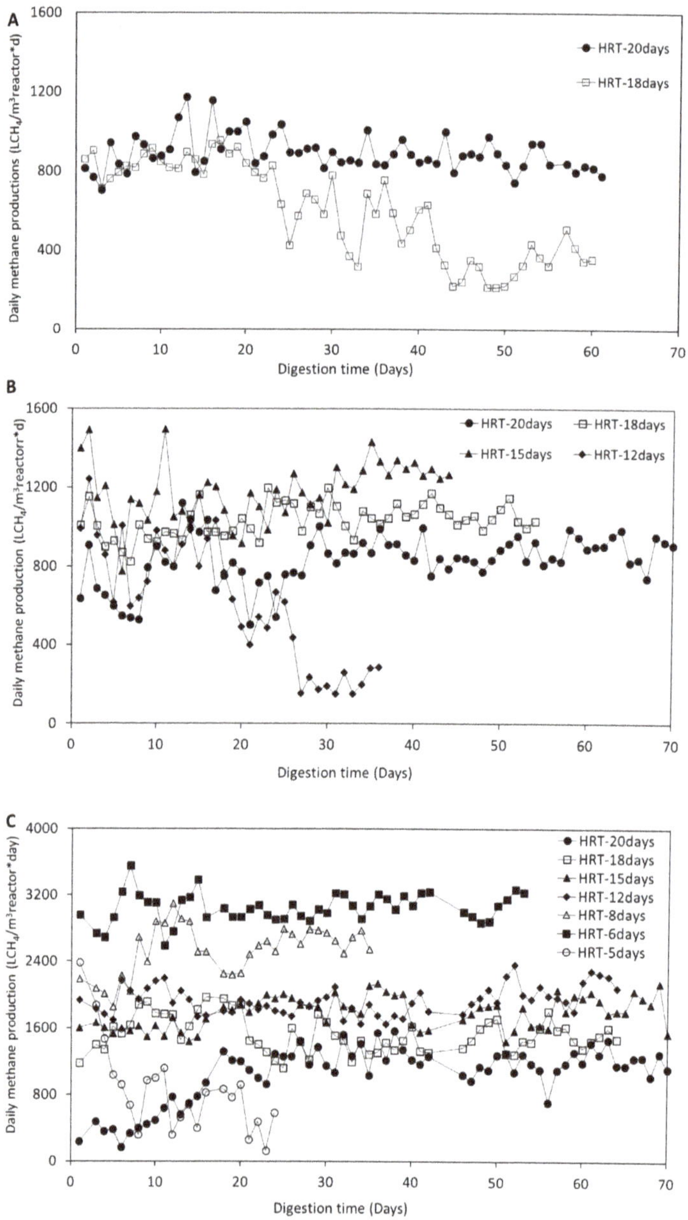

Figure 1. Daily methane productions (production rates as LCH_4/m^3 reactor*day) in the three SSTR digesters at different hydraulic retention times. (**A**) SBB, (**B**) SBB + CM, and (**C**) SBB + PM.

Nevertheless, the operation at 5-day HRT (12.8 gVS/Lreactor*d as OLR) was critical, leading to an abrupt decrease in methane generation, obtaining only 800 LCH_4/m^3reactor*d (a drop of 28%).

In both co-digestion assays, manure addition was shown to significantly increase methane production, as well as allowing increasing the organic load supplied to the reactor. In previous studies, Li et al. [32] reported that chicken manure or pig manure co-digested with apple pulp allowed the increase of OLR from 2.4 to 7.2 gVS/Lreactor*d, being 4.8 gVS/Lreactor*d the best OLR studied concerning the increase in methane generation. The daily methane productions increased by 56.9% and 47.4%, in comparison with individual digestion of chicken manure and pig manure, respectively. Kapajaru and Rinatal [33] studied anaerobic co-digestion of potato tuber (including stillage and peels) with pig manure in mesophilic SSTR. The authors reported that methane productions increased by 54.5% in comparison with the single digestion of PM. Similarly, Damaceno et al. [36] studied sweet potato co-digestion with sludge cake from poultry slaughtering. They found that the balanced nutrient content provided an optimum environment for the microorganism activity in co-digestion reactors, with 80% of poultry slaughtering and 20% of sweet potato. In another attempt, Panichnumcin et al. [37] reported that waste proportion in the mixture is a very important factor in co-digestion assays. They studied co-digestion of cassava pulp with PM in mesophilic SSTR at a constant OLR of 3.5 gVS/Lreactor*d and a 15-day HRT. In their study, cassava pulp up to 60% has shown the highest biogas generation. These findings are in agreement with those observed by Aboudi et al. [26] in batch studies about the optimization of the substrates mixture ratios for SBB and manure.

The methane content in the biogas produced by the SBB reactor decreased by 14.4% when decreasing the HRT to 18 days (OLR of 3.6 gVS/Lreactor*d). In co-digestion assays, the methane content in biogas from the SBB+CM reactor decreased by 46% at the HRT of 12 days (OLR of 6.2 gVS/Lreactor*d), while in reactors containing SBB+PM the methane content was almost maintained above 40% in all stages, with the highest content at the HRT of 20 and 18 days.

Figure 2 shows the obtained biomethanation degree in each assay.

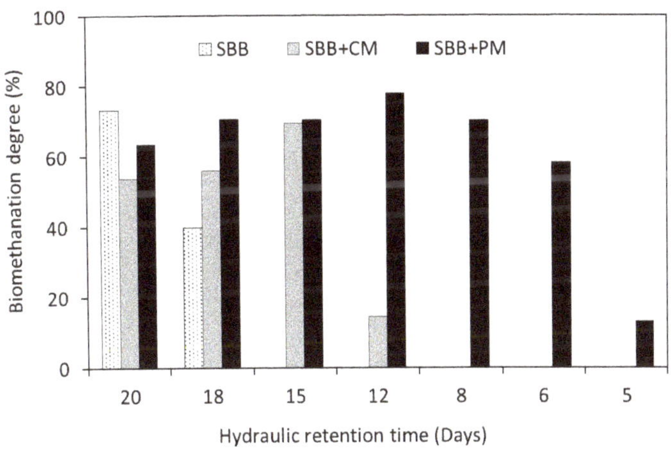

Figure 2. Biomethanation degree in the three SSTR digesters.

It can be observed that for SBB single digestion and at the optimum HRT of 20 days, a high biomethanation degree of 73% was obtained. The SBB substrate required a long time for its degradation (high HRT), and decreasing this time led to reaching low substrate biodegradability. Moreover, in both co-digestion assays, above 60% of biomethanation degree was achievable at the different HRT studied, excepting at the critical HRT tested in each experiment: 5-day HRT in SBB + PM (BD was only 13%) and 12-day HRT in SBB + CM (BD was only 14%). The highest value of the BD was around 78% in the reactor of SBB+PM at the HRT of 12 days. These results demonstrated that co-digestion played an important factor in wastes degradation likely due to the synergistic effect created in digesters, being PM more suitable to improve the biodegradability of the digester content than CM.

According to Smith and Mah [38] and from the results of their research using C^{14} tracers in an AD study of sludge, authors observed that the 73% of the total methane produced in an anaerobic process comes from the acetoclastic pathway, while the 27% usually comes from the hydrogen-utilizing pathway. Nevertheless, recent studies found that in the anaerobic digestion of nitrogen-rich substrates, the methanogenesis shift from acetoclastic pathways to the syntrophic acetate oxidation-hydrogenotrophic (SAO-HM) route [39–41]. In a recent study by Yin et al. [41], the authors attempted to explain methanogenic pathways in anaerobic digestion of nitrogen-rich substrates. The authors studied acetate conversion by C^{13} isotope under mesophilic and thermophilic conditions.

Results showed that in both temperatures, the syntrophic acetate oxidation-hydrogenotrophic (SAO-HM) pathway was dominant and that methanogenic pathway shifting, induced by high ammonia levels, closely correlated to the process performance. In this sense, it has been deduced that the substrate characteristics condition the predominance of one pathway or another.

In the present research, the combination of SBB as a high content carbonaceous material and manure as high content nitrogenous substrates led to balance the nutrients content of the anaerobic process and hence, the inhibition by ammonia was avoided.

Aiming for a comprehensive comparison of the obtained methane productions in the present research, the criterion developed by Smith and Mah [38] has been applied to the results. Thus, specific methane productions (SMPs) for the three semi-continuous digesters are depicted in Figure 3. The highest SMP from SBB+PM (HRT from 18 to 8 days) and the highest SMP from SBB+CM (HRT of 15 days) have been selected as representatives of the optimal operation area (the blue box). The average value of the five data of SMPs was 331 $LCH_4/kgVS_{added}$. The black discontinuous line in the figure corresponded to the SMP value of 27% of the average SMP obtained (89.5 $LLCH_4/kgVS_{added}$). The green line corresponded to the SMP value of 73% of the average SMP obtained (241.8 $LCH_4/kgVS_{added}$).

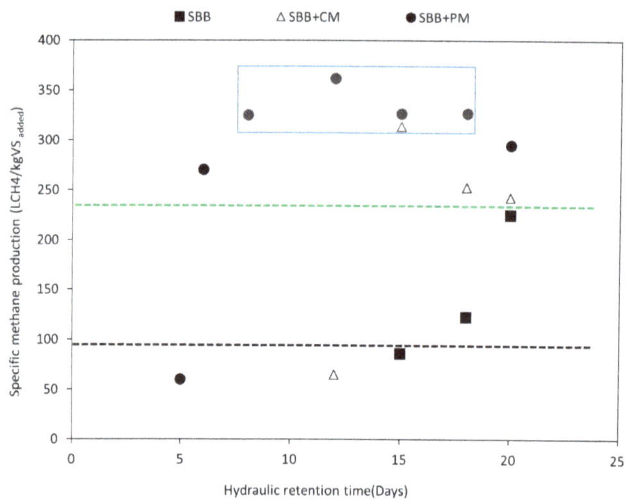

Figure 3. Specific methane productions at the different hydraulic retention time in the three SSTR digesters. Greenline: the specific methane production (SMP) value of 73% of the average SMP obtained. Blackline: the value of 27% of the average SMP obtained. Blue box: the highest SMP values in anaerobic digesters.

As can be observed, the critical operational condition in each assay (lower HRT/Higher OLR) has shown SMP values slightly below the limit value of 89.5 $LCH_4/kgVS_{added}$, indicating that acetoclastic methanogens were affected in failure periods. In critical HRTs, the accumulation of VFAs and the drop in pH to values ranged from 5.6 to 6.3 occurred indicating the inhibition of the systems. It has been reported that acetotrophic methanogens are strongly inhibited below a pH of 6.2 [41]. Figure 4 shows the average values of pH in each stage (HRTs) for the three SSTRs.

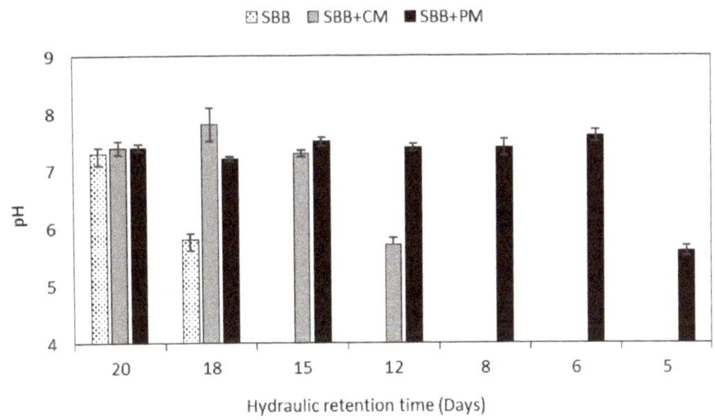

Figure 4. Average pH values in each stage in the three semi-continuous reactors.

Figure 5 shows the organic matter removal expressed as the percentage of volatile solids removed from the systems.

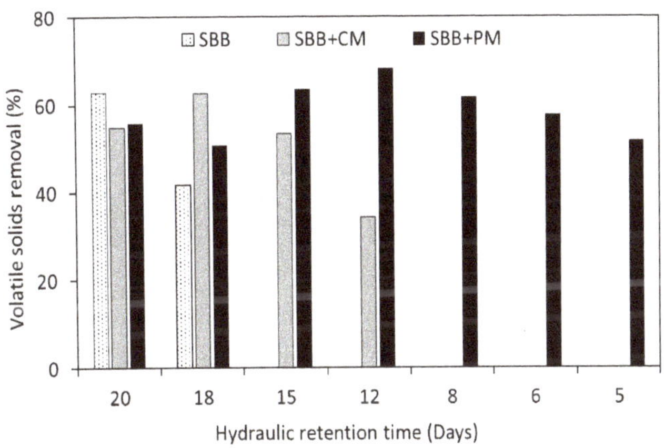

Figure 5. Volatile solids removal in the three semi-continuous reactors.

In the SBB digester, operation at a lower HRT (higher OLRs) of 18 days led to a decrease in the efficiency of the organic matter removal. In the SBB+PM reactor, the decrease in VS removal was gradual from 12-day HRT to 5-day HRT. The same behavior was observed in the SBB+CM reactor, with a gradual decrease from 18-day HRT until 12-day HRT.

3.2. Analysis of the Process Stability Based on the Indirect Carbon-Related Parameters

Fdez-Güelfo et al. [34] analyzed the indirect parameters of a semi-continuous anaerobic digester fed with the OFMSW. The authors reported that decreasing the HRT from 15 to 10 days affected the hydrolysis stage, which induced a sequential failure of the acidogenesis stage, and therefore, the methanogenesis acetoclastic pathway was disturbed. Moreover, the further decrease of HRT to 8 days has resulted in the wash-out of microbial populations involved in the anaerobic process with complete inhibition of both acetoclastic and H_2-utilizing archaea.

Figure 6 shows the evolution of the indirect parameters, together with the dissolved organic carbon (DOC) evolution in the three SSTR at the different OLRs applied.

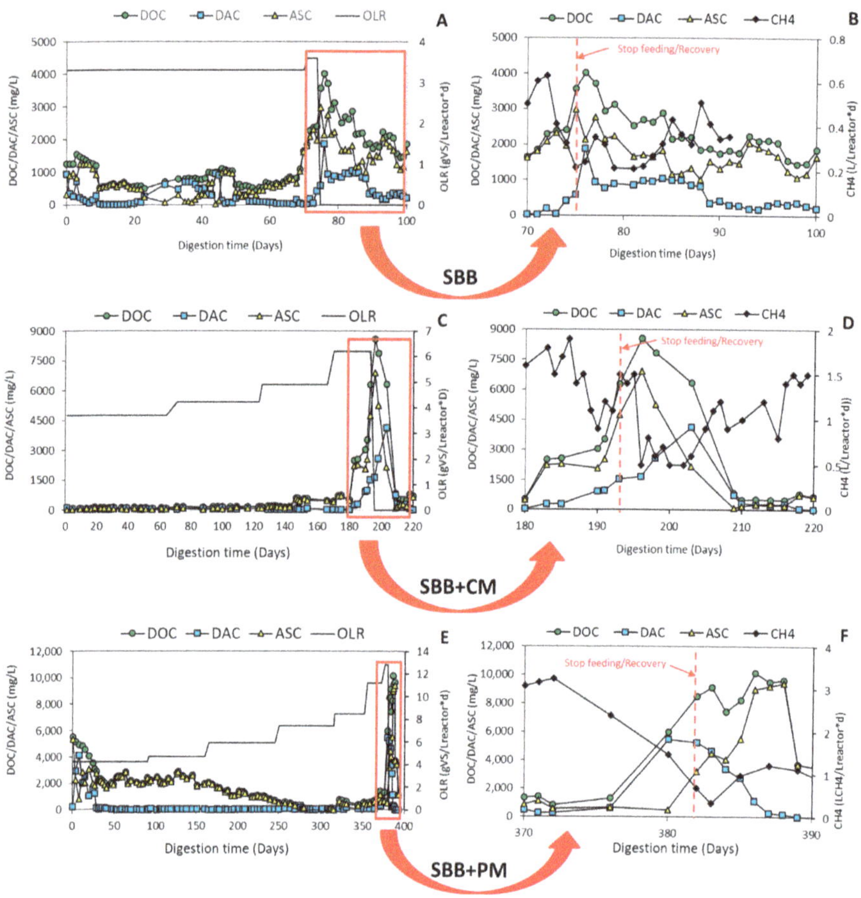

Figure 6. Evolution of the dissolved organic carbon (DOC) and indirect parameters in the three SSTR. The detailed figures of failure operation periods for each digester are illustrated on the right. (**A**,**B**) SBB digester; (**C**,**D**) SBB+CM digester; (**E**,**F**) SBB+PM digester.

At the startup period of the SBB digester and for a short period, the DOC parameter increased showing slight acidification due to the DAC increase, this behavior is usual at the inception period indicating the adaptation of microorganisms to the substrate and the operational conditions. Consecutively, the DOC and DAC values decreased gradually leading to methane generation (see Figure S1 in the supplementary material). It can be observed that in the stable operation period, the DAC values were low, being very similar to DOC values and, hence, no inhibition was observed.

The OLR increase to concentrations of 3.6 gVS/Lr*d, led to an initial increase in methane productions increased for the first few days, as a response to the organic material rise in the feedstock. The DOC and the ASC also showed a gradual increase during these few days. Nevertheless, once the DAC started increasing and despite the low proportion of the DAC with regard to the ASC, methane production dropped sharply indicating a distortion of the activity of methanogens. It has been observed that the lowest methane production (data not shown) matches the highest DAC value. The feeding was stopped to avoid further DAC accumulation and after several days, the SBB reactor has begun to recover as the DAC decreased in the medium.

In the SBB + CM reactor, operation at OLRs from 3.7 to 4.9 gVS/Lr*d showed low values of the DOC and the indirect parameters (DAC and ASC) with high methane production. It was observed that in the mid-operation period at OLR of 4.9 gVS/Lr*d (HRT of 15 days), these parameters showed a slight increase.

The next step using a further high OLR showed an increase in the three carbon-related parameters with the predominance of the ASC. It has been observed that during the failure period, the DOC increased with the rise of both the DAC and the ASC, however, a lag period between these two parameters has been detected. Hence, the disturbance period started with fluctuations in methane productions with lower values, but the highest content of the DOC was as ASC and only when the methane decrease was pronounced, the DAC proportion in the DOC was higher. This behavior could be likely due to failure in the hydrolysis and acidogenesis firstly, followed by the inhibition of methanogenesis. In this sense, an imbalance between the AD steps occurred. In this co-digestion digester, only the increase in OLR affected the process, however, CM shows the advantage that in the startup period, no accumulation of DAC or ASC was observed, indicating that a mixture of SBB and CM as complementary substrates has benefited the process. Furthermore, CM comes from the digestive tract of cows, which are herbivores ruminants and their rumen contains microorganisms able to degrade lignocellulosic substrates [42]. Nevertheless, the characteristics of CM showed that this manure provides more cellulosic material than PM.

In the SBB+PM digester, the startup period showed high DOC and DAC concentrations, which illustrate a necessary adaption period for microorganisms in this reactor. In this case, the DAC proportion in the DOC was very low in comparison to the ASC parameter. Moreover, it can be observed that as the OLR increased, the DOC and ASC decreased gradually to reach very low values when operating at OLR between 7.4 and 11.2 gVS/Lr*d, corresponding to HRTs from 12 to 6 days. An immediate decrease in methane productions accompanied by an increase in the DAC parameter was observed at the OLR of 12.8 gVS/Lr*d (HRT of 5 days). In this case, and unlike what was observed in the two previous reactors, the ASC parameter was very low and all the DOC was in the acids form, indicating that only methanogenesis was affected when the reactor was overloaded. This behavior could likely be related to the wash-out of microbial populations involved in the anaerobic process, as previously reported by Fdz-Güelfo et al. [34]. The cease of feeding led to the DAC decrease and hence the reactor has begun to recover. In all reactors, the recovery of the process has only been possible when the DAC decreased significantly.

3.3. *Analysis of the Process Stability Based on the Classical Parameters*

The stability of the three SSTRs has been assessed by using the parameters alkalinity and total volatile fatty acidity, among others. Figure 7 shows the total alkalinity, the acidity to alkalinity ratio, and the propionic acid concentrations, and the propionic to acetic acids ratio (A, B, C, and D, respectively) in each operational condition.

The alkalinity of both co-digestion reactors was significantly higher than that observed in the single digestion of SBB. This fact is expected due to the characteristics of the substrates involved in each assay. The SBB is rich in carbonaceous material but is deficient in nitrogen and alkalinity. Besides, it can be observed that alkalinity in co-digestion with PM was higher than for CM. This fact is likely due to the high buffer capacity provided by pig manure, which allows neutralizing organic acids in the system, and therefore, the highest OLRs applied were reached in this reactor. Nevertheless, the critical OLR and HRT in each digester led to a decrease in alkalinity. The criteria acidity/alkalinity ratio is an indicator of systems failure due to inhibition by acidification [43–45].

Furthermore, it has been observed that when reactors failed, the propionic acid was the predominant VFAs in the medium, which justify disturbances of systems due to propionic acid, a well-known strong inhibitor in the anaerobic process [46,47]. The ratio between acetic and propionic acids (HPr/HAC) has been designed as a useful tool for AD systems failure due to acidification [46,48], being considered critical the values above 1.4. In the present study, the individual digestion of SBB at the 18-days-HRT led to a higher HPr/HAC value than the tolerated limit by the system.

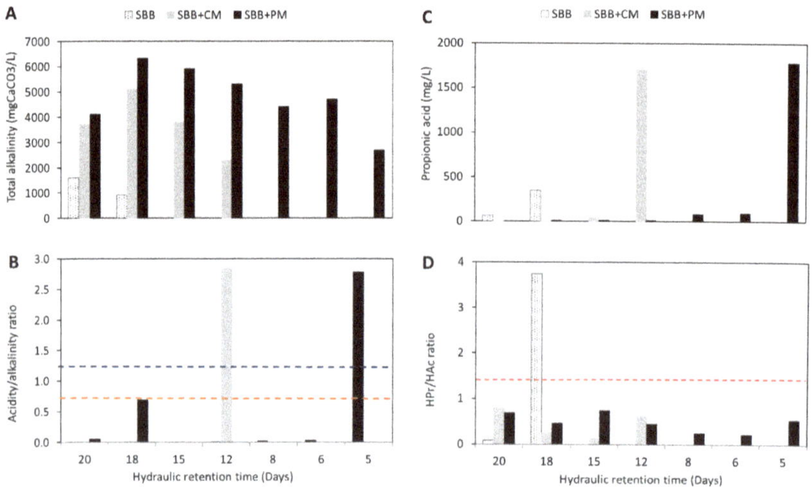

Figure 7. The classical parameters related to the process stability in the three SSTR. (**A**) The total alkalinity, (**B**) the acidity/alkalinity ratio, (**C**) the propionic acid concentrations, and (**D**) the propionic to acetic acid ratio. Values below the blue line refer to suitable values, values above the orange line indicate the critical threshold for proper operation, and values between both lines refer to alarming values that could distort the digesters. The red line refers to the threshold limit for the HPr/HAc ratio, as values above 1.4 were considered as critical for proper operation.

The characteristics of each manure in terms of alkalinity, ammonia, and the variety of microorganisms provided from the digestive tract of the animals may be the main reasons for the observed behaviors.

It has been reported that the ratio acidity to alkalinity is useful as an indicator of anaerobic reactor stability [43–45]. Thus, values above 0.8 predict the reactor instability. Values below 0.4 indicate proper operation and values with a range of 0.4–0.8 predict that the operating conditions are at the tolerance limit for microorganisms in which the destabilization of the system started. In this sense, it has been corroborated that at HRTs of 18 days, 12 days, and 5 days, respectively for SBB, SBB+CM, and SBB+PM reactors, the tolerance limit of microorganisms was altered and a lack of buffer capacity to neutralize the VFAs released occurred.

Previous studies have reported that SBB is rich carbonaceous material, suitable to be converted into biogas. However, limitations due to the high acidification resulted as a handicap for the process efficiency. Alkaya et al. [17] observed the increase of VFAs in the AD of sugar beet pulp at higher OLRs and lower HRTs. However, the use of an external source of alkalinity, decreased acidification of reactors, maintaining an adequate pH. Similarly, in a study of Demirel and Scherrer [49], the individual AD of sugar SBB (sugar beet silage) in SSTR at six different HRTs ranged from 95 to 15 days, showed better performance at the HRT of 25 days, with higher biogas production. The authors reported that SBB is a poor substrate in terms of nutrients such as nitrogen, phosphorus, and buffering capacity. In their study, authors regularly used an external source of nutrients to avoid system failure. In the present research, the use of manure as a co-substrate has a similar effect on the process, contributing to adjust the nutrients and stabilizing the system by the alkalinity supply.

Chen and Zhong [23] reported that despite the high VFAs concentration in co-digestion reactors of the cotton stalk with swine manure in comparison with single digestion of a cotton stalk, the anaerobic co-digestion assay showed higher stability with high biogas generation, indicating the role of swine manure on neutralizing VFAs in the medium. In the same way, Pagéz-Díaz et al. [50] studied semi-continuous anaerobic digestion of the solid cattle slaughterhouse (SB) and SB co-digestion with manure and various crops. They reported a failure of the individual digestion of SB. However, stable performance, operating with higher loading rates, was observed for co-digestion with different manure sources. The authors highlighted the established synergy effect by mixing substrates with different characteristics. In a study of Li et al. [18], it has been found that AD of PM has given the highest methane yield in comparison with dairy, chicken, and rabbit manure at the same operating conditions. These findings are in agreement with the results of the present study.

In addition to macronutrients such as carbon, nitrogen, phosphorus, and sulfur, trace elements are crucial for effective biogas production due to the microbial demand for these elements [51]. In a study of Schmidt et al. [52] using continuous mesophilic digesters of slaughterhouse wastewater, the authors reported that the addition of trace elements to the digester resulted in higher biogas production and higher degradation efficiency, by improving the process stability. Moreover, higher OLR and lower HRT could be used in comparison to digesters with deficiencies in trace elements. Similarly, Demirer and Scherer [49] highlighted that micronutrients availability plays a crucial role in the process performance and stability of the AD of agricultural substrates.

Moreover, authors have pointed out that the lack of trace elements is likely the main reason for poor process efficiency. In this sense, the supply of trace elements from the animal manure studied in this research could justify the high performance of reactors in comparison with the reactor of SBB digested individually.

4. Conclusions

- ➢ Non-classical parameters could give more insight into the coupling/uncoupling of the AD phases and the involved microorganisms, revealing that process failure was mainly due to methanogenesis inhibition in co-digestion with PM, while for co-digestion with CM or individual digestion of SBB, both hydrolysis–acidogenesis and methanogenesis phases were affected.
- ➢ Co-digestion with manure contributed to reducing the inhibitory effect of volatile fatty acids at high organic loading rates (OLRs), leading to increases in methane production by 70% and 31% in comparison with AD of SBB, for co-digestion with pig and cow manure, respectively.
- ➢ Biomethanation degree (BD) refers to the maximum methane potential that can be obtained from organic wastes under specific operating conditions. SBB required a long digestion-time to achieve high biodegradability. However, short digestion-times for co-digestion assays led to high BD.

Supplementary Materials: The following is available online at http://www.mdpi.com/2076-3417/10/15/5126/s1, Figure S1: Evolution of the DOC, DAC, ASC, and methane productions in the three SSTR. A: SBB reactor; B: SBB+CM reactor; C: SBB+PM reactor.

Author Contributions: Conceptualization, K.A., C.J.Á.-G., and L.I.R.-G.; Methodology, K.A., C.J.Á.-G., and L.I.R.-G.; Validation, K.A., C.J.Á.-G., and L.I.R.-G.; Investigation, K.A. and X.G.-Q.; Resources, C.J.Á.-G., and L.I.R.-G.; Data Curation, C.J.Á.-G., and L.I.R.-G.; Writing—Original Draft Preparation, K.A. and X.G.-Q.; Writing—Review & Editing, K.A. and X.G.-Q.; Visualization, K.A., C.J.Á.-G., and L.I.R.-G.; Supervision, C.J.Á.-G. and L.I.R.-G.; Project Administration, C.J.Á.-G., and L.I.R.-G.; Funding Acquisition, L.I.R.-G. All authors have read and agreed to the published version of the manuscript.

Funding: This research was supported by the project CTM2016-79071-R funded by the Spanish "Agencia Estatal de Investigación-AEI" and by the European Regional Development Fund (ERDF).

Acknowledgments: Authors acknowledge the sugar beet company (Azucarera) for the provision of the sugar beet by-products, and the livestock facilities for the supply of cow and pig manure.

Conflicts of Interest: The authors declare no conflict of interest.

References

1. Stanchev, P.; Vasilaki, V.; Egas, D.; Colon, J.; Ponsá, S.; Katsou, E. Multilevel environmental assessment of the anaerobic treatment of dairy processing effluents in the context of circular economy. *J. Clean. Prod.* **2020**, *261*, 121139. [CrossRef]
2. FAO. 2019. Available online: http://www.fao.org/news/story/en/item/196402/icode/ (accessed on 15 June 2020).

3. Vasco-Correa, J.; Khanal, S.; Manandhar, A.; Shah, A. Anaerobic digestion for bioenergy production: Global status, environmental and techno-economic implications, and government policies. *Bioresour. Technol.* **2018**, *247*, 1015–1026. [CrossRef]
4. De Corato, U.; De Bari, I.; Viola, E.; Pugliese, M. Assessing the main opportunities of integrated biorefining from agro-bioenergy co/by-products and agroindustrial residues into high-value added products associated to some emerging markets: A review. *Renew. Sustain. Energy Rev.* **2018**, *88*, 326–346. [CrossRef]
5. Joanna, B.; Michal, B.; Piotr, D.; Agnieszka, W.; Dorota, K.; Izabela, W. Sugar beet pulp as a source of valuable biotechnological products. In *Advances in Biotechnology for Food Industry*; Holban, A.M., Grumezescu, A.M., Eds.; Elsevier Science Publishers: Amsterdam, The Netherlands, 2018; pp. 359–392. [CrossRef]
6. Ileleji, K.E.; Martin, C.; Jones, D. Basics of energy production through anaerobic digestion of livestock manure. In *Bioenergy*; Dahiya, A., Ed.; Elsevier Science Publishers: Amsterdam, The Netherlands, 2015; pp. 287–295. [CrossRef]
7. Eurostat. 2017. Available online: http://ec.europa.eu/eurostat/statistics-explained/index.php/Agricultural_production_-_animals#Livestock_population (accessed on 15 June 2020).
8. Wainaina, S.; Awasthi, K.M.; Sarsaiya, S.; Chen, H.; Singh, E.; Kumar, A.; Ravindran, B.; Kumar-Awasthi, S.; Liu, T.; Duan, Y.; et al. Resource recovery and circular economy from organic solid waste using aerobic and anaerobic digestion technologies. *Bioresour. Technol.* **2020**, *301*, 122778. [CrossRef]
9. Zhang, Y.; Angelidaki, I. Counteracting ammonia inhibition during anaerobic digestion by recovery using submersible microbial desalination cell. *Biotechnol. Bioeng.* **2015**, *112*, 1478–1482. [CrossRef] [PubMed]
10. Moestedt, J.; Müller, B.; Westerholm, M.; Schnürer, A. Ammonia threshold for inhibition of anaerobic digestion of thin stillage and the importance of organic loading rate. *Microb. Biotechnol.* **2015**, *9*, 180–194. [CrossRef] [PubMed]
11. Aguirre-Villegas, H.A.; Larson, R.A. Evaluating greenhouse gas emissions from dairy manure management practices using survey data and lifecycle tools. *J. Clean. Prod.* **2017**, *143*, 169–179. [CrossRef]
12. Shou, Z.; Yuan, H.; Shen, Y.; Liang, J.; Zhua, N.; Gu, L. Mitigating inhibition of undissociated volatile fatty acids (VFAs) for enhanced sludge-rice bran composting with ferric nitrate amendment. *Bioresour. Technol.* **2017**, *244*, 672–678. [CrossRef] [PubMed]
13. Moreno, R.; Martínez, J.E.; Escapa, A.; Martínez, O.; Díez-Antolínez, R.; Gómez, X. Mitigation of volatile fatty acid build-up by the use of soft carbon felt electrodes: Evaluation of anaerobic digestion in acidic conditions. *Fermentation* **2018**, *4*, 2. [CrossRef]
14. Boldrin, A.; Baral, K.R.; Fitamo, T.; Vazifehkhoran, A.H.; Jensen, I.G.; Kjærgaard, I.; Lyng, K.-A.; van Nguyen, Q.; Nielsen, L.S.; Triolo, J.M. Optimised biogas production from the co-digestion of sugar beet with pig slurry: Integrating energy, GHG and economic accounting. *Energy* **2016**, *112*, 606–617. [CrossRef]
15. Aboudi, K.; Álvarez-Gallego, C.J.; Romero-García, L.I. Evaluation of methane generation and process stability from anaerobic co-digestion of sugar beet by-product and cow manure. *J. Biosci. Bioeng.* **2016**, *121*, 566–572. [CrossRef] [PubMed]
16. Neshat, S.A.; Mohammadi, M.; Najafpour, G.D.; Lahijani, P. Anaerobic co-digestion of animal manures and lignocellulosic residues as a potent approach for sustainable biogas production. *Renew. Sustain. Energy Rev.* **2017**, *79*, 308–322. [CrossRef]
17. Alkaya, E.; Demirer, G.N. Anaerobic acidification of sugar-beet processing wastes: Effect of operational parameters. *Biomass Bioenergy* **2011**, *35*, 32–39. [CrossRef]
18. Li, K.; Liu, R.; Su, C. Comparison of anaerobic digestion characteristics and kinetics of four livestock manures with different substrate concentrations. *Bioresour. Technol.* **2015**, *198*, 133–140. [CrossRef]
19. Vivekanand, V.; Mulat, D.G.; Eijsink, V.G.H.; Horn, S.J. Synergistic effects of anaerobic co-digestion of whey, manure and fish ensilage. *Bioresour. Technol.* **2018**, *249*, 35–41. [CrossRef] [PubMed]
20. Pagés-Díaz, J.; Sárvári-Horváth, I.; Pérez-Olmo, J.; Pereda-Reyes, I. Co-Digestion of bovine slaughterhouse wastes, cow manure, various crops and municipal solid waste at thermophilic conditions: A comparison with specific case running at mesophilic conditions. *Water Sci. Technol.* **2013**, *67*, 989–995. [CrossRef]
21. Lehtomäki, A.; Huttunen, S.; Rintala, J.A. Laboratory investigations on co-digestion of energy crops and crop residues with cow manure for methane production: Effect of crop to manure ratio. *Resour. Conserv. Recycl.* **2007**, *51*, 591–609. [CrossRef]
22. Xie, T.; Xie, S.; Sivakumar, M.; Nghiem, L.D. Relationship between the synergistic/antagonistic effect of anaerobic co-digestion and organic loading. *Int. Biodeterior. Biodegrad.* **2017**, *124*, 155–161. [CrossRef]

23. Cheng, X.-Y.; Zhong, C. Effects of feed to inoculum ratio, co-digestion, and pretreatment on biogas production from anaerobic digestion of cotton stalk. *Energy Fuels* **2014**, *28*, 3157–3166. [CrossRef]
24. Borowski, S.; Kucner, M.; Czyżowska, A.; Berłowska, J. Co-Digestion of poultry manure and residues from enzymatic saccharification and dewatering of sugar beet pulp. *Renew. Energy* **2016**, *99*, 492–500. [CrossRef]
25. Fernández-Rodríguez, J.; Pérez, M.; Romero, L.I. Mesophilic anaerobic digestion of the organic fraction of municipal solid waste: Optimisation of the semicontinuous process. *Chem. Eng. J.* **2012**, *193–194*, 10–15. [CrossRef]
26. Aboudi, K.; Álvarez-Gallego, C.J.; Romero-García, L.I. Improvement of exhausted sugar beet cossettes anaerobic digestion process by Co-digestion with pig manure. *Energy Fuels* **2015**, *29*, 754–762. [CrossRef]
27. Aboudi, K.; Álvarez-Gallego, C.J.; Romero-García, L.I. Influence of total solids concentration on the anaerobic co-digestion of sugar beet by-products and livestock manures. *Sci. Total Environ.* **2017**, *586*, 438–445. [CrossRef] [PubMed]
28. Wintsche, B.; Glaser, K.; Sträuber, H.; Centler, F.; Liebetrau, J.; Harms, H.; Kleinsteuber, S. Trace elements induce predominance among methanogenic activity in anaerobic digestion. *Front. Microbiol.* **2016**, *7*, 2034. [CrossRef] [PubMed]
29. Demirel, B.; Scherer, P. Production of methane from sugar beet silage without manure addition by a single-stage anaerobic digestion process. *Biomass Bioenergy* **2008**, *32*, 203–209. [CrossRef]
30. APHA-AWWA. *Standards Methods for the Examination of Water and Wastewater*, 20th ed.; American Public Health Association: Washington, DC, USA, 2005.
31. Van Soest, P.J.; Robertson, J.B.; Lewis, B.A. Methods for dietary fiber, neutral detergent fiber, and nonstarch polysaccharides in relation to animal nutrition. *J. Dairy Sci.* **1991**, *74*, 3583–3597. [CrossRef]
32. Li, K.; Liu, R.; Cui, S.; Yu, Q.; Ma, R. Anaerobic co-digestion of animal manures with corn stover or apple pulp for enhanced biogas production. *Renew. Energy* **2018**, *118*, 335–342. [CrossRef]
33. Kaparaju, P.; Rintala, J. Anaerobic co-digestion of potato tuber and its industrial by-products with pig manure. *Resour. Conserv. Recycl.* **2005**, *43*, 175–188. [CrossRef]
34. Fdez-Güelfo, L.A.; Álvarez-Gallego, C.; Sales, D.; Romero, L.I. New indirect parameters for interpreting a destabilization episode in an anaerobic reactor. *Chem. Eng. J.* **2012**, *180*, 32–38. [CrossRef]
35. Gómez-Quiroga, X.; Aboudi, K.; Álvarez-Gallego, C.J.; Romero-García, L.I. Enhancement of methane production in thermophilic anaerobic co-digestion of exhausted sugar beet pulp and pig manure. *Appl. Sci.* **2019**, *9*, 1791. [CrossRef]
36. Damaceno, F.M.; Chiarelotto, M.; Pires-Salcedo-Restrepo, J.C.; Buligon, E.L.; Costa, L.A.d.M.; de Lucas-Junior, J.; Costa-SilvadeMendonça, M.S. Anaerobic co-digestion of sludge cake from poultry slaughtering wastewater treatment and sweet potato: Energy and nutrient recovery. *Renew. Energy* **2019**, *133*, 489–499. [CrossRef]
37. Panichnumsin, P.; Nopharatana, A.; Ahring, B.; Chaiprasert, P. Production of methane by co-digestion of cassava pulp with various concentrations of pig manure. *Biomass Bioenergy* **2010**, *34*, 1117–1124. [CrossRef]
38. Smith, P.H.; Mah, R.A. Kinetics of acetate metabolism during sludge digestion. *Appl. Microbiol.* **1966**, *14*, 368–371. [CrossRef] [PubMed]
39. Gehring, T.; Klang, J.; Niedermayr, A.; Berzio, S.; Immenhauser, A.; Klocke, M.; Wichern, M.; Lübken, M. Determination of methanogenic pathways through carbon isotope (δ13C) analysis for the two-stage anaerobic digestion of high-solids substrates. *Environ. Sci. Technol.* **2015**, *49*, 4705–4714. [CrossRef]
40. Siegrist, H.; Vogt, D.; García-Heras, J.L.; Gujer, W. Mathematical model for meso- and thermophilic anaerobic sewage sludge digestion. *Environ. Sci. Technol.* **2002**, *36*, 1113–1123. [CrossRef] [PubMed]
41. Yin, D.M.; Westerholmc, M.; Qiaoa, W.; Bia, S.J.; Wanderaa, S.M.; Fana, R.; Jianga, M.M.; Dong, R.J. An explanation of the methanogenic pathway for methane production in anaerobic digestion of nitrogen-rich materials under mesophilic and thermophilic conditions. *Bioresour. Technol.* **2018**, *264*, 42–50. [CrossRef]
42. Jin, W.; Xu, X.; Yang, F.; Li, C.; Zhou, M. Performance enhancement by rumen cultures in anaerobic co-digestion of corn straw with pig manure. *Biomass Bioenergy* **2018**, *115*, 120–129. [CrossRef]
43. Callaghan, F.J.; Wase, D.A.J.; Thayanithy, K.; Forster, C.F. Continuous co-digestion of cattle slurry with fruit and vegetable wastes and chicken manure. *Biomass Bioenergy* **2002**, *22*, 71–77. [CrossRef]
44. Switzenbaum, M.S.; Giraldo-Gomez, E.; Hickeyt, R.F. Monitoring of the anaerobic methane fermentation process. *Enzym. Microb. Technol.* **1990**, *12*, 722–730. [CrossRef]

45. Zickefoose, C.; Hayes, R.B.J.; Bryant, J.O. *Anaerobic Sludge Digestion: Operations Manual*; National Technical Information Service, Ed.; EPA 430=9-76-001; United States Environmental Protection Agency: Washington, DC, USA, 1976.
46. Marchaim, U.; Krause, C. Propionic to acetic acid ratios in overloaded anaerobic digestion. *Bioresour. Technol.* **1993**, *43*, 195–203. [CrossRef]
47. Pullammanappallil, P.C.; Chynoweth, D.P.; Lyberatos, G.; Svoronos, S.A. Stable performance of anaerobic digestion in the presence of a high concentration of propionic acid. *Bioresour. Technol.* **2001**, *78*, 165–169. [CrossRef]
48. Hill, D.T.; Cobb, S.A.; Bolte, J.P. Using Volatile Fatty Acid Relationships to Predict Anaerobic Digester Failure. *Trans. ASAE* **1987**, *30*, 496. [CrossRef]
49. Demirel, B.; Scherer, P. Trace element requirements of agricultural biogas digesters during biological conversion of renewable biomass to methane. *Biomass Bioenergy* **2011**, *35*, 992–998. [CrossRef]
50. Pagés-Díaz, J.; Westman, J.; Taherzadeh, M.J.; Pereda-Reyes, I.; Sárvári-Horváth, I. Semi-Continuous co-digestion of solid cattle slaughterhouse wastes with other waste streams: Interactions within the mixtures and methanogenic community structure. *Chem. Eng. J.* **2015**, *273*, 28–36. [CrossRef]
51. González-Suárez, A.; Pereda-Reyes, I.; Oliva-Merencio, D.; Suárez-Quiñones, T.; José da Silva, A.; Zaiat, M. Bioavailability and dosing strategies of mineral in anaerobic mono-digestion of maize straw. *Eng. Life Sci.* **2018**, *18*, 562–569. [CrossRef]
52. Schmidt, T.; McCabe, B.K.; Harris, P.W.; Lee, S. Effect of trace element addition and increasing organic loading rates on the anaerobic digestion of cattle slaughterhouse wastewater. *Bioresour. Technol.* **2018**, *264*, 51–57. [CrossRef]

© 2020 by the authors. Licensee MDPI, Basel, Switzerland. This article is an open access article distributed under the terms and conditions of the Creative Commons Attribution (CC BY) license (http://creativecommons.org/licenses/by/4.0/).

applied sciences

MDPI

Article

Enhancement of Methane Production in Thermophilic Anaerobic Co-Digestion of Exhausted Sugar Beet Pulp and Pig Manure

Xiomara Gómez-Quiroga, Kaoutar Aboudi, Carlos José Álvarez-Gallego and Luis Isidoro Romero-García *

Department of Chemical Engineering and Food Technology, Faculty of Sciences (Wine and Agri-Food Research Institute-IVAGRO and International Campus of Excellence-ceiA3), University of Cádiz, Republic Saharawi Avenue, P.O. Box No. 40, 11510 Puerto Real (Cádiz), Spain; xiomara.gomez@uca.es (X.G.-Q.); kaoutar.aboudi@uca.es (K.A.); carlosjose.alvarez@uca.es (C.J.Á.-G.)

* Correspondence: luisisidoro.romero@uca.es; Tel.: +34-956-016-379

Received: 29 March 2019; Accepted: 26 April 2019; Published: 29 April 2019

Abstract: In this paper, the viability of thermophilic anaerobic co-digestion of exhausted sugar beet pulp (ESBP) and pig manure (PM) was evaluated. The effect of the proportion of ESBP on biogas production was investigated by using a series of lab-scale batch assays, in duplicates. The following five ESBP:PM mixture ratios were studied: 0:100, 10:90, 25:75, 50:50, and 100:0. The highest cumulative methane production (212.4 mL CH_4/g VS_{added}) was reached for the mixture 25:75. The experimental results showed that the increase in the proportion of ESBP in the mixture led to the distortion of the process, due to acidification by the volatile fatty acids generated. Acetic acid was the predominant acid in all the cases, representing more than 78% of the total acidity. Moreover, the results obtained by operating at thermophilic temperatures have been compared with those obtained in a previous study conducted at mesophilic temperatures. The results have shown that in the individual digestion of ESBP, the activity of acetoclastic methanogens was affected in both temperatures, but especially in thermophilic conditions. Thus, the methane produced in the individual thermophilic digestion of ESBP came almost entirely from the activity of hydrogen-utilizing methanogenic archaea.

Keywords: exhausted sugar beet pulp; pig manure; anaerobic co-digestion; thermophilic; lignocellulosic waste

1. Introduction

The main component of agro-industrial waste is lignocellulose, which is constituted by three fractions—lignin, hemicellulose and cellulose [1]. Lignocellulosic biomass can be converted into different products, including biofuels and fine chemicals with a high added-value, and can also be used as a cheap substrate source in microbial fermentation and enzyme production [2].

An interesting by-product generated in the process of sugar extraction from sugar beet (*Beta vulgaris*), is the exhausted sugar beet pulp (ESBP), which has the following composition—24%–32% of hemicellulose, 22%–30% of cellulose, 3%–4% of lignin, and 24%–32% of pectin [3]. Therefore, ESBP is a suitable substrate for biogas production, through the anaerobic digestion process (AD) [4–6]. The use of biogas produced from ESBP could generate economic and environmental benefits for energy-intensive consumers, such as sugar beet industries and, hence, contribute to the circular economy.

AD is a mature, highly efficient, and economically viable technology for transforming organic wastes into methane [7–9]. The process occurs under anaerobic conditions, requiring considerably less energy with respect to the aerobic treatment processes [10].

Agro-industrial by-products could release, depending on their nature and origin, toxic or inhibitory compounds for microorganisms, such as ammonia, sulphides, heavy metals, long chain fatty acids (LCFAs), and other organic compounds. The intensification of the microorganisms activity in the system, especially the increase of their hydrolytic ability, could be counterproductive if the breakdown of the lignocellulosic material resulted in the release of intermediate inhibitory compounds. The accumulation of these compounds would lead to the destabilization of the AD process [11]. Therefore, in order to improve the efficiency and stability of the AD process, a deep knowledge of the critical process parameters, affecting the different microbial groups is required [12,13].

According to the literature, operations in thermophilic temperature ranges (50–60 °C) lead to a higher biogas production than in mesophilic range (30–40 °C), but compromises the stability of the process [14,15].

However, the decomposition of agro-industrial by-products could present limitations due to the presence of non-biodegradable lignocellulose and the deficit in the nitrogen source, with respect to the high content in organic matter [1,16].

Aboudi et al. [11] have reported that thermophilic or hyperthermophilic acidogenic anaerobic digestion of ESBP led to the accumulation of LCFAs, such as palmitate and lignocerate. The use of the effluents from these reactors for the feeding of two subsequent mesophilic reactors led to the inhibition of methanogenic archaea population. LCFAs are considered inhibitory compounds for the AD process, especially for methanogenic microorganisms [17].

With respect to the deficit of nitrogen in ESBP, several authors [18,19] have shown that the anaerobic co-digestion (AcoD) of ESBP with animal manure, solves this problem. Thus, animal manure supplies the necessary nutrients, but it also provides alkalinity and specific microorganisms, originating from the digestive tract of animals, which are capable of degrading vegetal fibers [20,21]. Therefore, animal manure improves, both, the biodegradability of the lignocellulosic material and the stability of the AD process. Aboudi et al. [22], have studied the AcoD of ESBP with pig manure (PM) in continuous stirred tank reactor (CSTR) reactors operating in the mesophilic range of temperature. Other authors [23–25] have studied the AcoD of different agricultural wastes with animal manure and have shown that co-digestion improves the anaerobic process and waste digestibility.

The main objective of this study was to determine the optimal mixture ratio of ESBP and PM for the thermophilic anaerobic co-digestion of both wastes. In addition, the effect of temperature on the process performance was studied by performing a comparison with the previous results reported in the literature, for mesophilic AcoD, using similar wastes.

2. Materials and Methods

2.1. Substrate and Inoculum

The ESBP samples came from a high-size sugar beet factory (AB Sugar^{TM}) located in Jerez de la Frontera (Cádiz, Spain). The ESBP samples have a very high solid content (TS range 80%–90%) that facilitated their conservation. The samples of PM were collected in a local farm placed in San José del Valle (Cádiz, Spain). The farm did not have any device for the separation of urine and feces. The samples were frozen at −20 °C, unfrozen, and stored at 4 °C, when they were close to being used.

The mixtures of ESBP and PM were prepared, considering the proportions of each waste, on a wet weight basis, and a total solid content of 8% was maintained. ESBP samples were previously rehydrated for 24 h, in deionized water [11]. ESBP is highly hygroscopic and, hence, the total solids percentage must be limited to 8%, to avoid rheological problems into the reactors. Indeed, previous studies determined that the increase in solids content impeded an adequate mixing in the reactor, due to the high water holding capacity of ESBP [26–30]. The following mixtures of ESBP:PM (weight basis) were studied: 0:100, 10:90, 25:75, 50:50, and 100:0.

The inoculum used in the experiments came from a lab-scale semi-continuous reactor for the AcoD of ESBP and PM. The reactor has been working in the thermophilic temperature range (55 °C) and was fed with an ESBP:PM mixture of (10:90), also with a total solids content of 8%. When the inoculum samples were taken, the reactor was operating in stable conditions with a hydraulic retention time of 30 days, a methane yield around 105–110 mL CH_4/gVS_{added}, and a productivity of 0.22–0.24 $LCH_4/L_{reactor}$·day. The use of an adapted inoculum to carry out the AD assays guaranteed an efficient start-up of the process [30]. The inoculum to substrate ratio was at 1:1 (volume basis).

The physicochemical characteristics of the substrates and the inoculum are shown in Table 1.

Table 1. Physicochemical characteristics of the substrates and inoculum.

Component	Units	ESBP	PM	Inoculum
pH	-	6.22 ± 0.03	8.26 ± 0.10	7.96 ± 0.08
TS	g/kg	857.7 ± 0.04	276.3 ± 0.27	55.67 ± 0.15
VS	g/kg	758.6 ± 0.06	156.8 ± 0.30	28.43 ± 0.11
sCOD	g/kg	21.0 ± 0.42	7.0 ± 0.28	3.4 ± 0.39
TVFA	g/kg	1.99 ± 0.31	1.2 ± 0.02	0.72 ± 0.10
Alkalinity	g/kg	2.19 ± 0.10	33.1 ± 0.50	11.8 ± 0.77
C/N ratio	-	37.4 ± 0.22	3.03 ± 0.27	-
Pectins	%	55.54	-	69.31
Hemicellulose	%	22.52	-	11.29
Cellulose	%	21.14	-	11.90
Lignin	%	3.50	-	5.61

All results were calculated in a wet basis. ESBP: Exhausted Sugar Beet Pulp, PM: Pig Manure, TS: Total Solids, VS: Volatile Solids, sCOD: Soluble Chemical Oxygen Demand, TVFA: Total Volatile Fatty Acid, -: No unit/Not analyzed.

2.2. Experimental Design

A series of 10 stainless steel reactors with a working volume of 1.7 L were operated in batch mode. Each reactor had independent devices for heating and stirring. The biogas production was continuously measured using a MilliGas counter (Ritter®) and a series of specific infrared on-line sensors for CH_4 and H_2 determinations (Calibrated and certified by Bluesens®). Figure 1 shows an illustration of the equipment used for the biogas measurement and the characterization.

Figure 1. Equipment used for the biogas measurement and characterization. From left to right—infrared on-line sensors, the MilliGas counter, and the data interface module.

Subsequently, the reactors were hermetically sealed and purged with N_2 to remove the residual O_2. The temperature was continuously measured by an inner sensor and was maintained at 55 °C. The mixing was performed by paddles (20 rpm).

2.3. Analytical Methods

All analytical determinations were performed according to the Standard Methods [31]. The following reference methods were used—Total Solids (TS) by 2540B, Volatile Solids (VS) by 2540E, total Chemical Oxygen Demand (tCOD), and soluble Chemical Oxygen Demand (sCOD) by 5220C,

Dissolved Organic Carbon (DOC) by 5310B, using a carbon/nitrogen analyzer (Analytic-Jena multi N/C 3100 with chemiluminescence detector, Endress+Hauser Company®, Waldheim, Germany), alkalinity by 2320B and pH by 4500H$^+$. In addition, the concentration of the main individual volatile fatty acids (VFAs), from C2 to C7, were measured by using a gas chromatograph GC-2010 (Shimadzu Corporation®, Kyoto, Japan) with flame ionization detector and a Nukol® capillary with a diameter of 0.25 mm and 30 m of length column (Merck KGaA®, Darmstadt, Germany). The hydrogen was used as carrier gas with a flow of 50 mL/min and 75.5 kPa; synthetic air (400 mL/min and 50 kPa) and hydrogen (40 mL/min and 60 kPa) were used for the flame ignition, and nitrogen (30 mL/min and 75 kPa) was used as the make-up gas. All the gases were provided by Abello-Linde® (Barcelona, Spain).

The pH was measured daily and directly into the reactor. The remaining parameters were analyzed three times per week. Prior to the analysis of sCOD, DOC, and VFA, the samples were centrifuged in a Consul-21 Ortoalresa® equipment (Madrid, Spain) at 3220× g during 15 min and filtered with a 0.47 µm glass microfiber filter (ref. GF52047, Hahnemühle®, Dassel, Germany). For the analysis of VFAs, an additional filtration was carried out through a 0.22 µm PTFE filter manufactured by Rephile Bioscience® (Shanghai, China) [32].

To determine the lignocellulosic fractions of the samples (cellulose, hemicellulose, and lignin), the FIBERTEC™ 8000 equipment (FOSS®, Hilleroed, Denmark) was used by applying the Van Soest method [33].

2.4. Indirect Parameters

Fdez-Güelfo et al. [8] have proposed a series of indirect parameters that can be used to analyze the performance of the different stages of the anaerobic digestion process and to determine the rate-limiting stage for specific operating conditions.

These parameters include the Dissolved Acid Carbon (DAC) and the Acidogenic Substrate as Carbon (ASC). The DAC is the carbon contained in the volatile fatty acids (C2 to C7) present in the medium. An increase in DAC is related to a failure in the methanogenic stage. The ASC corresponds to the difference (expressed in carbon units) between the Dissolved Organic Carbon (analytical value of DOC) and the DAC (calculated value). The ASC represents the solubilized organic carbon that has not been transformed into acids, and an increase in this parameter could be interpreted as a distortion of the acidogenic stage.

Angeriz-Campoy et al. [34] have used the indirect parameter ASC and have corroborated that the failure in the AcoD of organic fraction of municipal solid waste and sewage sludge, was due to a distortion in the hydrolytic phase. Aboudi et al. [35] have used the indirect parameters to analyze the AcoD of exhausted sugar beet pulp and cow manure, in the mesophilic range of temperature as well.

3. Results and Discussion

3.1. Waste Characteristics

As can be seen in the Table 1, the ESBP is a lignocellulosic material with a very high organic matter content (high values of VS y COD). However, the alkalinity of ESBP is relatively low with respect to its organic content. Therefore, a mixture with PM could be favorable, since PM has a 12 times higher alkalinity. In addition, the C/N ratio indicates that ESBP is deficient in nitrogen and, hence, co-digestion with PM might adjust the nutrient content.

3.2. Process Stability

The pH is the result of the acid–base equilibrium in the process and, therefore, is related to the production of VFAs and alkalinity. The pH is considered a key parameter in the stability of the anaerobic digestion process, since a low pH value implies that the organic acids are mainly in their undissociated form, which is the most inhibitory for the anaerobic microbiota [19]. In fact, if the pKa of the acetic acid (pKa = 4.76 at 25 °C) is considered, approximately 15% of the acid would be present in its undissociated form at pH = 5.5, while at pH = 4.76 the undissociated form could reach a 50%. The low alkalinity of the lignocellulosic waste leads to an eventual acidification of the system, because VFAs accumulation significantly diminishes the buffer capacity, affecting the growth rate of the microorganisms capable of transforming the organic matter into biogas [36].

The buffer capacity of the system could be increased by the addition of strong bases or carbonate salts, even though the CO_2 percentage in biogas could be affected [37]. Figure 2 shows the pH and the acidity/alkalinity ratio evolution in the experiments.

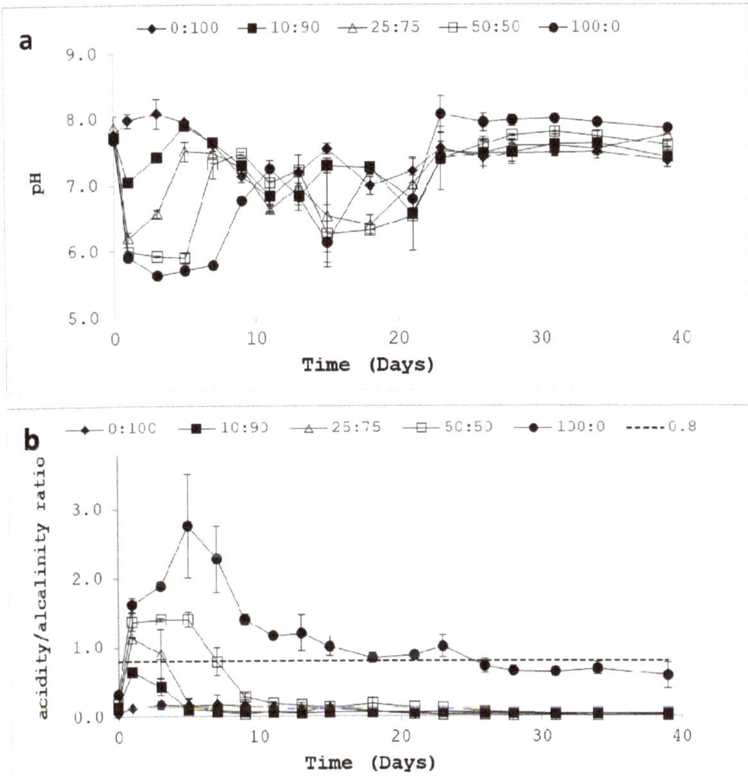

Figure 2. pH evolution (**a**) and acidity/alkalinity ratio evolution (**b**) in all ESBP:PM ratios.

In this work, during the first days of operation, the pH diminished below the required range, especially in the reactors with high content in ESBP, as can be seen in Figure 2a. The daily correction of pH with potassium carbonate, permitted all reactors to operate in the appropriate pH range, within a few days [38]. However, for reactors with the mixture ratios 0:100, 10:90, and 25:75, the pH was stabilized rapidly and the required additions of the alkaline reagent were very low. Specifically, the additions of potassium carbonate were 3, 5, and 8 g for the tests 0:100, 10:90, and 25:75, respectively, whereas for reactors 50:50 and 100:0 were necessary 12 and 46 g, respectively. To the end of the process, the pH in all reactors were in the required range.

The acidity/alkalinity ratio has been used as an indicator of acidification problems in anaerobic reactors. In the literature, three different ranges have been identified for this parameter values below 0.4 indicate an optimal process performance; values between 0.4 and 0.8 are evidence of the beginning of process distortion and, finally, values higher than 0.8 indicate that a total or a partial inhibition of the methanogenic activity occurred [39–41].

Figure 2b shows that for the mixture 0:100, the ratio was in the optimum range, throughout the whole assay, while for the mixture 10:90, this ratio was over 0.4 for the first days, and then rapidly reached the optimum range. However, for the mixtures 25:75 and 50:50, the acidity/alkalinity ratio initially exceeded 0.8 and a longer period was required (which increased with an in increase in the ESBP content), to reach the optimum level. For the reactor containing only ESBP (100:0), the ratio remained above 0.8 for most of the test, and only reached values between 0.4 and 0.8 at the end of the assay.

Therefore, according to the acidity/alkalinity ratio, the increase in the percentage of animal manure (PM) in the mixture had a beneficial effect on the stability of the AcoD of ESBP and PM.

Figure 3a shows the evolution of the total volatile fatty acidity (TVFA) in all tests. The initial acidification was due to the rapid release of VFAs and it seemed to be a characteristic of the ESBP, as a substrate in the AD process [4]. Indeed, as the content of ESBP in the mixtures increased, the release of VFAs was more abrupt, reaching higher maximum concentrations of TVFA. Moreover, the time needed to reduce the TVFA concentration from the maximum value was longer for reactors with a high fraction of ESBP in the mixture, showing a decoupling between the acidogenic and acetoclastic-methanogenic activities. In the case of the mixture 100:0, the higher values of VFAs seemed to have inhibited the activity of the acetoclastic-methanogenic archaea population, since the VFAs had not degraded over the 40 days of operation.

Figure 3b shows the main individual VFAs observed in the different tests. Acetic (HAc), propionic (HPr), and butyric (HBu) acids together represented more than 80%–90% of the TVFA (when the maximum of TVFA was reached). Acetic acid was the predominant in all the cases, representing more than 78% of TVFA. It could be observed that an increase in the concentrations of the three main organic acids occurred when the content in ESBP was increased in the mixtures. In the test 100:0, the maximum concentration of acetic acid was 19.5 g/L, whereas the concentrations of propionic and butyric acids were 1.4 and 2.5 g/L, respectively. Probably, these high concentrations of VFAs had inhibited the acetoclastic-methanogenic stage [11]. Thus, despite the slight decrease in the acetic acid concentration at the end of the test, the methanogenic activity was clearly affected and, hence, the observed methane production must have come mainly from the activity of the hydrogen-utilizing methanogenic archaea. As can be seen later, in Section 3.5, there was a high fraction of organic matter as VFAs, which was not transformed into methane at the end of the tests.

The relationship between the concentrations of propionic and acetic acids (HPr/HAc ratio) was another very useful parameter to study the stability of the AD process. According to the literature, when the HPr/HAc ratio exceeds 1.4, the destabilization of the process occurs [42]. However, other studies have determined that the threshold value of this relationship depends on the characteristics of the substrates used in co-digestion [35,43]. Figure 3c shows the evolution of the HPr/HAc ratio in all the mixtures. In this study, despite certain initial high values, the HPr/HAc ratio was usually less than 2. For the mixture 50:50, values above 1.4 were recorded for 7 days, but a decrease was observed from day 20 of the test. However, for the mixture 100:0, the HPr/HAc ratio remained much lower with respect to the mixtures 25:75 and 50:50 throughout the assay. This behavior was due to the fact that the initial production of acetic acid in the mixture 100:0 was very high. Thus, although the highest concentration of propionic acid was obtained in this test, the HPr/HAc ratio remained at low level.

It can be pointed out that the microbiota coming from the inoculum was very effective to remove propionic acid. Therefore, the transitory accumulation of propionic acid could be interpreted as a decoupling between the acidogenic and acetogenic stages for a specific period.

In the literature, different threshold values of HPr have been reported as causing inhibition. Barredo and Evison [44] have indicated that concentrations of propionic acid higher than 1500–2220 mg/L could lead to inhibition. However, in the mesophilic AcoD of ESBP and PM, concentrations of HPr in the range 1099–5500 mg/L were observed and the system was not inhibited [35].

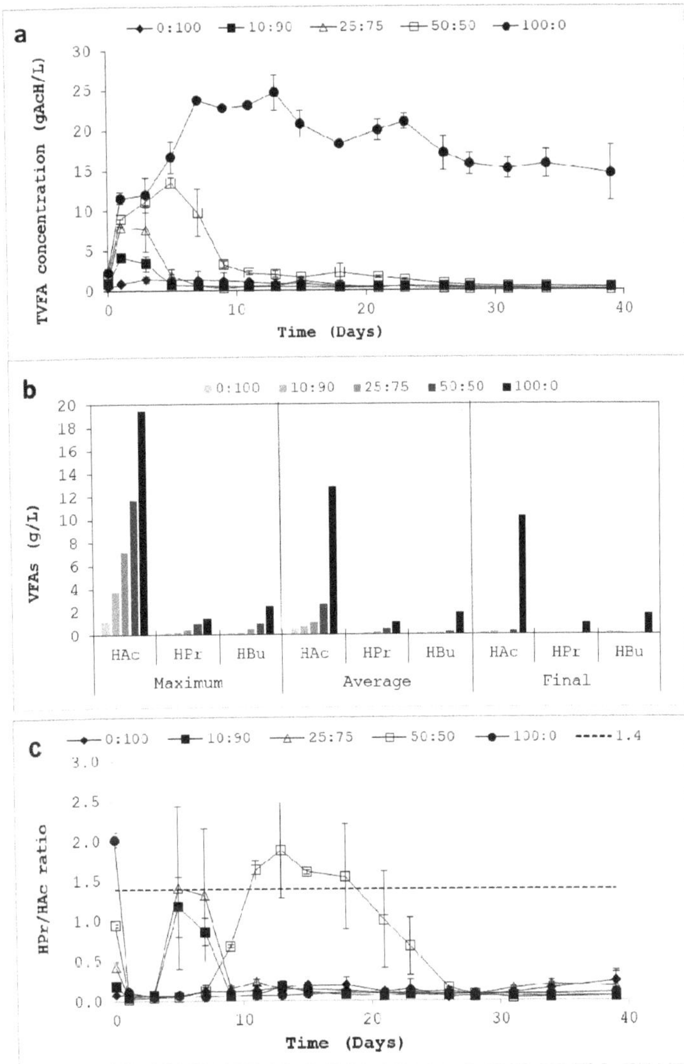

Figure 3. Evolution of total volatile fatty acidity (TVFA) concentration (**a**), the individual volatile fatty acids (VFAs) (**b**), and the ratio propionic (HPr)/ Acetic (HAc) acid (**c**) in all the ESBP:PM ratios.

In this work, the maximum concentrations of propionic acid detected were: 125.42 mg/L, 277.02 mg/L, 447.75 mg/L, 972.88 mg/L, and 1401.26 mg/L for the mixtures 0:100, 10:90, 25:75, 50:50, and 100:0, respectively. Therefore, the concentrations of propionic acid for the mixture 100:0 could not be considered as the only responsible factor for the destabilization of the process.

3.3. Methane Production

The reactors were maintained in operation until no significant methane production was detected. Figure 4 shows the specific methane production (SMP) obtained in all the digesters. It can be seen that the SMP was improved in all co-digestion mixtures, with respect to the individual digestion of ESBP. The following data were obtained for the different tests: 19.2; 108.1; 212.4; 141.4, and 47.9 mL CH_4/gVS_{added} for the mixtures 0:100; 10:90; 25:75; 50:50, and 100:0, respectively.

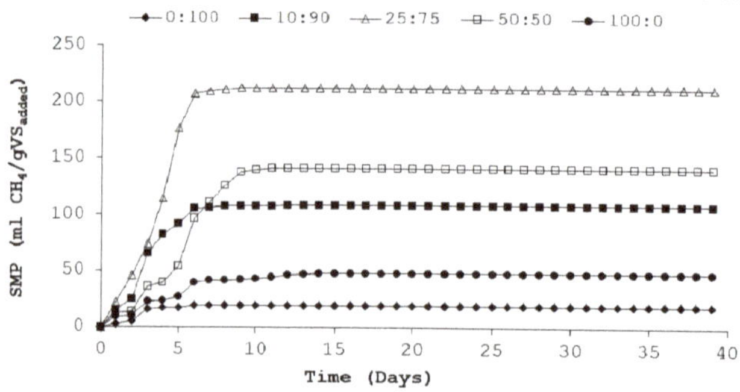

Figure 4. Specific accumulated methane production for the different mixtures tested.

In terms of methane production, the best mixture was 25:75. In fact, the reactor with the mixture 25:75 had rapidly reached the maximum SMP of 212.4 mL CH_4/gVS_{added} with a very short lag phase (2 days). Aboudi et al. [40] found an optimal ratio of 32:68 with an SMP of 494 mL CH_4/gVS_{added} at the mesophilic range. Ohuchi et al. [6] obtained an SMP of 377–422 mL CH_4/gVS_{added} for anaerobic co-digestion of cow manure with ensiled sugar beet tops at the proportion of 40%. The maximum SMP found for the mixture 25:75 in this work was approximately 11 times higher than for the mixture 0:100 (PM alone) and 5 times higher than 100:0 (ESBP alone), highlighting the synergy of the co-digestion of both substrates.

Therefore, it could be admitted that the co-digestion of ESBP and PM in the thermophilic range showed a synergistic effect in the methane production and the rate of the organic matter degradation, with respect to the individual anaerobic digestion of PM and ESBP. The synergy observed was due to the fact that the mixture of both wastes generated more appropriate environmental conditions for the development of the process; that is, both the nutrient balance (C/N ratio) and the available alkalinity were improved. In addition, pig manure added microorganisms from its digestive tract which could enhance the hydrolysis of the lignocellulosic material present in the ESBP. These results were in line with what has been reported by several authors studying AD of agro-industrial wastes [45,46].

Nevertheless, it must be noted that the SMP obtained for the PM used in this study was clearly lower than that obtained in a previous study, conducted under mesophilic conditions [46], where a value of 468 mL/VS_{added} was reached. Moreover, this effect seemed to extend to the assay with the ratio 10:90, due to its high PM proportion. This aspect could be due to the different characteristics of the manures used in both tests. Another possible explanation for this difference could be related to the presence of typical animal manure toxics (ammonia nitrogen, drugs, and antibiotics). The inhibition by ammonia nitrogen was especially relevant in the thermophilic anaerobic digestion [47] and several authors reported that animal manures were a reservoir of antibiotic-resistant genes, outlining that for PM, around 149 antibiotic-resistant genes were found [48–50]. However, to support this hypothesis, additional specific analytical determinations should be performed.

3.4. Acidogenic Substrate as Carbon (ASC)

In order to obtain a more detailed interpretation of when the process failed (see Section 2.4), the indirect parameters ASC and DAC were calculated since they allowed a better understanding of the origin of the distortions [8]. The study of the ASC parameter, which represented the hydrolyzed and solubilized organic matter that had not been degraded to VFAs, permitted to interpret the appearance of the inhibitory effects in acidogenesis stage [34]. A transitory accumulation of ASC denoted the existence of some type of limitations in acidogenesis because the hydrolysis stage was normally the rate-limiting step of the anaerobic degradation of solid wastes [32]. Figure 5 shows the evolution of these parameters in all reactors.

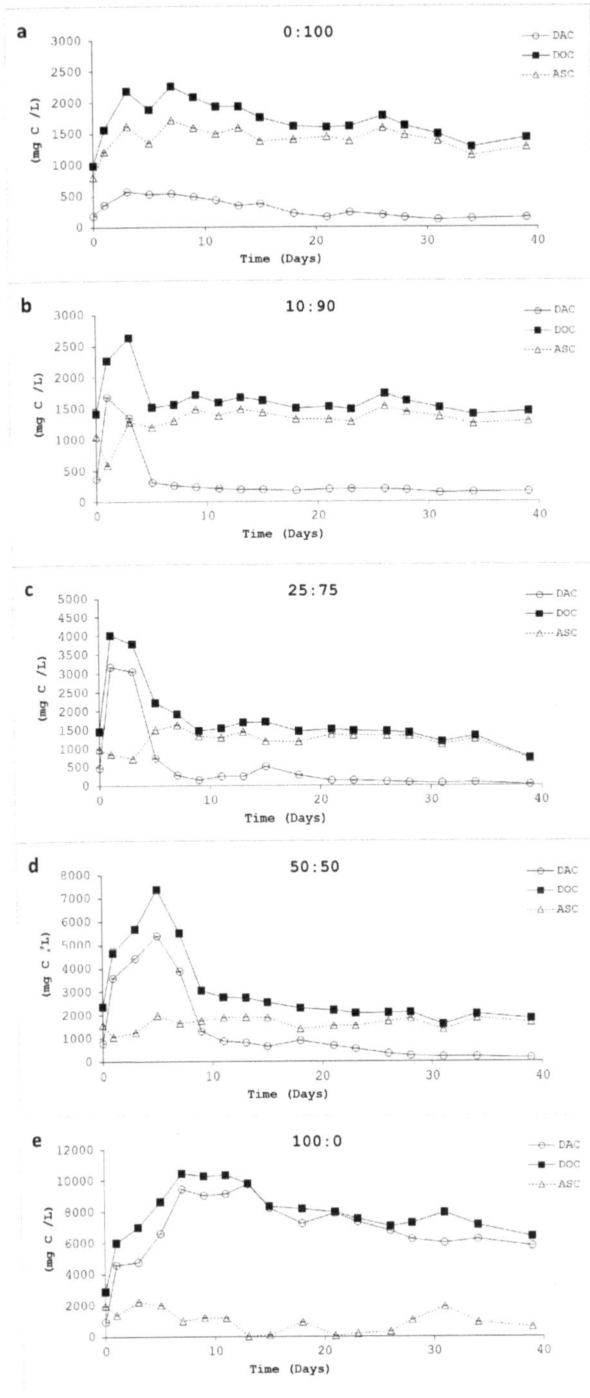

Figure 5. Evolution of dissolved organic carbon (DOC), dissolved acid carbon (DAC), and acidogenic substrate as carbon (ASC) in each reactor. (**a**) Reactor 0:100; (**b**) Reactor 10:90; (**c**) Reactor 25:75; (**d**) Reactor 50:50, and (**e**) Reactor 100:0.

It could be observed that the ASC was not accumulated in any test. In fact, ASC reached rapidly the final level for this parameter, indicating that this value represented the concentration of the organic carbon that could not be transformed into VFAs. This behavior was observed for all tests, including the mixture 100:0 (ESBP alone). Moreover, the increase of organic matter added to the medium, when the ESBP proportion in the mixture increased, led to an increase in the maximum TVFA obtained. Therefore, it could be deduced that the inhibition of the acidogenic stage did not occur and that the hydrolytic and acidogenic stages were well-coupled in all reactors. However, the acidogenic and methanogenic stages were not coupled in the determined periods in each test. In these periods, the acidogenesis rate was higher than that of methanogenesis and, hence, an accumulation of VFA was observed. The decoupling periods were longer in reactors with high proportions of ESBP in the mixtures. Indeed, the decoupling period covered the entire test for the mixture 100:0 (ESBP alone), due to the inhibition of the acetoclastic metanogenic population, as a response of the excessively high level of TVFA.

3.5. Influence of the Temperature Range

Temperature is a key variable in all biological processes. Although it is generally admitted that thermophilic anaerobic digestion improves the kinetics and the methane yield of the process, in comparison to the mesophilic range, there are some possible drawbacks in the function of the substrate and the operational conditions. Therefore, when the hydrolytic and acidogenic stages are predominant in a thermophilic system, a significant accumulation of VFAs occurs due to the inhibition of acetoclastic methanogenic microorganisms, which are the most sensitive microorganisms to temperature [51].

As was previously pointed out in Section 3.3, a significant improvement in methane production was obtained by applying the co-digestion option with respect to the anaerobic digestion of the individual substrates. However, the calculated SMPs were markedly lower than those obtained in the mesophilic process [46].

Figure 6 depicts the DOC evolution in this study, together with data from a previous mesophilic study of Aboudi et al. [46]. Maximum values of DOC have been taken into account for this comparison, since they represent the situation of maximum unbalance between solubilization and consumption of organic matter in batch processes. At the thermophilic temperature conditions, the evolution of DOC was very similar to that previously commented on TVFA and the maximum of DOC achieved in each test was directly related to the ESBP content. In fact, and for the mixture 100:0, the removal of the dissolved organic matter (as DOC) was very low, since an important fraction remained in the medium as VFAs, mainly as acetic acid.

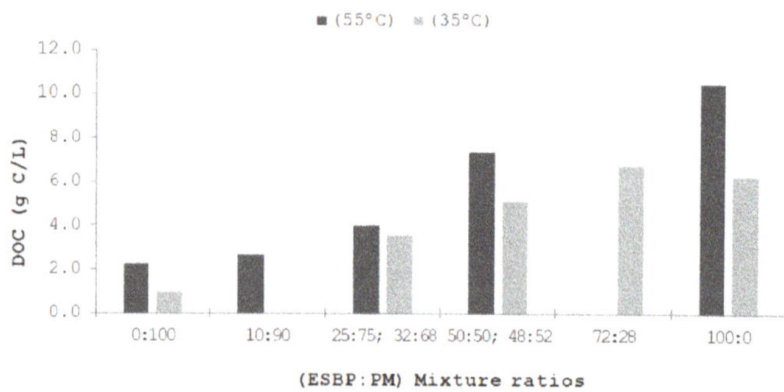

Figure 6. Maximum values of DOC in thermophilic assays (this study) and mesophilic assays [46].

Aboudi et al. [46] have found the same trend in the increase of the maximum values of DOC when the ESBP proportion in the mixture increased. However, as can be seen in Figure 6, these maximum values were generally lower with respect to those observed in this study. The reason is that hydrolysis rate was normally sharper in the thermophilic process than in the mesophilic one [14,15].

Likewise, the PM degradation was not complete in the reactors with a high proportion of PM, as has been discussed previously. Considering all these issues, the comparison between the mesophilic and thermophilic assays needed a new indicator, allowing the estimation of the theoretical methane that could be obtained from the non-degraded organic matter. This variable would be the percentage of the non-produced methane from ASC or VFA in the system, with respect to the total theoretical methane (non-produced and produced) that could be generated if all the organic matter was converted.

VFAs and ASC concentrations refer to the organic material which has not been converted into biogas. For this, the final levels of VFAs and the ASC were considered for the theoretical calculations. Only concentrations of the predominant VFAs (HAc, HPr, and HBu) were used. Moreover, the ASC was considered to be completely converted to VFAs in the same proportion observed at the end of the tests.

The following equations were considered for the calculations:

$$CH_3COO^- + H_2O \rightarrow CH_4 + HCO_3^- \quad (1)$$

$$CH_3CH_2COO^- + 3H_2O \rightarrow CH_3COO^- + HCO_3^- + H^+ + 3H_2 \quad (2)$$

$$CH_3CH_2CH_2COO^- + 2H_2O \rightarrow 2CH_3COO^- + H^+ + 2H_2 \quad (3)$$

$$CO_2 + 2H_2 \rightarrow CH_4 + 2H_2O \quad (4)$$

In an efficient AD system, the combination between the above-mentioned equations enables the calculation of the carbon proportions converted into methane for each substrate. These carbon proportions are 1/2, 7/12, and 5/8 for acetic, propionic, and butyric acids, respectively. Table 2 shows the obtained results of this estimation.

Table 2. Estimation of the theoretical methane from the non-converted Volatile Fatty Acids (VFAs) and Acidogenic Substrate as Carbon (ASC).

Assay	Reference	CH_4 (HAc) ($L/L_{reactor}$)	CH_4 (HPr) ($L/L_{reactor}$)	CH_4 (HBu) ($L/L_{reactor}$)	CH_4 (ASC) ($L/L_{reactor}$)	CH_4 (VFAs+ASC) (L)	CH_4 (Produced) (L)	CH_4 (Non-Produced) (%)
0:100T	This study	0.052	0.010	0.077	0.106	0.42	0.86	32.7
10:90T		0.100	0.006	0.042	0.100	0.42	4.03	9.40
25:75T		0.039	0.011	0.012	0.100	0.27	9.16	2.90
50:50T		0.064	0.009	0.051	0.103	0.39	8.43	4.40
100:0T		4.056	0.499	1.121	0.098	9.82	2.92	77.1
0:100M	Aboudi et al. [44]	0.004	0.002	0.002	0.102	0.22	7.63	2.8
32:68M		0.003	0.000	0.002	0.100	0.21	8.48	2.4
48:52M		0.011	0.001	0.004	0.099	0.23	9.34	2.4
72:28M		0.041	0.025	0.003	0.099	0.34	10.19	3.2
100:0M		0.266	0.200	0.022	0.107	5.19	11.04	32.0

T—thermophilic; M—mesophilic. It can be observed that the methane production associated with the theoretical ASC conversion was very low and was similar in all the assays. This finding enforces the fact that the hydrolysis and acidogenesis stages have not been limited in any operational condition.

As can be observed, for all co-digestion assays, the organic matter was almost completely converted into methane, and the percentage of the non-produced methane ranged from 2.4%–4.4%. However, a different behavior was observed for the two thermophilic reactors with high PM content, in which a noticeable fraction of the methane, associated with the non-converted ASC and VFAs (9.4% and 32.7% in 10:90T and 0:100T, respectively), was observed. In terms of the estimated methane from ASC and VFAs, both data were similar (0.42 L) and the final difference was attributed to the additional organic matter provided by the ESBP in the reactor 10:90T. In addition, the differences with respect to the mesophilic tests could be attributed to the PM characteristics, as previously commented.

The results from the reactors containing only the ESBP as the substrate have demonstrated that the methanogenesis stage was clearly affected, in both 100:0T and 100:0M reactors.

In the reactor 100:0T, the fast release of VFAs had induced a maximum level of acetic acid of 25 g/L around day 13 (Figure 3), which had prevented a normal development of the methanogenesis, leaving a final concentration of acetic acid equivalent to 4 L $CH_4/L_{reactor}$. The estimation of the methane fraction from the non-converted ASC and VFAs was 77.1%. This percentage was similar to that proposed by Smith and Mah [52], who estimated that 73% of the total methane was produced by acetoclastic methanogenesis, while the remaining 27% was produced by the hydrogen-utilizing methanogens.

However, the estimation of the methane fraction from the non-converted ASC and VFAs in the reactor 100:0M was 32.0%. Thus, the acetic acid was degraded in a higher proportion than in the thermophilic reactor, while the propionic acid had slightly increased with respect to the other mesophilic co-digestion tests. The different behavior of methanogenesis in both cases could be likely related to a possible accumulation of intermediate inhibitory compounds in the thermophilic reactor, such as long chain fatty acids (LCFAs) [17,53,54].

In a recent work, Aboudi et al. [11] had studied the anaerobic digestion of ESBP in a temperature phased CSTR reactors (hyperthermophilic at 65 °C–mesophilic at 35 °C and thermophilic at 55 °C–mesophilic at 35 °C). Authors reported an LCFAs accumulation in the hyperthermophilic and thermophilic reactors. Hence, palmitic acid concentrations of 140 and 210 mg/L and lignoceric acid concentrations of 80 and 125 mg/L were detected in thermophilic and hyperthermophilic reactors, respectively. The concentration of these acids was significantly reduced in the subsequent mesophilic reactors. Therefore, the inhibition of methanogenic acetoclastic archaea in the anaerobic digestion of ESBP, could be linked to the operations at high temperatures.

4. Conclusions

The thermophilic co-digestion of the exhausted sugar beet pulp with pig manure allowed for the achievement of an increase in the methane yield, with respect to the anaerobic digestion of ESBP, individually. The addition of PM had a beneficial effect, as it provided additional alkalinity and nitrogen and, consequently, increased the stability of the process. Moreover, the ESBP provided carbon and had contributed to a better C/N balance in the process. Therefore, the mixture of both wastes had a synergistic effect on the process. Thus, the ratio 25:75 (ESBP:PM) was the best proportion tested with a specific methane production of 212 mL/VS_{added}.

Despite the thermophilic range being normally considered favorable for increasing the process rate and product yield, the comparison of the results obtained with previous mesophilic studies from the literature indicated that the temperature had a detrimental effect on the process. Thus, in this study, the population of methanogenic acetoclastic archaea seems to have been affected and the methane production would have been mostly generated by the pathway of the hydrogen-utilizing microorganisms.

Author Contributions: The authors contribute equally to this work.

Funding: This research was supported by the projects CTM2013-43938-R and CTM2016-79071-R (Spanish Ministry of Economy, Industry and Competitiveness) and financed by the Spanish State Research Agency ("Agencia Estatal de Investigación"-AEI), and by the European Regional Development Fund (ERDF). The Authors also acknowledge the University of Cádiz (Spain) for the Scholarship UCA-2014-043/PU/EPIF-FPI-CT/CP, the Agri-food Campus of International Excellence (Ceia3) and the Azucarera (an AB Sugar Company) for the supply of the samples of exhausted sugar beet pulp.

Conflicts of Interest: The authors declare no conflict of interest.

References

1. Anwar, Z.; Gulfraz, M.; Irshad, M. Agro-industrial lignocellulosic biomass a key to unlock the future bio-energy: A brief review. *J. Radiat. Res. Appl. Sci.* **2014**, *7*, 163–173. [CrossRef]

2. Singh, R.; Krishna, B.B.; Mishra, G.; Kumar, J.; Bhaskar, T. Strategies for selection of thermo-chemical processes for the valorisation of biomass. *Renew. Energy* **2016**, *98*, 226–237. [CrossRef]
3. Hutnan, M.; Drtil, M.; Mrafkova, L. Anaerobic biodegradation of sugar beet pulp. *Biodegradation* **2000**, *11*, 203–211. [CrossRef]
4. Aboudi, K.; Álvarez-Gallego, C.J.; Romero-García, L.I. Biomethanization of sugar beet byproduct by semi-continuous single digestion and co-digestion with cow manure. *Bioresour. Technol.* **2016**, *200*, 311–319. [CrossRef]
5. Montañés, R.; Solera, R.; Pérez, M. Anaerobic co-digestion of sewage sludge and sugar beet pulp lixiviation in batch reactors: effect of temperature. *Bioresour. Technol.* **2015**, *180*, 177–184. [CrossRef]
6. Ohuchi, Y.; Ying, C.; Lateef, S.A.; Ihara, I.; Iwasaki, M.; Inoue, R.; Umetsu, K. Anaerobic co-digestion of sugar beet tops silage and dairy cow manure under thermophilic condition. *J. Mater. Cycles Waste Manag.* **2015**, *17*, 540–546. [CrossRef]
7. Ward, A.J.; Hobbs, P.J.; Holliman, P.J.; Jones, D.L. Optimisation of the anaerobic digestion of agricultural resources. *Bioresour. Technol.* **2008**, *99*, 7928–7940. [CrossRef]
8. Fdez-Güelfo, L.A.; Álvarez-Gallego, C.; Sales, D.; Romero, L.I. New indirect parameters for interpreting a destabilization episode in an anaerobic reactor. *Chem. Eng. J.* **2012**, *180*, 32–38. [CrossRef]
9. Velásquez Piñas, J.A.; Venturini, O.J.; Silva Lora, E.E.; Calle Roalcaba, O.D. Technical assessment of mono-digestion and co-digestion systems for the production of biogas from anaerobic digestion in Brazil. *Renew. Energy* **2018**, *117*, 447–458. [CrossRef]
10. Salsabil, M.R.; Laurent, J.; Casellas, M.; Dagot, C. Techno-economic evaluation of thermal treatment, ozonation and sonication for the reduction of wastewater biomass volume before aerobic or anaerobic digestion. *J. Hazard. Mater.* **2010**, *174*, 323–333. [CrossRef]
11. Aboudi, K.; Gómez-Quiroga, X.; Álvarez-Gallego, C.J.; Quiroga-Alonso, J.M.; Romero-García, L.I. Effect of Temperature on Biohydrogen and Biomethane Productions by Anaerobic Digestion of Sugar Beet by-Products. *Int. J. Environ. Sci. Dev.* **2018**, *8*, 762–766. [CrossRef]
12. Zahedi, S. Energy efficiency: Importance of indigenous microorganisms contained in the municipal solid wastes. *Waste Manag.* **2018**, *78*, 763–769. [CrossRef]
13. Zahedi, S.; Dahunsi, S.O.; Perez, M.; Solera, R. Assessment of chemical inhibitor addition to improve the gas production from biowaste. *Waste Biomass Valoriz.* **2017**, *10*, 1091–1099. [CrossRef]
14. Fernández-Rodríguez, J.; Pérez, M.; Romero, L.I. Comparison of mesophilic and thermophilic dry anaerobic digestion of OFMSW: Kinetic analysis. *Chem. Eng. J.* **2013**, *232*, 59–64. [CrossRef]
15. Fernández-Rodríguez, J.; Preze, M.; Romero, L.I. Bio-methanization of organic fraction from municipal solid waste: Temperature effects. *Polish J. Chem. Technol.* **2013**, *15*, 99–106. [CrossRef]
16. Sawatdeenarunat, C.; Surendra, K.C.; Takara, D.; Oechsner, H.; Khanal, S.K. Anaerobic digestion of lignocellulosic biomass: Challenges and opportunities. *Bioresour. Technol.* **2015**, *178*, 178–186. [CrossRef]
17. Palatsi, J.; Illa, J.; Prenafeta-Boldú, F.X.; Laureni, M.; Fernandez, B.; Angelidaki, I.; Flotats, X. Long-chain fatty acids inhibition and adaptation process in anaerobic thermophilic digestion: Batch tests, microbial community structure and mathematical modelling. *Bioresour. Technol.* **2010**, *101*, 2243–2251. [CrossRef]
18. Mata-Alvarez, J.; Dosta, J.; Romero-Güiza, M.S.S.; Fonoll, X.; Peces, M.; Astals, S. A critical review on anaerobic co-digestion achievements between 2010 and 2013. *Renew. Sustain. Energy Rev.* **2014**, *36*, 412–427. [CrossRef]
19. Neshat, S.A.; Mohammadi, M.; Najafpour, G.D.; Lahijani, P. Anaerobic co-digestion of animal manures and lignocellulosic residues as a potent approach for sustainable biogas production. *Renew. Sustain. Energy Rev.* **2017**, *79*, 308–322. [CrossRef]
20. Sanaei-Moghadam, A.; Abbaspour-Fard, M.H.; Aghel, H.; Aghkhani, M.H.; Abedini-Torghabeh, J. Enhancement of Biogas Production by Co-digestion of Potato Pulp with Cow Manure in a CSTR System. *Appl. Biochem. Biotechnol.* **2014**, *173*, 1858–1869. [CrossRef]
21. Rodriguez-Verde, I.; Regueiro, L.; Carballa, M.; Hospido, A.; Lema, J.M. Assessing anaerobic co-digestion of pig manure with agroindustrial wastes: the link between environmental impacts and operational parameters. *Sci. Total Environ.* **2014**, *497–498*, 475–483. [CrossRef]
22. Aboudi, K.; Álvarez-Gallego, C.J.; Romero-García, L.I. Semi-continuous anaerobic co-digestion of sugar beet byproduct and pig manure: Effect of the organic loading rate (OLR) on process performance. *Bioresour. Technol.* **2015**, *194*, 283–290. [CrossRef]

23. Fang, C.; Boe, K.; Angelidaki, I. Anaerobic co-digestion of by-products from sugar production with cow manure. *Water Res.* **2011**, *45*, 3473–3480. [CrossRef]
24. Umetsu, K.; Yamazaki, S.; Kishimoto, T.; Takahashi, J.; Shibata, Y.; Zhang, C.; Misaki, T.; Hamamoto, O.; Ihara, I.; Komiyama, M. Anaerobic co-digestion of dairy manure and sugar beets. *Int. Congr. Ser.* **2006**, *1293*, 307–310. [CrossRef]
25. Wu, X.; Yao, W.; Zhu, J.; Miller, C. Biogas and CH(4) productivity by co-digesting swine manure with three crop residues as an external carbon source. *Bioresour. Technol.* **2010**, *101*, 4042–4047. [CrossRef]
26. Aboudi, K.; Álvarez-Gallego, C.J.; Romero-García, L.I. Influence of total solids concentration on the anaerobic co-digestion of sugar beet by-products and livestock manures. *Sci. Total Environ.* **2017**, *586*, 438–445. [CrossRef]
27. Ge, X.; Xu, F.; Li, Y. Solid-state anaerobic digestion of lignocellulosic biomass: Recent progress and perspectives. *Bioresour. Technol.* **2016**, *205*, 239–249. [CrossRef]
28. Chandra, R.; Takeuchi, H.; Hasegawa, T. Methane production from lignocellulosic agricultural crop wastes: A review in context to second generation of biofuel production. *Renew. Sustain. Energy Rev.* **2012**, *16*, 1462–1476. [CrossRef]
29. Liotta, F.; d'Antonio, G.; Esposito, G.; Fabbricino, M.; Frunzo, L.; van Hullebusch, E.D.; Lens, P.N.L.; Pirozzi, F. Effect of moisture on disintegration kinetics during anaerobic digestion of complex organic substrates. *Waste Manag. Res.* **2014**, *32*, 40–48. [CrossRef]
30. Holliger, C.; Alves, M.; Andrade, D.; Angelidaki, I.; Astals, S.; Baier, U.; Bougrier, C.; Buffière, P.; Carballa, M.; de Wilde, V.; et al. Towards a standardization of biomethane potential tests. *Water Sci. Technol.* **2016**, *74*, 2515–2522. [CrossRef]
31. APHA-AWWA-WPCF. *Standard Methods for the Examination of Water and Wastewater*, 20th ed.; American Public Health Association: Washington, DC, USA, 2005.
32. Álvarez-Gallego, C.J. Testing Different Procedures for the Start-Up of a Dry Anaerobic Co-Digestion Process of OFMSW and Sewage Sludge at Thermophilic Range, University of Cadiz. 2005. Available online: https://dialnet.unirioja.es/servlet/tesis?codigo=52224 (accessed on 30 September 2018).
33. Van Soest, P.J.; Robertson, J.B.; Lewis, B.A. Methods for Dietary Fiber, Neutral Detergent Fiber, and Nonstarch Polysaccharides in Relation to Animal Nutrition. *J. Dairy Sci.* **1991**, *74*, 3583–3597. [CrossRef]
34. Angeriz-Campoy, R.; Fdez-Güelfo, L.A.; Tyagi, V.K.; Álvarez-Gallego, C.J.; Romero-García, L.I. New criteria to determine the destabilization of the acidogenic anaerobic co-digestion of organic fraction of municipal solid waste (OFMSW) with mixed sludge (MS). *Bioresour. Technol.* **2018**, *248*, 174–179. [CrossRef]
35. Aboudi, K.; Álvarez-Gallego, C.J.; Romero-García, L.I. Evaluation of methane generation and process stability from anaerobic co-digestion of sugar beet by-product and cow manure. *J. Biosci. Bioeng.* **2016**, *121*, 566–572. [CrossRef]
36. Wang, M.; Zhou, J.; Yuan, Y.-X.; Dai, Y.-M.; Li, D.; Li, Z.-D.; Liu, X.-F.; Zhang, X.-Y.; Yan, Z.-Y. Methane production characteristics and microbial community dynamics of mono-digestion and co-digestion using corn stalk and pig manure. *Int. J. Hydrogen Energy* **2017**, *42*, 4893–4901. [CrossRef]
37. Guwy, A.J.; Hawkes, F.R.; Wilcox, S.J.; Hawkes, D.L. Neural network and on-off control of bicarbonate alkalinity in a fluidised-bed anaerobic digester. *Water Res.* **1997**, *31*, 2019–2025. [CrossRef]
38. Angelidaki, I.; Ahring, B.K. Thermophilic anaerobic digestion of livestock waste: the effect of ammonia. *Appl. Microbiol. Biotechnol.* **1993**, *38*, 560–564. [CrossRef]
39. Callaghan, F.J.; Wase, D.A.J.; Thayanithy, K.; Forster, C.F. Continuous co-digestion of cattle slurry with fruit and vegetable wastes and chicken manure. *Biomass Bioenergy* **2002**, *22*, 71–77. [CrossRef]
40. Switzenbaum, M.S.; Giraldo-Gomez, E.; Hickeyt, R.F. Monitoring of the anaerobic methane fermentation process. *Enzyme Microb. Technol.* **1990**, *12*, 722–730. [CrossRef]
41. Zickefoose, C.; Hayes, R.B.J.; Bryant, J.O. *Anaerobic Sludge Digestion: Operations Manual. EPA 430=9-76-001*; National Technical Information Service, Ed.; United States Environmental Protection Agency: Washington, DC, USA, 1976.
42. Hill, D.T.; Cobb, S.A.; Bolte, J.P. Using Volatile Fatty Acid Relationships to Predict Anaerobic Digester Failure. *Trans. ASAE* **1987**, *30*, 496. [CrossRef]
43. Pullammanappallil, P.C.; Chynoweth, D.P.; Lyberatos, G.; Svoronos, S.A. Stable performance of anaerobic digestion in the presence of a high concentration of propionic acid. *Bioresour. Technol.* **2001**, *78*, 165–169. [CrossRef]

44. Barredo, M.S.; Evison, L.M. Effect of propionate toxicity on methanogen-enriched sludge, Methanobrevibacter smithii, and Methanospirillum hungatii at different pH values. *Appl. Environ. Microbiol.* **1991**, *57*, 1764–1769. [PubMed]
45. Wang, Y.; Li, G.; Chi, M.; Sun, Y.; Zhang, J.; Jiang, S.; Cui, Z. Effects of co-digestion of cucumber residues to corn stover and pig manure ratio on methane production in solid state anaerobic digestion. *Bioresour. Technol.* **2018**, *250*, 328–336. [CrossRef] [PubMed]
46. Aboudi, K.; Álvarez-Gallego, C.J.; Romero-García, L.I. Improvement of Exhausted Sugar Beet Cossettes Anaerobic Digestion Process by Co-Digestion with Pig Manure. *Energy Fuels* **2015**, *29*, 754–762.
47. Yang, Z.; Wang, W.; He, Y.; Zhang, R.; Liu, G. Effect of ammonia on methane production, methanogenesis pathway, microbial community and reactor performance under mesophilic and thermophilic conditions. *Renew. Energy* **2018**, *125*, 915–925. [CrossRef]
48. Selvam, A.; Xu, D.; Zhao, Z. Fate of tetracycline, sulfonamide and fluoroquinolone resistance genes and the changes in bacterial diversity during composting of swine manure. *Bioresour. Technol.* **2012**, *126*, 383–390. [CrossRef]
49. Zhu, Y.-G.; Johnson, T.A.; Su, J.-Q.; Qiao, M.; Guo, G.-X.; Stedtfeld, R.D.; Hashsham, S.A.; Tiedje, J.M. Diverse and abundant antibiotic resistance genes in Chinese swine farms. *Proc. Natl. Acad. Sci. USA* **2013**, *110*, 3435–3440. [CrossRef]
50. Zhang, J.; Sui, Q.; Zhong, H.; Meng, X.; Wang, Z.; Wang, Y.; Wei, Y. Impacts of zero valent iron, natural zeolite and Dnase on the fate of antibiotic resistance genes during thermophilic and mesophilic anaerobic digestion of swine manure. *Bioresour. Technol.* **2018**, *258*, 135–141. [CrossRef]
51. Montañés Alonso, R.; Solera del Río, R.; Pérez García, M. Thermophilic and mesophilic temperature phase anaerobic co-digestion (TPAcD) compared with single-stage co-digestion of sewage sludge and sugar beet pulp lixiviation. *Biomass Bioenergy* **2016**, *93*, 107–115. [CrossRef]
52. Smith, P.H.; Mah, R.A. Kinetics of acetate metabolism during sludge digestion. *Appl. Microbiol.* **1966**, *14*, 368–371.
53. Chan, P.C.; de Toledo, R.A.; Shim, H. Anaerobic co-digestion of food waste and domestic wastewater. Effect of intermittent feeding on short and long chain fatty acids accumulation. *Renew. Energy* **2018**, *124*, 129–135. [CrossRef]
54. Rasit, N.; Idris, A.; Harun, R.; Wan Ab Karim Ghani, W.A. Effects of lipid inhibition on biogas production of anaerobic digestion from oily effluents and sludges: An overview. *Renew. Sustain. Energy Rev.* **2015**, *45*, 351–358. [CrossRef]

© 2019 by the authors. Licensee MDPI, Basel, Switzerland. This article is an open access article distributed under the terms and conditions of the Creative Commons Attribution (CC BY) license (http://creativecommons.org/licenses/by/4.0/).

applied sciences

MDPI

Review

A Review of the Role of Critical Parameters in the Design and Operation of Biogas Production Plants

Shiplu Sarker [1,*], Jacob J. Lamb [2,3], Dag R. Hjelme [2] and Kristian M. Lien [3]

1 Department of Manufacturing and Civil Engineering, Norwegian University of Science and Technology, 2815 Gjøvik, Norway
2 Department of Electronics Systems, Norwegian University of Science and Technology, 7491 Trondheim, Norway; jacob.j.lamb@ntnu.no (J.J.L.); dag.hjelme@ntnu.no (D.R.H.)
3 Department of Energy and Process Engineering, Norwegian University of Science and Technology, 7491 Trondheim, Norway; kristian.m.lien@ntnu.no
* Correspondence: shiplu.sarker@ntnu.no; Tel.: +47-9127-1077

Received: 4 April 2019; Accepted: 3 May 2019; Published: 9 May 2019

Abstract: Many operating parameters, individually or together, may influence the performance of anaerobic digestion towards biogas or digestate yield and quality maximization. The most preferred method of optimizing an anaerobic digestion plant often relies on how carefully the crucial parameters, such as pH, temperature, organic loading rate, hydraulic retention time, and pressure, are chosen. There is a large amount of literature available on optimization of anaerobic digestion; however, given the continued development and implementation of innovative technologies, together with the introduction of increasingly complex systems, it is necessary to update present knowledge on process parameters and their role on operational ranges and flexibilities in real-life anaerobic digestion system. Accordingly, the present review discusses the importance of the selection of operational parameters in existing technologies and their impact on biogas yield. Notably, the four broad areas of feedstock utilization (substrate, inoculum, codigestion and pretreatment), process condition (pH, temperature, pressure, and reactor design), reactor control (HRT and OLR) and inhibition (Ammonia and VFAs) are covered in this review. In addition, particular emphasis is placed on the most recent innovations that have been or may be implemented in current or future biogas plants.

Keywords: anaerobic digestion; biogas; optimization; operating parameters; review

1. Introduction

Among all the forms of renewable energy, biomass-based technologies are foreseen to play a critical role in fulfilling the continuously increasing demand for future energy. Because biomass resources are abundant and easily transformable into various forms (solids, liquids, and gases), a vast range of energy applications, i.e., heat, power, chemicals, and liquid biofuels for transportation vehicles, are suited to this source of energy. The global carbon footprint reduction in a long-term future as a result of deploying biomass driven renewable energy technologies is promising. Today, the final energy consumption using biomass-derived fuels have reached 50 EJ globally with potential growth of 150–400 EJ/year in 2100 [1]. Simultaneously, the conversion of fossil fuel vehicles to biofuel vehicles is accelerating rapidly.

Anaerobic digestion (AD), one class of biomass conversion technology, mediated by the syntrophic association of bacteria and archaea in the absence of oxygen [2], has been considered a promising option of treating various types of biomass and wastes, e.g., energy crops, agricultural residues, bio and municipal wastes, industrial wastes (food & beverage), livestock and poultry wastes, sewage, and algae. The products of AD include energy-rich biogas, a potential candidate for diverse downstream conversion technology, and a liquid residue enriched with macro and micronutrients, suitable as organic fertilizer or soil amendments to agricultural lands. As a result of these manifold output opportunities, AD has been regarded as an excellent biomass conversion alternative, stabilizing utilization of biomass-derived wastes as well as offering the benefit of achieving the circular economy [3]—preserving the concept: 'reduce', 'reuse', and 'recycle'.

With the objectives of addressing future energy needs and the global warming reduction, the implementation of commercial and industrial scale biogas plants via AD have been evolving massively in all parts of the world. In Germany alone, the country that has the highest number of biogas plants in the EU, the installed capacity of biogas to electricity plants in the year 2015 exceeded the ~8900 MW capacity [4]. China, one of the biggest biogas producers in Asia, has been experiencing a boom in the biogas generation units predominantly for household-scale applications for the last few decades. However, the considerable investment and public initiatives on innovative measures and technologies have promoted the transformation of many of these household-scale applications to large-scale technologies including enormous growth towards biogas to electricity. Consequently, the biogas to electricity installation capacity in China reached 5500 MW in 2015, which is estimated to escalate up to 30,000 MW by the year 2020 [5]. Among the Nordic countries, the progress in biogas development in Sweden, Denmark, and Finland is reasonably steady [6]. Additionally, Norway has recently made an exemplary stride by commercializing the world's biggest liquid biogas plant with production capacity estimated to be 25 million Nm^3 of biofuel annually (Biokraft, Skogn, Norway). Once commenced, this will have a direct contribution in reducing 60,000 ton of CO_2 equivalent emission annually, and 25 million L of fossil fuels replacement in transportation application. The biogas integration to the energy sector in other countries like UK, USA, and Australia is also progressing.

Despite the diverse applicability and rapid expansion globally, some factors including process complexity, poor stability, inefficient biodegradability, substrate complexity, and low productivity impede methane production from AD. Numerous ways to overcome operational shortcomings suppressing methane yield have been suggested in previous studies, where the innovative approaches like three-stage digester [7], novel enzyme addition [8] and continuous microbial growth analysis [9] have been developed and implemented successfully. In parallel, optimization of the process performance by manipulating operational variables [10] such as feedstock choice, pretreatment, codigestion, reactor type, temperature, pH and HRT (Hydraulic retention time) [11] have been widely considered. An example of the optimum and conventional AD reactor comparison is shown in Figure 1.

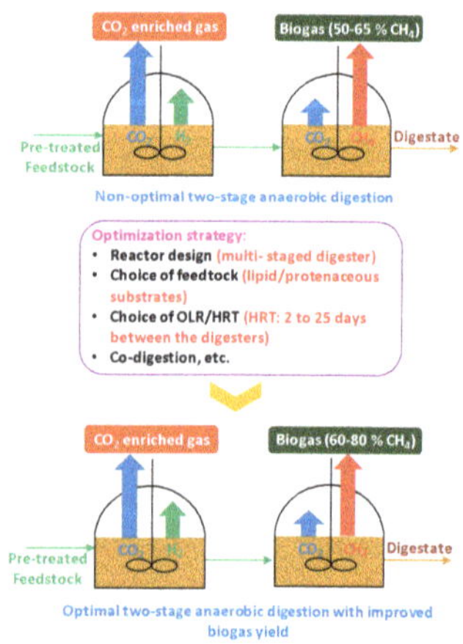

Figure 1. Methane content in biogas between conventional and optimum anaerobic digestion (AD).

Despite this, the complex microbiological interactions influencing the performance of operational parameters remain to be extensively explored. In this vein, the overview of the critical parameters and their interdependence to anaerobic digestion efficiency would always be an interesting field of study. This review intends to discuss this relationship and emphasizes on the innovations that may potentially bring significant prospects for future applications.

2. Anaerobic Digestion Process and Microbial Communities

Anaerobic digestion offers a valuable option for converting biodegradable feedstock into renewable energy. In this process, the conversion occurs by bacteria and archaea in the absence of oxygen, wherein a series of complex biochemical reactions determine the product output [12]. Generally, the organic part of the biomass is utilized by the microorganism consortia and yielded to methane and carbon dioxide, while the rest is transformed into other minorities [13]. The complete digestion process takes place in four different stages called hydrolysis, acidogenesis, acetogenesis, and methanogenesis [14], which are briefly overviewed in Figure 2.

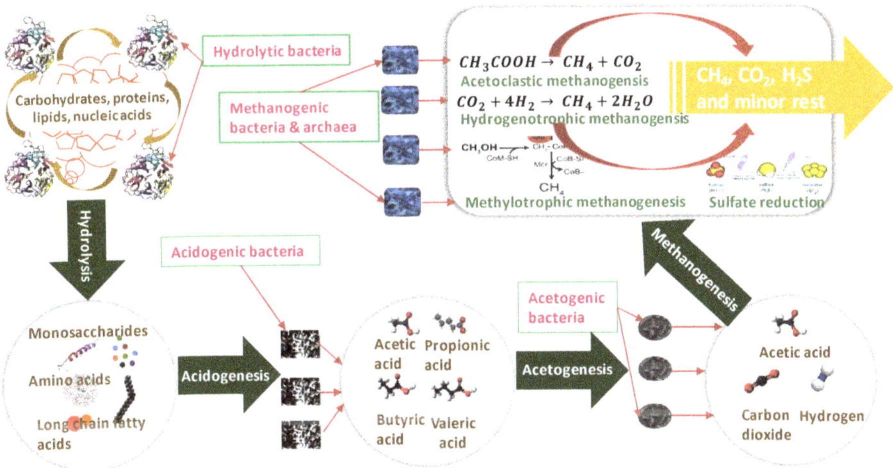

Figure 2. Simple schematic representation of anaerobic digestion steps.

Hydrolysis involves depolymerization of insoluble complex organic hydrocarbons into soluble monomers where the principle substrate compounds (i.e., carbohydrates, lipids, and proteins) are broken down into corresponding low molecular weight monosaccharides, long chain fatty acids, and amino acids that are favorable for bacterial degradation. Hydrolysis is a complex multistep process mediated by extracellular enzymes. The enzymes required for hydrolysis can either be attached to microbial cells or secreted to the solution [15]. Several groups of hydrolytic microorganisms are involved in the degradation of several substrate compositions, where the bacteria Bacteriods, Clostridium and Staphylococcus are significant drivers [16] (see Table 1).

Table 1. Bacterial groups involved in hydrolysis of different substrate components [16].

Primary Substrate Components	Hydrolyzed Products	Bacterial Group
Carbohydrates	Soluble sugars	*Clostridium, Acitovibrio celluliticus, Staphylococcus, Bacteriodes*
Lipids	Higher fatty acids or alcohols and glycerol	*Clostridium, Staphylococcus, Micrococcus*
Proteins	Soluble peptides and amino acids	*Clostridium, Proteus vulgaris, Peptococcus, Bacteriods, Bacillus, Vibrio*

In the acidogenesis stage, acidogenic bacteria such as Lactobacillus, Streptococcus, and Clostridium [16] (Table 1) transform hydrolysis products (amino acids and sugars) into volatile fatty acids (VFAs) (acetic acid, butyric acid, and propionic acid), organic acids (succinic acid and lactic acid), ammonia (NH_3), hydrogen gas (H_2), carbon dioxide (CO_2), hydrogen sulfide (H_2S), and low alcohols [17]. The concentration of hydrogen produced at this stage affects the final product after digestion and the resulting organic matters such as VFAs are not suited for direct conversion to methane by the methanogens.

Hence, the third stage, namely acetogenesis, converts the VFAs, especially acetic acids and butyric acids, into acetate, H_2 and CO_2. Among the VFAs, 65–95% of methane is directly produced from acetic acid, while propionic acid remains mainly unconverted because its degradation is thermodynamically less favorable (based on the relationship between hydrogen partial pressure and VFA degradation), in comparison to butyric acid [18].

In the final stage, methane is generated by the function of three groups of methanogens, namely acetotrophic, hydrogenotrophic, and methylotrophic [19]. The majority of the methane is produced by acetotrophic methanogens, which transform the acetate (resulting from acetogenesis) into CH_4 and CO_2 [20]. In this process, the principle reaction can be interpreted as shown by Equation (1).

$$CH_3COOH \rightarrow CH_4 + CO_2 \quad (1)$$

The hydrogenotrophic group converts hydrogen and carbon dioxide into methane through the reactions stated in Equations (2) and (3) [20]. From this route, around 30% of methane may be produced.

$$CO_2 + 4H_2 \rightarrow CH_4 + 2H_2O \quad (2)$$

$$4CO + 2H_2O \rightarrow CH_4 + 3CO_2 \quad (3)$$

Besides the above two groups, some methane can also be produced by the methylotrophic methanogens [19]. Through this pathway, the methyl or trimethylamine component of a given feedstock is transformed into methane following the chemical reactions given by Equations (4) and (5).

$$3CH_3OH + 3H_2 \rightarrow 3CH_4 + 3H_2O \quad (4)$$

$$4(CH_3)_3N + 6H_2O \rightarrow 9CH_4 + 3CO_2 + 4NH_3 \quad (5)$$

3. Process Parameters Involved in a Biogas Production Plant

As mentioned earlier, the characteristics of various process parameters impact heavily on the quantity and quality of biogas production. Variables that are broadly relevant to AD include feedstock type, reactor type, pH, temperature, retention time, organic loading, pressure, and inhibitory components. Properties such as inoculum type, codigestion, and pretreatment type are directly linked to feedstock, while temperature, mixing, retention time, organic loading, volatile fatty acids, pH, and pressure are greatly influenced by the type of reactor used and the biochemical interactions involved. The effect of all these parameters on various aspects of anaerobic digestion in light of the present state-of-the-art is widely covered in the following subsections.

3.1. Feedstock

3.1.1. Substrate

Diverse substrates originated from agricultural wastes, municipal solid wastes, industrial wastes, wastewater, aquatic biomass, and energy crops may be included as feedstock for anaerobic digestion [21]. Extracting biogas from these materials depends on the physical and chemical compositions that favor biological degradation. Not all the substrate components (carbohydrates—cellulose, hemicellulose, and lignin; lipids—fats, oil, and glycerols; and proteins) are readily degradable. For instance, lignin is highly nondegradable, whereas cellulose breaks down in several weeks; hemicelluloses, fats, and proteins within a few days; and volatile fatty acids, alcohols in several hours [21]. Thus, the choice of appropriate feedstock concerning the amount of energy production, reactor design and waste disposal is essential.

Theoretically, the biogas potential of feedstock constituents can be predicted by employing stoichiometry, as shown in Table 2, or using the empirical formula, proposed by Boyle [22], given in Equation (6).

$$C_aH_bO_cN_dS_e + \left(\frac{4a-b-2c+3d+2e}{4}\right)H_2O \rightarrow \left(\frac{4a+b-2c-3d-2e}{8}\right)CH_4 + \left(\frac{4a\pm b\mp 2c+3d+2e}{8}\right)CO_2 + dNH_3 + eH_2S \quad (6)$$

Among the main components, lipids have the highest methane potential (see Table 2); however, they can cause process problem [23] and inhibition when hydrolyzed into long chain fatty acids (LCFA) and accumulated. Accumulated LCFA due to the adsorption on to anaerobic sludge damages bacterial cell walls, hinders nutrients and metabolic transports, and eventually results in process inhibition [24].

Table 2. Stoichiometry of biogas potential determination from various feedstock components [25].

Feedstock	Methane Formation Stoichiometry	Methane Concentration, %
Carbohydrate	$(C_6H_{10}O_5)_n+nH_2O \rightarrow 3nCH_4+3nCO_2$	50
Lipid	$C_{50}H_{90}O_6 + 24.5H_2O \rightarrow 34.75CH_4 + 15.25CO_2$	69.5
Protein	$C_{16}H_{24}O_5 + 14.5H_2O \rightarrow 8.25CH_4 + 3.75CO_2+4NH_4^+ +4HCO_3^-$	68.8

Food wastes generated from slaughterhouses, food processing industries, and partly from municipal solid waste (MSW) contain a high level of proteins and fats, which, when utilized for AD, influence the production level of ammonia and sulfide. High levels of ammonia are toxic to methanogens and likely to cause process imbalances in terms of pH buffering [26]. Additionally, substrates with a high lipid content may also lead to other process problems associated with the substrate and product transport limitations, sludge floating, foaming, blockage of pipes and pumps, clogging of gas collectors and gas transport system [27]. Nevertheless, due to the high methane potential, lipid-rich feedstocks can offer an enhanced methane production if codigested in combination with substrates like municipal solid wastes [27], sewage sludge [28], paper waste [29], and rice husk [30]

Protein-rich substrates like food waste, fish waste, algae, and energy crops have also been utilized for biogas production via AD. The major problem associated with anaerobic degradation of protein is the development of a high concentration of ammonia that often leads to process instability or inhibition. Some of the process problems related to protein-rich feedstock are suggested to be overcome by means of various approaches where reducing the feedstock particle size [31], keeping operating temperature to mesophilic range [32], increasing HRT [32], ammonia stripping by addition of CaOH, KOH or NaOH [33], and adjusting pH of bioreactor liquid by adding acidic iron and acid [32,34] are commonly used options.

Besides, substrates rich in carbohydrate content, such as lignocellulosic biomass, can be good sources of biogas production. Theoretical biomethane potential of lignocellulosic biomass, such as grass, wheat, straw, and sorghum [35], is relatively high, but due to the high level of lignin, this class of feedstock opposes microbial hydrolysis [36], and hence recalcitrant to anaerobic conversion. A lot of research has suggested pretreatment to maximize utilization of lignocellulosic biomass (see Section 3.1.3).

While lignocellulosic biomass contains a rather high level of lignin, marine feedstock algae (micro and macroalgae) tend to have very low or no lignin, which makes them suitable for AD [37]. Algae as a substrate can circumvent the issue of food vs. fuel, provide an excellent alternative to the biogas industry, and can offer a synergistic benefit of coproducing various chemicals and value-added products including biogas [38]. So far, the biomethane potential of various algae species has been documented in a number of past studies [39–41]. However, a host of problems are identified, including cell wall resistance for degradation, production of toxic substances, increasing level of pH in case of high C:N ratio algae species, LCFA inhibition [42], and NH_3 inhibition [37,43]. To overcome part of these challenges, pretreatment, i.e., washing [44], maceration [44], thermal [45], sonic [45], and mechanical [46] treatment along with codigestion [47] may be implemented.

Animal manures are a common source of organic material used as feedstock for AD. Using manures for biogas production reduces anthropogenic greenhouse gas emissions, which would have otherwise released during storage [48]. Depending on factors such as animal species, breed, growth stage, feed, amount, and type of bedding, the biomethane potential (BMP) of animal manure can vary widely. Manure provides essential nutrients (micro and macro) for bacterial proliferation [49], buffer capacity for degradation of low nitrogen substrates, and controlling the level of VFA during AD. Due to its many advantages, most of the anaerobic digesters in Europe [50] are based on animal manure as a substrate. However, the biogas potential of manure is somewhat reduced by its recalcitrant material content, especially the biofibers or the embedded bed materials (e.g., straw). To achieve accelerated conversion of the recalcitrant fraction, manure is often suggested to undergo pretreatment before utilization for biogas production.

3.1.2. Inoculum

The choice of right inoculum type in combination with the right temperature is the key to set a desired anaerobic start-up condition [51,52]. Generally, biogas production and the substrate utilization rate in an anaerobic digester show an inverse trend with increasing input of substrate to inoculum (S:I) [53,54]. The usage of too little inoculum can result in incomplete feedstock degradation and process problems associated with VFA accumulation, inhibition and slower methane production rate. For example, depending on the type of inoculum, operating temperature, and the volatile solids content of the substrate, a four-fold increase in S:I ratio could contribute to decreasing as much as 40% of biogas production from AD of cattle manure [55]. Additionally, a higher S:I ratio can lead to longer HRT, resulting in larger reactor working volumes [56]. However, the use of a high amount of inoculum or a low S:I ratio might induce process instability as well as low biogas yield [57]. Thus, a careful compromise is necessary. The effect of the type and amount of inoculum on biogas production from various substrates was studied previously; for example, by Gu et al. [58].

3.1.3. Pretreatment

Pretreatment increases feedstock utilization towards AD. Substrates composed of high recalcitrant matter, or not readily biodegradable matter, are considered for pretreatment. The technique used for pretreatment depends on the type of substrate and varies to a wide degree of methods, but, generally, thermal [59], chemical [60], physical/mechanical [61], ultrasound [62], microwave [63], biological [64], and metal addition [11] methods are used to perform pretreatment. The primary feedstock pretreatment categories are shown in Figure 3.

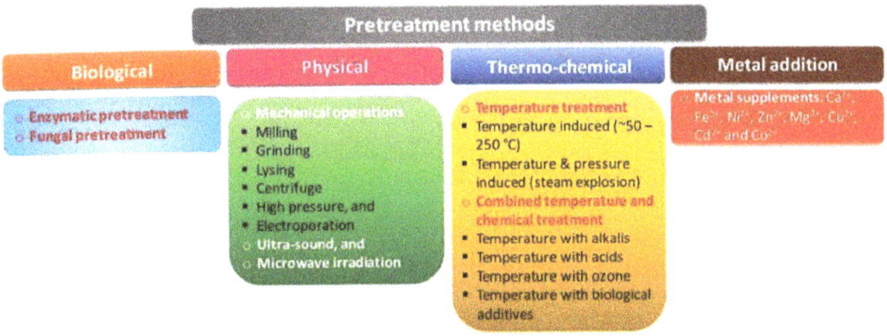

Figure 3. Primary feedstock pretreatment categories—modified from Ref. [65].

Thermal pretreatment occurs at a vast range of temperature, which across various studies spans between 70 °C and 275 °C [59,66–68]. Before anaerobic digestion, the thermal treatment can enhance feedstock hydrolysis resulting in increased anaerobic digestion rate and extent. Substrates like waste activated sludge [66], food waste [69], municipal solid waste [70], agricultural byproducts [71], grass [72], and algae [73] showed a positive effect on methane yield as a result of thermal pretreatment. However, the methane yield from thermally pretreated substrates is not always higher than those that are untreated. Rafique et al. [74] showed no improvement in methane yield from the solid fraction of pig manure at a pretreatment temperature of over 100 °C, while Carrère et al. [75] observed that a pretreatment temperature of 190 °C was found favorable for improved biogas yield from the total liquid and solid fractions of a similar substrate.

Steam explosion is another thermal pretreatment technique mostly applied for treating lignocellulosic biomass. Through steam explosion, the fibers in lignocellulosic biomass open-up and become more accessible for biological degradation. According to the literature [76], biogas yield from lignocellulosic biomass such as Salix woodchips can be maximized with the steam explosion method at 210 °C for about 10 min. Steam explosion pretreatment is usually defined by a severity factor that is calculated from the temperature and duration of the process. The relation between the steam explosion severity, duration, and temperature are expressed by Equation (7) below [77]. A similar trend to the Salix woodchips was observed for birch wood chips [78], where an approximately two-fold increase in methane yield was achieved compared to untreated woodchips due to the steam explosion with the severity of 4.5 at temperature 220 °C. For agricultural biomass such as wheat straw, the different severity of steam explosion had shown no positive impact on methane yield, but the degradation rate was found to be increased [79]. The severity factor of steam explosion for thet majority of feedstocks usually lies within the range of 3.14–3.56 [77].

$$\log R_o = \log\left(t \cdot e^{\left(\frac{(T-100)}{14.75}\right)}\right) \quad (7)$$

where,

$\log R_o$: the severity factor as a function of treatment time;

T: the temperature in °C;

t: is the residence time in (min); and

14.75: the activation energy where the process obeys first-order kinetics and the Arrhenius temperature dependence

Apart from thermal pretreatment, thermochemical pretreatment combining heat and chemicals/additives can contribute to enhancing AD productivity by increasing the COD (chemical oxygen demand) solubility [80], reducing feedstock particle size [81], and increasing volatile solid reduction [82]. The chemicals predominantly used for thermochemical pretreatment include alkali [83], acids [84], biological additives [85], and ozone [86]. Acid pretreatment, performed with acids such as HCl, HNO_3, H_2SO_4, and H_3PO_4, accelerates hemicellulose solubilization into oligomers (i.e., molecules that consist of a small and specifiable number of monomers; usually less than five), while alkali pretreatment, involving NaOH and $Ca(OH)_2$, helps to break lignin together with the solubilization of hemicellulose [65]. Alkali pretreatment (also called saponification) is a commonly adopted technique to help accelerate the hydrolysis of the lipid-like substrate [87]. Compared to acid pretreatment, alkali pretreatment was demonstrated to be more effective in enhancing methane production [65,88], although some issues concerning digestate land application and methanogen toxicity were identified [89]. Despite the certain advantages, the cost associated with the use of additives/chemicals may offset the benefit earned from the revenue of the biogas plants. Therefore, a careful trade-off between cost and biogas yield optimization needs to be practiced. The pros and cons of using a selective list of additives in an anaerobic digestion application are given in Table 3.

Table 3. The advantages and disadvantages of using additives in AD [90].

Additive	Element or Compound	Benefit	Adverse Effect
Macro-nutrients	P, N and S	Methane production improvement and enhanced process stability	Methane or biomass inhibition by overdosing
Micro-nutrients, heavy metals	Cu^{2+}, Zn^{2+}, Cr^{3+}, Cd, Ni, Pb^{4+} and Hg^{2+}	Promoting various enzymatic reactions	Inhibition to the acetogens
Micro-nutrients, light metals	Na^{+}, K^{+}, Mg^{2+}, Ca^{2+} and Al^{3+}	Enhancing microbial growth	Restricting production of double cells (Mg^{2+}), Inhibition of acetoclastic methanogens (Na^{+}), Destabilizing buffering system (Ca^{2+}), etc.
Iron	Zero valent iron, Clean scrap, rusty scrap and iron additives	Sulfide fixation, biomass stimulation, etc.	Precipitation and clogging risk
Ash	Bottom ash and fly ash		
Inorganic absorbent materials	$MgCl_2$, $MgCl_2.6H_2O$, $MgPO_4.3H_2O$	Ammonia sequestration through struvite formation	
Inorganic nitrogen		Availability of nitrogen as nutrient	Inhibition of methane producing enzymes, VFA accumulation
Nano particles	Ag, Au, Fe, Al_2O_3, SiO_2, TiO, ZnO	Methane production improvement	Inhibition of hydrolytic and methanogenic biomass
Biological additives (bioaugmentation and enzyme)	Compost, C. proteolyticus, SAO co-culture and Methanoculleus bourgensis MS2	Increased methanogenic activity, increased hydrolytic activity	Cost, process control
Others	Biochar, activated carbon, sand, zeolite, Ni-Zeo, Co-Zeo, Mg-Zeo, rockwool, membrane, molecular sieve, polyurethane foam and loofah	Biomass immobilization, buffering agents, enhanced VFA degradation	

Unlike thermochemical pretreatment, physical pretreatment does not require the addition of any external compounds but uses physical techniques such as physical operations (milling, grinding, lysing, centrifuge, high pressure, electroporation, ultrasound, and microwave irradiation). Mechanical pretreatment is a widely used physical method that improves bioconversion effectivity, particle densification and distribution, flow properties, porosity, bulk density, and the overall conversion of lignocellulosic biomass into biogas without producing any toxic side streams [91]. With mechanical milling, commonly achieved by the different mills such as attrition mills, ball mills, centrifugal mills, colloid mills, hammer mills, extruders, knife mills, pin mills, and vibratory mills [92], the crystallinity of cellulose, particle size, and the degree of polymerization is reduced and consequently the surface area and digestibility of feedstock are enhanced. Employing milling, such as screw press extrusion, results in an increase of up to 30% in the methane yield from dip litter (a combination of a solid fraction and straw) manure was accomplished [93]. The combination of milling ($\leq$2 mm) with fungal pretreatment was observed to achieve a profound methane yield increase of up to 160% [94].

Microwave irradiation is another physical pretreatment option, or a heating method, which directly applies electric and magnetic field components to the molecular structure of the substrate, stimulating physical, chemical, or biological reactions due to the heat and extensive collisions by the vibrations of the polar molecules and ion movements [95,96]. As a result of microwave radiation, a number of pretreatment benefits can be acquired, which includes increased substrate surface area and accessibility to enzyme attack, decreased polymerization and crystallinity of cellulose and lignin depolymerization [97]. Microwave-assisted pretreatment, using microwave irradiation alone or in combination with various solvents (i.e., water, alkali, acid, ionic liquid, salt, and other organic components), has been studied. With the use of microwave pretreatment alone, a 60% methane yield increase for a continuous reactor was observed [65]. Other evidence suggests that microwave irradiation with a solvent combination is either able to rapidly hydrolyze more sugars [98] or to remove more hemicellulose and lignin from the lignocellulosic substrate [99] compared to the conventional thermal and thermochemical treatment. However, the application of this technique to date is limited to the lab-scale, as technical difficulties to expand it to the industrial scale still exist.

Sonication, involving a sound frequency greater than 20 kHz (ultrasound), can be applied to physical pretreatment of biomass [100]. Brief exposure to ultrasound can cause the thinning of microbial cell walls, resulting in the release of cytoplasm (material within a living cell excluding cell nucleus) and thereby facilitating the intercellular matter to be available for further degradation to CH_4 and CO_2. Although a large variety of feedstock can be successfully pretreated using ultrasound, until now, this method is widely considered for sewage or waste activated sludge [101], as this material, compared to other substrates, requires a lower sonication time and energy. The past study found that longer sonication time and energy may lead to a lower conversion efficiency of solubilized matter into methane [65]. Based on the literature, the threshold of specific sonication energy ranges between 1000 and 16,000 kJ/kg TS (total solids) and is strongly correlated to substrate solid content, which for sludge was reported to be optimum at 20–30 g/L [65]. With use of sonication, the BMP of waste activated sludge and dairy cattle manure (TS of 5.8%) was improved by 140% [102] and 19% [102], respectively. With high TS substrates, such as mixed sludge and MSW (~9% TS), a very high specific energy requirement of about 90,000 kJ/kgTS resulted in an increase of 24% in the methane yield [103]. Sonication of high solid substrates also caused high cavitation, increased dewaterability and disturbance to the homogeneity of acoustic waves [65]. However, the energy applied to treat high TS (9%) sludge in a full-scale plant (sonix) was found energetically feasible as the benefit obtained as a result of revenue earned from increased biogas production (35–55%) overcompensated the cost spent on sonication [104], making this method suitable for a wide range of feedstock. Despite this, the use of electric energy in comparison to waste heat utilization for thermal pretreatment could be energetically ineffective and a significant drawback to the sonication technique.

Enzymatic pretreatment has been investigated in many previous studies. The enzymes used during or prior to AD include cellulases, cellobiases, endoglucanase, xylanases, pectinases, laccases, versatile peroxidases, as well as α-amylases and proteases [65]. Deconstruction of lignin and cellulose by enzymes in different stages of hydrolysis can result in the conversion of oligomers and various monomer sugars [105]. Although many enzymes do exist, the application of these in most cases was not practical, and biogas yield in many instances either remained unaffected or dropped [106]. Enzymes can be applied in different ways at different stages of AD. When used upstream during hydrolysis, the sugars released can be potentially consumed by endogenous microorganisms before being further converted into biogas [65]. For cases like this, the endogenous microorganisms are typically eliminated by sterilization (an extra processing step), which incurs additional cost and affects the overall plant economics [65]. Since enzymes for AD are produced from different forms of fungi (e.g., Aspergillus and Trichoderma genus) [107], they can be used alternatively, so the process cost for producing enzymes can be avoided. Further, an auxiliary enzyme, such as LPMO (Lytic polysaccharide mono-oxygenases) in combination with fungi can also be used for enhancing cellulose degradation [8]. The cost and selectivity are apparently the major drawbacks concerning enzyme application to pretreatment, which, in the context of optimization of AD, calls for further research in future.

Adding certain metals, such as Ca^{2+}, Fe^{2+}, Ni^{2+}, Zn^{2+}, Mg^{2+}, Cu^{2+}, Cd^{2+}, Co^{2+}, and W^{6+} [108,109] at specific concentrations, enhances the biogas yield from AD. The enzymes required for essential bacterial growth are chemically linked with metals (such as those stated above). By supplementing appropriate amounts of these metals, the appropriate levels of nutrients in the anaerobic digester can be maintained, resulting in enhanced substrate degradation and ultimately increased biogas yield [110]. For example, the enzyme complex acetyl-CoA decarboxylase/synthase has an essential role in degrading acetate to methane, which can be further accelerated by adding the metal Ni^{2+} [111]. More enzymes were found to be linked to micronutrients or metals [112]. Augmenting heavy metals Ni^{2+}, Zn^{2+}, and Cd^{2+} at a concentration of 2.5 mg/L to a mesophilic reactor (37 ± 1 °C) operating on potato waste and cattle manure (50:50), the biogas production was improved [113]. Methane yield was also reported to be increased from a lab-scale digester after adding Fe, Zn, Mn, Cu, and Mo [114]. Using trace elements, the conversion of organic solids can be enhanced allowing stable digester operation at a low organic loading rate (OLR) [115]. However, an excess amount of heavy metals can reduce the methanogen's activity, leading to process failure or collapse [113]. Therefore, the right amount of metal supplementation is essential.

3.1.4. Codigestion

Codigesting multiple substrates can improve biogas yield during AD at various combinations and fractions. For instance, methane production increase of up to 65% [116] can be achieved by codigestion of cattle manure with organic wastes (C5 molasses) [117,118], seaweed [47], energy crops [116], food wastes [119], agrowastes [120], and fruit and vegetable wastes [121]. A considerable increase in methane yield from codigestion of many other substrates has also been detected and documented in a comprehensive work by Poulsen T. et al. [122].

Generally, substrate in the AD process is degraded to biomethane by the synergistic interaction of a bacterial consortium, consisting of three functional group of microorganisms involved into the processing steps: hydrolysis, acidogenesis, and methanogenesis. Due to the insufficient buffering capacity, the products of each of these steps can ultimately cause inhibition of methanogenesis. For example, the reduction in pH due to the accumulation of VFAs from acidogenesis caused by the insufficient buffering capacity can result in substantial drop in production of biomethane and thereby loss in AD efficiency.

By deploying codigestion, the digester can be designed for operating at an optimum C:N ratio [123] under which enhanced nutrient balance with improved pH buffering can be achieved. Theoretically, with a C:N ratio of about 25–30:1 [11,124], methanogens use nitrogen to fulfil their protein requirements and optimize the production of methane. At higher ratios, the nitrogen availability depletes quicker, resulting in incomplete methanogenesis, leading to reduced biogas productivity. Conversely, lower C:N ratios give excess nitrogen causing ammonia formation together with an increase in pH and consequently drop in methane yield [125].

Besides C:N ratio enhancement, codigestion can also improve overall macronutrients balance by adjusting the proportion of carbon, nitrogen, phosphorous and sulphur content. The requirement of macronutrients in AD is usually not so high, with a typical ratio of C:N:P:S as 600:15:5:1 [126]. Nonetheless, this depends on the characteristics of feedstock and an individual AD process. Gil et al. [127] showed that substrates with simultaneous high concentrations of N and P, resulting in a combined C:(N + P) ratio between 53 and 40, have a pronounced effect in an increased methane content in the produced biogas. Many other combinations and proportions of macronutrient contents can also lead to possible conditions for enhanced biogas production for a codigestion process, and research into this area is still ongoing.

3.2. Reactor

3.2.1. Configuration

The way feedstock is converted in AD is greatly dependent on the way the reactor is configured. Generally, two broad classes of substrates, i.e., wet substrate (TS < 15%) and high-solid substrate

(15% < TS < 40%), are treated [128] in a wide range of reactor designs, including anaerobic baffled stacking reactor (ABSR) [129], anaerobic contact process (ACP) [130], anaerobic filter (AF) [131], anaerobic fluidized bed reactor (AFBR) [132], agitated granular sludge bed reactor (AGSBR) [133], anaerobic hybrid reactor (AH) [131], anaerobic membrane reactor (AnMBR) [134], anaerobic sequencing batch reactor (ASBR) [135], anaerobic migrating blanket reactor (AMBR) [136], batch system anaerobic reactor (BSAR) [137], carrier-induced granular sludge bed reactor (CIGSBR) [138], continuously stirred anaerobic bioreactor (CSABR) [139], continuously stirred tank reactor (CSTR) [47], expanded granular sludge bed reactor (EGSBR) [140], fixed-bed reactors (FBR) [141], plug flow reactor (PFR) [142], submerged anaerobic membrane reactor (SAnMBR) [143], super-high-rate anaerobic bioreactor (SAB) [144], temperature phased anaerobic digestion reactor (TPAD) [145], tubular reactor [146], upflow static media anaerobic reactor (USMAR) [147], upflow anaerobic sludge blanket reactor (UASB) [148], and solid-state digester [149]. Selected operational parameters used in these reactors based on existing literature are summarized in Table 4.

Table 4. Selected reactor configurations and their operating parameters.

Reactor Configuration	Application	Operational Parameters (Reactor Size, Feedstock Type, Reactor Temperature, pH, OLR and HRT)	Results	Reference
ABSR	Biohydrogen production	3 L, Sucrose, 37 °C, 5.5 ± 0.2 (adjusted), 10–30 gCOD/L, 8 h	**HPR**: 10.9 ± 1.5 L/L-d, **HY**: 1.7 ± 0.2 mol/mol-sucrose	[129]
ACP	Methane production	10 L, Olive mill waste water and urea, 35 °C, 7.5 (adjusted), 2 gCOD/L/d, 15 days	**MY:** 0.15 L-CH_4/gCOD removed	[130]
AF	Pre-treatment & process performance	60 L, Domestic sewage, 13 °C, **N/G**, 234 mgCOD/L, 4 h	**MC**: 70.7 ± 2.9%; **MC** in **AF** was found higher than that of **AH**	[131]
AFBR	Biohydrogen production and waste water treatment	ca 4 L, Synthetic waste water, 37 °C, 4 (adjusted), 10 g/L, 0.5 to 4 h	Max. **HPR**: 2.36 L/L-h, Max. **HY**: 1.16 mol/mol-glucose	[132]
AGSB	Biohydrogen production	ca 0.9 L, Glucose, 40 °C, 6.5, 20 gCOD/L, 4, 2, 1 and 0.5 h	**HC**: 36–41%, **HY**: 1.4 to 31.5 mol/mol-glucose	[133]
AH	Pre-treatment & process performance of sewage treatment plant	88 L, Domestic sewage, 13 °C, **N/G**, 340 mgCOD/L, 4 h	**MC**: 58.9 ± 3.2% (see also **AF**, given above)	[131]
AnMBR	Biogas generation	2-phase (7 L & 20 L), Cheese whey, 37 ± 2 °C (both phases), 6.5 at start (acidogenic), Max. 19.78 gCOD/L-d (methanogenic), 1 d & 4 d	**MC**: Max. 70% (methanogenic); biogas production exceeded 10 times reactor volume increased with OLR	[134]
AMBR	Methane production & waste water treatment	12 L, Sucrose base synthetic wastewater, 35 ± 1 °C, 6.5 (adjusted), 30 g/L/d, 12 h	**MPR:** 6.5 L/L/d with 62.2% average methane based COD removal efficiency	[136]
ASBR	Biogas generation	**N/A**, Swine waste, 25 °C, 6.8 to 7.4, 0.9 to 5.5 g/L/d, 2 to 6 days	Biogas production rate: 0.9 to 1.8 L/L/d	[135]
BSAR	Biogas generation	1 L (5 units, equal volume), Pig manure (PM) and grass silage (GS), 35 °C, 6.5 to 8.0, 5 PM:GS (1:1, 1:3, 3:1, 1:0, 0:1), 90 days	Max. **MY**: 304.2 mL/gVS (at OLR 3:1 for PM:GS) Max. cumulative **MY**: 8517 L (at OLR 3:1 for PM:GS)	[137]
CIGBR	Biohydrogen production & waste water treatment	ca 1 L, Sucrose base waste water, 35 °C, 3 (adjusted), 2.5 to 5 gCOD/L/h, 4 to 8 h	Max. **HPR**: ~7.3 L/L-h, Max. **HY**: 3.03 mol/mol-sucrose	[138]
CSAB	Biohydrogen production	ca 1 L, Sucrose, inoculum heat shock, 40 °C, 6.6 ± 0.2, 30 to 40 gCOD/L/h, 0.5 to 6 h	Max. **HPR**: 15 L/L-h, Optimal **HY**: 3.5 mol/mol-sucrose	[139]
CSTR	Methane production	5 L, **CM** and *Laminaria digitata*, 35 ± 2 °C & 50 ± 1 °C, 8.0 ± 0.3, 2.5 to 2.9 gVS/L/d, 22 days	**MY avg.**: ca 225 L/kg VS (meso), ca 170 L/kg VS (thermo)	[118]
EGSB	Biogas generation	1 L, Potato-juice, 37 °C, 8 (adjusted to 4, 5, 6 & 7), 2.5 to 4.2 gCOD/L-d, 6 to 10 days	**MY avg.**: 385 mLCH_4/gVS; **MPR avg.**: 1496 mLCH_4/L-d	[140]
FBR	Biogas generation and biomass development	86 L, Skim milk, whole milk and oleate (variable feeding in 3 periods), 35 °C, 7 to 7.2 (adjusted), 12 g/L (2.4 - 4.15 g oleate/L skim. milk), 426 days	Max. **MPR**: 33 and 46 mLCH_4/gVS-d	[141]
PFR	Biogas generation	ca 5 m^3 (field scale plant), Terrestrial weeds and leafy biomass, 25 to 35 °C, **N/D**, 50 to 100 kg leafy biomass/day, 35 to 70 days	Average biogas yield: 50 L/kg fresh biomass (at OLR: 50 kg/day); 30 to 45 L/kg fresh biomass (at OLR: 100 kg/day)	[142]
SAnMBRs	Biogas generation & waste water treatment	6 L (3 units), Synthetic low strength waste water, 25 to 30 °C, 7.0 ± 0.5 (adjusted), 1.1 to 1.65 kg COD/m^3/day, 8 to 12 h	Max. **MPR avg.**: ca 2.9 L/d (HRT: 8, SRT: infinitive), Max. **MY average**: 0.29 L/gCOD (HRT: 8, SRT: infinitive), Max. **specific MY**: 0.068 L/**MLVSS**/d (HRT: 12, SRT: infinitive)	[143]
TPAD	Biogas generation and performance analysis	30 L (meso) & 20 L (thermo), **CM**, 38 °C (meso) and 58 °C (thermo), 7.00 to 7.75, 2 to 8 gVS/L/d, 14 days	**MY**: 0.21 to 0.22 L/gVS fed (thermo); 0.15 L/gVS (meso)	[145]
TR	Biogas generation	18 L (4 units), Fruit and vegetable waste, 35 ± 1 °C, 6.8 to 7.6, 2 to 8 gVS/L/d, 12 to 20 days	Max. **biogas production rate**: 2.62 L/L/d, Max. **biogas yield**: 707 L/kgVS fed, Max. **MC**: 65%	[146]
USMAR	Methane production	85 L (3 equal cylinders), Synthetic waste & dry milk, 35 °C, 4.5 to 7.2, 1 to 12 gCOD/L/d, 0.5 to 2 days	**MY**: 0.1 to 0.2 L/gVS fed with 5 to 13% increase in **MC** as a result of OLR at 10 gCOD/L/d	[147]
UASB	Hydrogen and methane production	24 L, Cassava waste water, 37 °C, 5.5 (with control), 10 to 30 kg/m^3-d (biohydrogen reactor) & 2 to 10 kg/m^3-d (methane reactor, Short (**N/G**)	Max. **HPR**: 0.39 L/L-d (at OLR: 25 kg/m^3-d), Max. **HY**: 39.83 L/kgCOD fed (at OLR: 25 kg/m^3-d), Max. **MPR**: 0.91 L/L-d (at OLR: 8 kg/m^3-d), Max. **MY:** 115.23 L/kgCOD (at OLR: 8 kg/m^3-d)	[148]

CM: Cattle manure, **HC**: Hydrogen content, **HPR**: Hydrogen production rate, **HY**: Hydrogen yield, **MC**: methane content, **MLVSS**: Mixed liquid volatile suspended solids, **MPR**: methane production rate, **N/A**: Not accessed, **N/D**: Not determined, **N/G**: Not given.

In recent years, the process efficiency and feedstock characteristics (i.e., dry matter content) and reactor systems have been modified to meet the specific criteria of product output, and further development is expected. Broadly, digesters can either be batch or continuous. Batch configurations are straightforward, require less moving parts and are inexpensive [17], but they are generally employed to determine methane potential of substrates where the reactors are subject to the anaerobic environment for a long period, until a degree of digestion close to the theoretical maximum is achieved. However, in a real life scenario, constant gas production is more desirable, requiring continuous digestion.

Continuous digestion system can be designed with single or multiple digesters. In a single stage digestion system, all steps in microbial degradation, i.e., hydrolysis, acidogenesis, acetogenesis, and methanogenesis, occur simultaneously in a single reactor vessel. For the majority of anaerobic digestion applications, single stage reactors are still dominant because of their simplicity in design and cost [150]. So far, an array of substrates such as food waste [151], vegetable waste [152], sludge [151], municipal solid waste (MSW), biowaste [150], and livestock manure [153] have been treated by this configuration and process performance is optimized in a vast number of cases. For example, optimum methane recovery using a single stage reactor (CSTR, PFR, UASB, ASBR, and TPAD) for livestock manure (i.e., cattle manure, pig manure, and poultry manure) was shown to be achieved and even maximized by recirculating the process liquid back into the reactor [154]. By digestate recirculation, liquid retention time increases and the microbe wash out from the process consequently decreases [155].

In a two-stage system, the first step involves feeding material into the first digestion tank (acidogenic reactor), where hydrolysis, acetogenesis, and acidogenesis take place. The partially digested materials from this reactor are then introduced into the second stage (methanogenic reactor), where the methane is produced. Compared to the single stage digestion, the two-stage process allows fast and efficient formation of biogas in the second stage where methane recovery of volatile solids can exceed 90% in exceptional cases [156,157]. However, due to the complexity in design, two-stage reactors have a higher cost [150].

The three-stage reactor concept was developed in the early 1990s [158]. In this configuration, the first phase involves semi-anaerobic hydrolysis of feedstock at low hydraulic retention time and the removal of un-degraded materials to the secondary reactor for acidogenisis. From the secondary reactor, output liquids and solids are transferred to a tertiary reactor where methane-rich biogas is produced. Likewise, in a two-stage system, a part, or all of the effluent from the tertiary reactor can be recirculated back to the digester depending on the targeted HRT and process efficiency [159]. According to the study by Kim et al. [159], using a three-stage reactor, COD removal efficiency of over 95% was achieved for food waste AD. In a more recent study [7], an 11–23% increase in methane yield for a three-stage codigestion of food waste and horse manure was reported.

In addition to the configurations mention above, AD systems have undergone several modifications over the last few decades to address the retention efficiency of slow-growing methanogens. With the development of the UASB reactor, where a dense sludge bed allows methanogenic sludge to maintain within the reactor [160], the improvement in retention efficiency was found to be significant. So far, numerous UASB reactor types are being used for treating waste and wastewater effluents of various characteristics. Among these, the internal circulation (IC) reactor was observed to exhibit high efficiencies in terms of considerable feasibility, robust resistance to outside accidents, high organic loading rate, and low investment cost [161]. Membrane bioreactors (MBR) are another technique quite successfully able to retain active biomass in the system. By employing a membrane, both cells and inhibitory components within the bioreactor can be separated and as such, an efficient biological process can be achieved [162]. Available membrane reactors and their application in the context of AD can be learned from a comprehensive review authored by Visvanathan and Abeynayaka [162]. Figure 4 illustrates two examples of novel reactor configurations that are used in both lab- and industrial-scale applications.

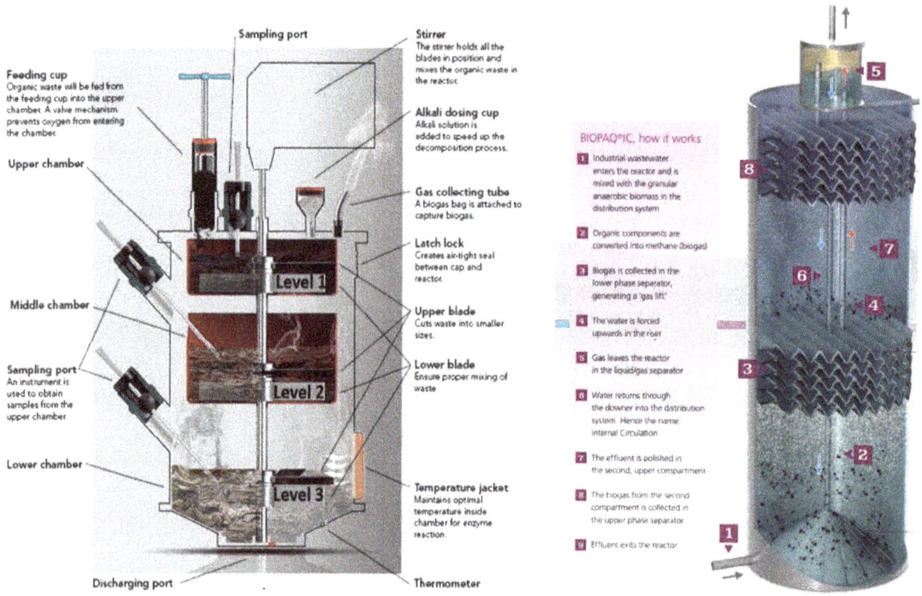

Three-stage reactor

Internal circulation (IC) reactor

Figure 4. Example of two novel anaerobic reactor types—reproduced with permission from the original authors [7,161].

Apart from bioreactor design, the developments in biosensor technologies and molecular biology techniques can enable understanding complex microbiological processes and potential disturbances much earlier than the conventional AD processes that do not involve any sensors for that matter. Reviews of the current progress in sensor technologies for controlling AD is documented excellently by Gaida et al. [163].

3.2.2. Mixing

Mixing, to a great extent, can determine the performance and cost of an AD system. Mixing promotes contact between microorganisms, substrates, and nutrients and provides uniform temperature distribution in the digester. Adequate mixing can reduce sedimentation and foaming caused by floating fat with adhering gas bubbles or by filamentous microorganisms, such as Microthrix or Norcardia [164]. Reactors equipped with mixing tend to produce more biogas [165] than those without, although contrasting results do exist [166,167]. Mixing can be accomplished by means of mechanical (employing a mixer), hydraulic (liquid recirculation), and pneumatic (recirculation of gases) techniques [168] at various frequencies (continuously or intermittently with several hours or several times in an hour during a day) [11] and intensities (gentle, intermittent, and rigorous rotation speed). Ong et al. [169] showed that employing intermittent mixing, mass transfer from the liquid phase to the gas phase greatly enhanced resulting in an increased gas release as much as 70% higher than the periods without mixing. In addition, among various mixing intensities, gentle mixing leads to the formation of aggregates and prevents methane-producing organisms from being washed out by the liquid and thereby proved to be more effective [170].

Dispite the benefits of increased gas release, mixing does have an extra cost and requires an energy input. Continuous mixing in a full-scale AD can demand as much as ~50% of the total plant energy [171], with mixer motor start-up consuming 2.5 % additional energy [172]. Therefore, mixing should be compromised or chosen carefully, given the feedstock type and AD technology. Strategies like switching from continuous to intermittent mixing mode could save over 25% of the energy demand and ensure higher plant productivity [173]. Additionaly, using innovative technologies such as a gas-lifting reactor (that works on biogas's rising tendency for partial recirculation and digester's agitation) can provide the effect of mixing without requiring a mixer, and has been shown to result in improved plant economies [174]. Two innovative mixing configurations operating with the airlifting mixing concept are shown in Figure 5.

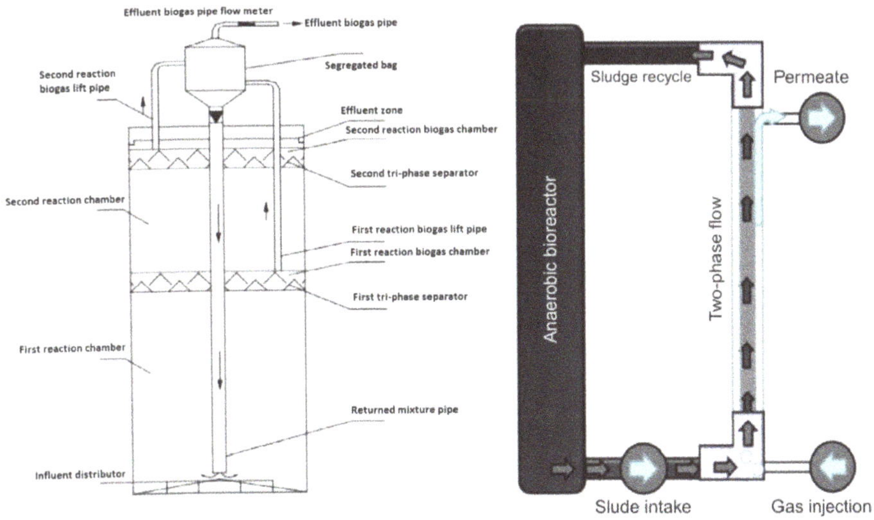

(a) (b)

Figure 5. Two novel configurations (**a**,**b**) of airlifting bioreactor agitating sludge without having an installed mixer—reproduced with permission from the original authors [175,176]. (**a**) biogas-lift reactor (BLR): right upward arrow- first biogas lift included with sludge and water; middle downward arrow—flow of sludge and water; left upward arrow—second biogas lift with separation of sludge and water. (**b**) gas-lift membrane bioreactor: the flow directions of materials are indicated with the arrows and texts in the body of the figure.

3.3. Temperature

The AD temperature depends on the tolerance of the microorganisms and is categorized as psychrophilic, mesophilic, thermophilic, or extremophilic [177] established at temperature ranges of 4–25 °C, 30–40 °C, 50–60 °C, and >65 °C, respectively [177,178]. Although anaerobic digestion can be achieved with temperatures lower than 20 °C [179], below 10 °C, the degradation is three times slower than the normal mesophilic process with methanogenesis becoming the rate-limiting step [180]. With higher temperatures, such as those between mesophilic and thermophilic regimes, anaerobic digestion tends to have a perfect environment for biological degradation resulting in high rates of hydrolysis and consequently high biogas yields. The correlation between the reaction rate and the biological process with temperature is typically represented by Equation (8) below [181]:

$$k_T = k_{20}\theta^{(T-20)} \tag{8}$$

where k_T denotes the reaction rate constant at temperature T, k_{20} represents the reaction rate constant at 20 °C, θ is the temperature activity constant, and T denotes the temperature.

The methane production amount in thermophilic AD is almost identical to mesophilic AD, but higher temperatures improve the production rate [182], and reduces the requirement of high operational

HRT and consequently reducing the reactor size [183]. However, thermophilic digestion system can be energy intensive, unstable, and sensitive to inhibition [184] (see Section 3.5), which, in addition to the robustness of the operation, is why mesophilic processes are presently the most preferred techniques implemented at industrial scale AD [185].

For a given operational temperature type, the fluctuation of a few degrees of temperature can have a severe impact on methane yield, as microorganisms adjust to one certain temperature and re-adaptation corresponding to a different temperature requires an alternated microbial structure. Noticeably, a variation in mesophilic temperature of ±4 °C and thermophilic temperature of ±1 °C was found to result in a sharp decrease in biogas yield [19,186].

Within physio-microbial activities, some AD processing phases are more influenced by the temperature than others. For example, the temperature requirements for optimizing the growth of methanogens bacteria, specifically the mesophilic methanogens species, may differ from the temperature requirements for optimizing hydrolysis or acidification [19]. Investigation of phase-specific temperature effects has been conducted and it was found that the temperature-phased (70 and 55 °C in the two successive stages) digesters, separating hydrolysis and methanogenesis, produced better results in terms of biogas yield and organic matter removal than the thermophilic or mesophilic digestion alone [187]. Similar results were demonstrated in previous research works by the authors Puchajda et al. [188,189]. Moreover, a staged digester combining thermophilic (55 °C) and extremophilic (68 °C) temperatures was reported to accelerate hydrolysis of recalcitrant organics, resulting in higher biogas yields compared to a thermophilic or extremophilic reactor alone [190]. Figure 6 illustrates the energy generation comparison between a single and staged digester operated at different temperature conditions.

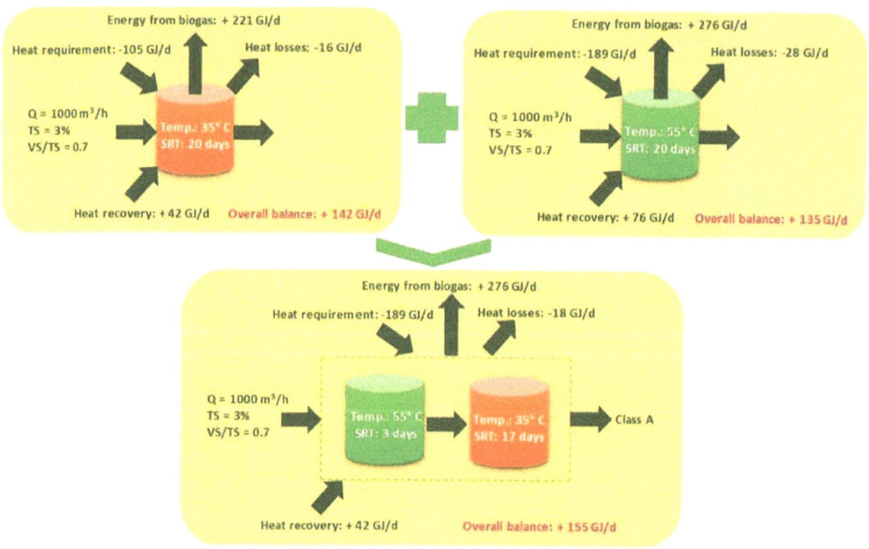

Figure 6. Higher energy generation from a two-stage thermophilic-mesophilic digestion compared to a single stage mesophilic or thermophilic digestion alone—modified from Puchajda et al. [189].

Temperature optimization by employing various technological options such as reactor design [187], OLR or HRT variation [191], pretreatment [192], and post-treatment [190] have been the major topics investigated in the existing literature. Moreover, incorporating computer simulation with the use of advanced modelling tools, such as ADM1 (Anaerobic Digestion Model No. 1), is gaining popularity in recent research [193].

3.4. pH

The pH is measured to indicate the health of anaerobic microorganisms and the performance of AD system [17]. A variation of pH in the magnitude of 0.5 can result in a substantial change in microbial metabolism influencing reaction kinetics and produced gas yield. Although feasible anaerobic digestion can be achieved within the pH of 5.5 and 8.5 [194], methanogens are highly sensitive to the pH change and generally optimize at a pH of close to 7 [177]. A pH under 6.3 or over 7.8 can adversely affect methanogenesis with a tendency of process failure to occur [195]. However, unlike methanogenesis, the other processing steps (e.g., hydrolysis and acidogenesis), can potentially optimize between pH of 5.5 and 6.5 [196]. With regard to the intermediates, specifically the VFAs produced during the acidogenesis if not metabolized into products, the pH can be significantly lowered (<3), eventually leading to process collapse. Process disturbances can also result due to high alkalinity caused by the formation of total ammonium nitrogen (TAN) during hydrolysis. Generally, the pH range of extreme acidic (~3 or less) or extreme alkaline (~12 or higher) can be inhibitory to the acidogenesis [197] and limit the rate of hydrolysis.

The development of various chemical species during biochemical interaction (e.g., NH_3, CO_3^{2-}, and CH_3COO^-) can cause a considerable impact in digestate pH variation. For example, the formation of ammonium carbonate ($(NH_4)_2CO_3$) or reduction of CO_2 from the digestate liquid can directly lead to a pH increase [198,199]. Also, the production of basic cations (e.g., Mg^{2+}, Ca^{2+}, K^+) or the reduction of multivalent anions (e.g., SO_4^{2-}, $Fe(OH)_3$) can cause electric charge imbalances in the digestate liquid resulting in a pH rise. Similar to observations in pH increases, the precipitation of carbonates (e.g., calcite $CaCO_3$) or the fatty acids as a consequence of high organic loading could be the critical factors influencing pH reduction. Kovács et al. [200] showed that the protein-rich substrates (i.e., meat extracts), tend to wane buffering capacity at an elevated organic loading input resulting in pH drop. The tendency of the various ions originating from various degradation pathways to influence pH is shown by Figure 7.

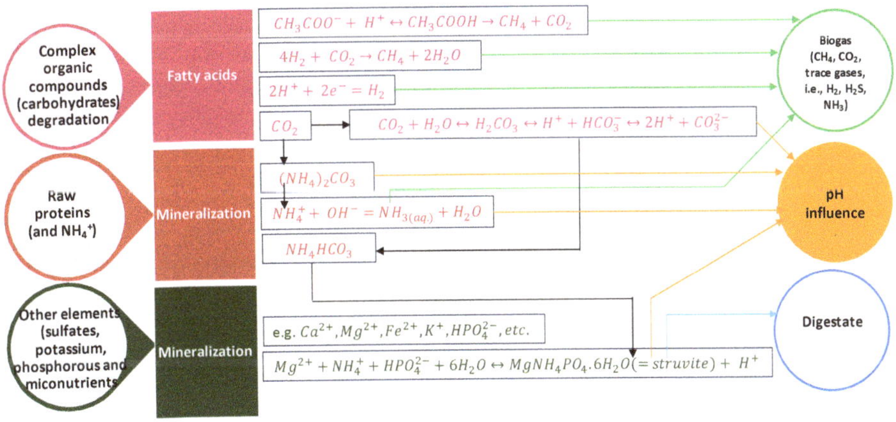

Figure 7. Interaction of degradation pathways and the formation of products influencing pH.

To control pH fluctuation and to operate AD process at the desired pH range, modern biogas plants are often installed with an automatic pH controller. The primary objective of the controller is to help balancing the system buffer (alkalinity) by adding an appropriate neutralizing agent at an appropriate concentration when necessary. Strong bases (e.g., NaOH) or carbonate salts (e.g., Na_2CO_3 and $NaHCO_3$) and acids (e.g., HCl) are primarily the components used as chemicals to achieve a pH increase or decrease. Moreover, indirect pH control can be applied by manipulating operating parameters including a reduction in organic loading rate [201], digestate recirculation [202], HRT variation [203] and codigestion [204].

The controlling of pH, whether achieved by the direct addition of a buffer or by the manipulation of the operating conditions, requires a careful and sophisticated application to avoid disturbances to methanogenesis and further to biogas yield. Robust controlling of pH by incorporating multiple parameters towards an optimum methane yield using sensor technologies is a growing field of research in AD.

3.5. HRT

Retention time is an important parameter used for design and optimization of anaerobic digestion. Retention time refers to both hydraulic retention time (HRT) and solid retention time (SRT). HRT represents the retention time of the liquid phase, whereas SRT denotes the retention of the microbial culture in the digester. In an anaerobic reactor system where the feedstock and microbial mixed cultures are present at the same phase, the HRT equals SRT. AD of food waste, kitchen waste, and municipal solid waste are the examples for which HRT is essentially SRT and vice-versa, while for the substrates like waste activated sludge and primary sludge, the interaction between solids and microbial cultures are biphasic making HRT and SRT different. HRT and SRT are typically represented by Equations (9) and (10) [205]:

$$\text{HRT} = \frac{V}{Q} \tag{9}$$

$$\text{SRT} = \frac{V \cdot X}{Q_X \cdot X_X} \tag{10}$$

where, V: Individual reactor volume [m^3]; Q: Influent flow rate [m^3/d]; X: Mixed liquid suspended solids in an individual reactor [mg/L]; Qx: Excess biosolids removal rate [m^3/d]; X_X: Mixed liquid suspended solids in excess biosolids flow [mg/L].

In general, the choice of HRT differs based on feedstock composition, reactor volume, processes, and temperature. Substrates rich in starch and sugar can be easily digested [177], and as such, hydrolysis is less or not required, resulting in shorter retention times, while longer retention times are required for fiber and cellulose plant matter as hydrolysis of these substrates is slow and rate limiting. High reactor temperature increases the decomposition rate and consequently lowers HRT, which is why most thermophilic reactors are generally operated at a lower HRT than mesophilic reactors. Comparing two HRT types, shorter HRT risks bacterial mobilization including the build-up of higher molecular weight VFAs and consequently elevated stress to the methanogens [206]. Conversely, a longer HRT increases the digester size [183]. Therefore, the optimal operational HRT is usually neither too long nor too short and in the majority of cases lies between 10 and 25 days, although a very high HRT in the order of 50–100 or more days may be needed for digesters operated in colder climates [206].

With regard to the different phases of AD, typically a longer HRT is preferred for methanogenesis to match the slower growth rate of methanogens compared to acidogens. This also validates the fact that many current digesters are designed in a multistaged manner so that acidogenesis and methanogenesis are technically separated into two different volumes allowing each group of microorganisms to operate at optimal conditions [207]. The correlation between the different bacterial species and different levels of HRT has been studied and depicted in a recent paper by Yang et al. [208].

In the case of anaerobic digestion of sludge type feedstock, consideration of both HRT and SRT are essential. For a given HRT, longer SRT usually provides sufficient time for methanogens to complete biological activities and result in a high biogas yield. With an operating HRT of 12 h, infinitive SRT for a submerged anaerobic membrane bioreactor (SAnMBR) was found to double the methane production rate (L/L/d) than that for SRT of 30 days [143]. However, contrasting results were also reported [209], where the shorter SRT (or increased OLR) was found to give a higher biogas production rate (L/L/d) [209]. This was explained by the fact that a raised OLR may support an increased methanogenic population and hence enhanced microbial activities leading to higher biogas productivity, but the higher VFA accumulation might indicate that the process is already overloaded. Therefore, the balance between the choice of SRT and OLR is critical for an optimum AD. At the start-up period, in general, a long acclimatization time under longer SRT and lower OLR are required for the satisfactory development of slow-growing populations [210], while in the stable period, a relatively shorter operational SRT may be needed allowing improved process performance including a higher methane yield. In a past study, shorter SRT was deliberately used to control the production of VFA by shifting the microbial communities from one route of production to another, for example, from acetogenotrophic to hydrogenotrophic [211]. A similar effect (but with a shift from methane to VFA production), was observed while the HRT was shortened [212].

3.6. Ammonia

Ammonia is produced from the nitrogen-containing matters in the feedstock (i.e., from the degradation of proteins and urea in the manure). In the aqueous phase, it is present in two forms, ammonium ions (NH_4^+) and free ammonia (FA) or unionized ammonia (NH_3). Together, this is called the total ammonia nitrogen (TAN) [213]. In anaerobic digestion, only a partial fraction of the organic nitrogen (Kjeldahl nitrogen) is biologically degraded to inorganic ammonia with the proportion spanning between 34 and 80% [214–216]. Most of the ammonia is usually generated during hydrolysis [217], and the type produced is influenced by the factors such as a change in temperature, pH, inoculum or microbial community. The FA and ionic ammonia with relation to TAN are typically expressed by Equation (11) below [218,219]:

$$NH_3{-}N = \frac{NH_4^+{-}N}{1 + \frac{10^{-pH}}{10^{-(0.09018 + \frac{2729.92}{T})}}} \tag{11}$$

FA is a strong candidate for methanogen inhibition. It is membrane permeable, and when it diffuses through the cells, proton imbalance or variation in intracellular pH occurs causing inhibition of enzymatic reactions [220]. High temperature combined with a high pH decreases the solubility of ammonia resulting in the dissociation equilibrium of aqueous ammonia to shift towards free ammonia leading to the production of this ammonia to be dominated over ionic ammonia (see Equations (11) and (12)).

$$NH_4^+(\text{aqueous ammonia}) \leftrightarrow NH_3(\text{free ammonia}) + H^+ \tag{12}$$

At mesophilic temperature with a pH of 7, approximately 1.25% of TAN converts to FA, while for the same temperature and a pH of 8, approximately 11.25% TAN transforms to FA; suggesting FA to be ten times more toxic to methanogens at a pH of 8 than at a pH of 7 [221]. Generally, a FA concentration between 1.7 and 14 gN/L is inhibitory to the methanogens, mainly to the acetoclastic species [219], which results in 50% or more reduction in methane yield [222], and in the worst case, the process will collapse. The degree at which methanogens are affected by the level of ammonia differs across several bacteria types and environmental conditions. An investigation by Niu et al. [223] observed that acetoclastic methanogens are more sensitive than hydrogenotrophic methanogens, while other researchers [224] observed the opposite, where hydrogenotrophic methanogens were found to be less tolerant. Some species of both acetoclastic and hydrogenotrophic methanogens were observed to tolerate TAN at levels higher than 10 g/L [225]. At elevated ammonia levels, for example, with a concentration greater than 3 g/L, the shift of acetoclastic methanogens to syntrophic acetate oxidation was observed that consequently led to a two-fold decrease in methane yield [226]. With regard to environmental conditions, thermophilic methanogens are generally deemed to tolerate higher levels of free ammonia than the mesophilic methanogens [227], but the process was found to be unstable and easily prone to inhibition [228]. Therefore, to obtain an optimum AD, a careful choice of process condition, particularly operating temperature, pH, type of inoculum, and feedstock, are necessary, so the level of free ammonia, as per recommendation from the published literature, is kept below 0.2 g/L [229]. Lowering FA and overcoming ammonia toxicity is achieved by employing various physical and chemical strategies including air stripping [230], use of zeolite [231], membrane filtration [232], diluting substrates [229], post-treatment of AD effluent [233], bioaugmentation [234], anaerobic ammonia oxidation [235], ultrasonication [235], microwave irradiation [236], struvite precipitation [237], and C/N ratio adjustment [238], all of which are comprehensively reviewed in a recent paper authored by Krakat et al. [213]. A novel ammonia recovery system using vacuum thermal stripping, an acid absorption process developed by the authors Ukwuani and Tao [239], is shown in Figure 8.

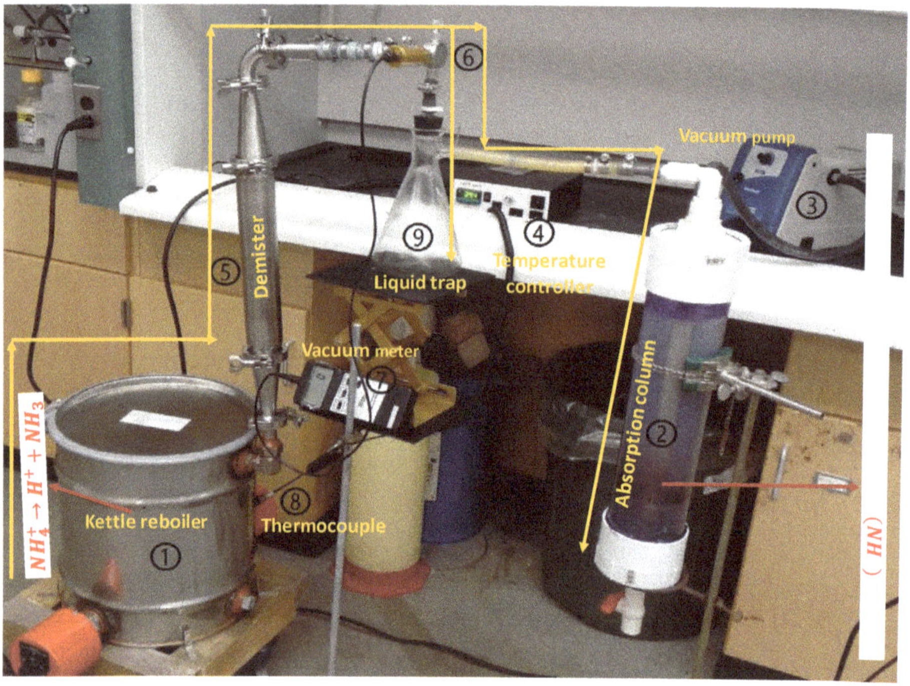

Figure 8. Innovative vacuum thermal stripping–acid absorption process for ammonia recovery—reproduced with permission from the original authors [239].

3.7. VFA

VFAs, mainly composed of acetic acid/acetate, propionic acid/propionate, butyric acid/butyrate, valeric acid/valerate, caproic acid/caproate, and enanthic acid/enanthate [240], are the essential intermediates during the anaerobic digestion process. The majority of the VFAs degrade to acetate and further to methane via methanogenesis through acetoclastic and hydrogenotrophic pathways (see Figure 2). In an AD process, a higher VFA conversion efficiency is usually desirable, as this allows the achievement of greater stability in methane production and balance in different production stages.

During the lifetime of an AD plant, it can suffer from nonoptimal VFA formation because of various operational disturbances caused by the feedstock organic loading rate, pH, temperature, and H_2 partial pressure change. Accumulated VFAs, depending on the levels and types, can be toxic and in extreme cases inhibitory to the process. For example, the development of propionic acid, in excess of levels as low as 1 g/L, can be extremely detrimental to methanogens, resulting in reduced or stagnated production of methane [241]. Other VFAs, such as butyrate and acetate, although not as potent as propionic acid, can cause process collapse when the concentration reaches to 10 g/L or more [241]. Therefore, for a well-controlled AD system, formation of a high concentration of VFAs of any type is not recommended, as this leads to decreases in the potential production of methane and consequently the diminished AD efficiency. Several measures have been proposed to enable accelerated conversion of VFAs into methane and discussed in the previously cited research articles. Among these, the selection of an appropriate reactor type, operating temperature, pH, hydrogen partial pressure, organic loading rate, and chemical additives are widely considered to be essential [242].

The selection of an appropriate reactor design can play a pivotal role in determining VFA conversion. For example, multistage reactors over single stage reactors can allow establishing operating conditions simultaneously suitable for optimal interaction of different groups of bacteria in individual AD phases, resulting in a balance in exchange of products and reactants, including increased conversion of VFAs. Despite this, due to the design complexity and high capital cost, the two-stage reactors are not preferred in many real applications and, alternatively, the design optimization of a single stage reactor towards maximizing VFA reduction have been given special attention. Installing baffles at various levels of a granular bed reactor [243] and employing membrane bioreactors [244] are some of the examples identified to have promising results in accelerated degradation of VFAs. Nevertheless, the future research on either reducing the cost of a multistage reactor or incorporating more advanced features to single stage reactors on optimal utilization of VFAs for improving AD efficiency is desirable.

Temperature is another variable having a substantial impact on VFA production and utilization. Since the presence and growth rate of microbial species are highly influenced by the level of temperature changes, their intercommunication that leads to the production and conversion of VFAs also changes. The effect of various temperature ranges (psychrophilic & mesophilic; 4–20 °C & 20–50 °C, respectively) on both production and composition of VFAs during hydrolysis have been studied, and it has been observed that with the rise in temperature, the rate of hydrolysis increases, resulting in increased solubility of carbohydrates and proteins, and consequently an accelerated production of VFAs. However, this temperature change did not have any noticeable impact on the VFA composition change with only some reductions in acetate production from 55–43% corresponding to the temperature rise between 4 and 14 °C [245]. Another study [246] investigated the change in temperature on the conversion of different VFA components. Here, the degree of conversion on specific VFA components (i.e., acetate, propionate and butyrate), as a result of sudden temperature change (from 63–55 °C) was examined, and the results suggested that the sudden increase in temperature influenced methanogenic populations resulting in unconverted (partial) propionate. However, as soon as the reactors were exposed back to the original temperature (55 °C), microorganisms adapted faster, giving increased methane yield followed by an improved propionate conversion extent. Some past researchers also studied temperature changes in combination with feedstock pretreatment on conversion of VFAs [247]. Their results, although inconsistent, reported that the extent of VFA conversion is a function of both temperature, feedstock and the given pretreatment technique, as one or all of these parameters directly influences the population and the type of microbial culture, dictating the way at which VFAs are produced and utilized towards biogas production. Identifying the correlation between microbial growth patterns and their mechanism towards VFA production and degradation will, therefore, be an interesting future research topic in this direction.

In addition to temperature, the pH in an anaerobic reactor can also be an indication of the status of VFAs. Varying pH over the range of extreme acidic (less than 3.0) and extreme alkaline (above 12.0) on accumulated VFAs was examined, and it was found that the VFA buildup reached their peak once the pH went to 10. A similar phenomenon was observed by another study where the SFAs (short-chain fatty acids) were found peaking when pH ranged between 8.0–10.0 compared to that of 3.0–7.0 [248]. This was thought to be caused by the fact that at an alkaline pH, there was higher availability of soluble proteins and carbohydrates than those at a neutral or acidic pH. In contrast, a study dealing with chicken manure with a high TS as a substrate found the VFA formation to be an inverse function of pH change. Here, the authors [216] performed further analysis on accumulated VFA types, and observed that the unutilized VFAs are mainly composed of unionized VFA types, called U-VFAs as calculated based on the total VFA concentration using Equation (13) below:

$$\mathrm{U-VFA} = \mathrm{VFA}\ \left(10^{(\mathrm{pK_a}\ -\ \mathrm{p^H})}\right) / \left(1 + 10^{(\mathrm{pK_a} - \mathrm{p^H})}\right) \tag{13}$$

where the dissociation constant values of acids in water (K_a) at 353 °C were taken from Weast (1987).

These VFAs, unlike ionized VFAs, are characteristically more toxic to methanogenesis and strongly linked with pH variation. The past study by [216] observed that U-VFA concentration as low as 10 mg/L in acetic acid was already a cause of inhibition to methanogenesis. However, the research on U-VFA and its role on various anaerobic digestion aspects are scarce, and its impact on the overall AD optimization and effectivity have no solid proof yet.

The partial pressure of H_2 has a direct impact on the interaction of several groups of bacteria that exchange VFAs. For example, the syntrophic acetate oxidation, symbiotically linked to the conversion of acetate to methane, is thermodynamically hindered (when Gibbs free energy of this reaction goes towards positive) if the partial pressure of H_2 exceeds the acceptable operating limit (i.e., 74 Pa [249]). Also, the propionate degradation can be severely reduced as a result of H_2 partial pressure rise [250], and, in the worst case, the process can stall when Gibbs free energy for this conversion becomes more positive [251]. The relationship between H_2 partial pressure and the interaction between H_2-producing and H_2-consuming anaerobes towards influencing individual VFA component are complex and are not clearly understood. Nonetheless, the research on H_2 partial pressure manipulation by employing various external techniques is ongoing [249], and some promising results in this line are already achieved.

As with the factors above, the organic loading rate (OLR) is another parameter strongly correlated with the VFA yield and its conversion. Independent of the feedstock type, a high OLR in general leads to VFA accumulation as a result of an imbalance between the growth rates of methanogenic archaea and VFA producing bacteria and speed of methane and VFA producing bacteria. Also, a high OLR can contribute to changes in VFA composition from low molecular weight ones to the high molecular weight ones, for example, from acetic acid to n-butyric acid [244], which eventually can promote methane inhibition. To achieve improved VFA utilization, a sophisticated OLR adjustment with reduced or withdrawn feeding were proved to be beneficial in previous research [118,244,252]. However, the range of optimal OLR values corresponding to the reactor and feedstock type, microbial adaptation pattern, and methane yield remain unexplored.

3.8. OLR

The amount of raw-material added to an anaerobic digester per day per unit volume is called the organic loading rate, and is typically expressed as Equation (14):

$$\mathrm{OLR} = \frac{\mathrm{C}}{\mathrm{HRT}} \tag{14}$$

where C is the feed concentration in g·VS/L, and HRT is the hydraulic retention time.

OLR is an essential controlling parameter in AD, and its deliberate variation can determine the degree of digestion for a broad level of influent input. In a typical anaerobic digester application, a high operating OLR is usually preferred, as this allows enriched bacterial species, reduced reactor sizes, lowered heating requirement and a reduced investment cost [253]. Several reactor configurations have been investigated to achieve a high OLR, wherein the primary aim was to reduce HRT either by adding water to the substrate or to recirculate digestate back to the main reactor [254]. By lowering HRT of a lab-scale spiral automatic circulation (SPAC) bioreactor, a high OLR equaling to 300 kg/m^3/d [144] was successfully accomplished. Another reactor type, up-flow anaerobic fixed bed (UAFB), was also reported to provide high operational OLR, but the maximum OLR level for these configurations was reported to be much lower than the SPAC type. However, a lowered HRT may lead to microbial washout and eventually digester failure. Also, there are possibilities of VFA accumulation [255] as the anaerobic digester operates either with higher OLR or forced to run with a lower HRT. Generally, methanogenic reactors were found to have higher stability at a somewhat higher HRT, typically between 10 and 25 days for CSTR type configurations. Nevertheless, Zhang et al. [254] recently used a novel feeding strategy where OLR was kept constant while HRT was simultaneously reduced. This results in a reduction in ammonia nitrogen inhibition with concomitant improvement in methane yield. Various other approaches of OLR optimization leading to enhanced biogas production are continually being researched. Among others, microbial management [256] and use of additives [90] are gaining constant popularity. Table 5 lists a range of OLR applied in recent AD publications (during the year 2018) on a number of feedstocks including agricultural residue [257], algae [258], energy crops [259], industrial wastes [260,261], fish waste [262], food waste [263–266], horticulture waste [267], manure [268], MSW [269], and sludge [270,271].

Table 5. A range of OLR variations used in selected recent publications on anaerobic digestion.

OLR	CH_4 Yield	Aim	Feedstock	Reference
0.4 to 3.1 kg COD/m^3	Maximum (0.46 $LCH_4/gCOD_{removed}$) at OLR of 2.5 kg COD/m^3	Co-digestion	Rice straw and pig manure	[257]
1 to 4 gVS/L/d for methane reactor; 3 to 12 gVS/L/d for H_2 reactor	Methane production maximized at OLR of 2 gVS/L/d and thereafter decreased. H_2 production maximized at OLR of 6 gVS/L/d	Co-production of H_2 and CH_4	Macro-algae *Laminaria digitata* and micro-algae *Arthrospira plantensis*	[258]
30, 60 and 90 gVS/L	Methane yield for all the co-digestion types maximized at OLR of 30 gvs/l and 60 gVS/L	Co-digestion performance	Sweet potato vine and animal manure	[259]
2.5 to 27.7 gVS/L	Ammonia inhibition at OLR > 20 gVS/L	Ammonia inhibition	Tannery fleshing, municipal solid waste, chrome shaving and others	[260]
0.4 to 0.7 gCOD/L/d	Methane yield decreased with increased OLR	Pilot scale two stage AD	Slaughter house waste	[261]
1.5 to 4.3 g/L/d	Maximum methane yield at OLR of 3.5 g/L/d	Methane production by ammonium tolerant microorganisms	Protein rich fish silage	[262]
1, 2 & 3 gVS/L/d	70% and 73% reduction of SMY and SCOD for OLR increment from 1 to 3 gVS/L/d	Semi-continuous AD at different psychrophilic range	Food waste	[263]
Various	Specific gas production (0.88 m^3biogas/kg_{vs})at two stage reactor was found higher than that of (0.75 m^3 biogas/kg_{vs})single stage reactor for an optimum OLR of 3.5 $kg_{vs}/m^3/d$	Comparison between single and two stage reactor performance	Food waste	[264]
2.4 and 3.6 gCOD/d	Higher OLR led to reactor's acidification problem and hence affected methane yield	Performance and metagenomics analyses of single and two stage thermophilic anaerobic digestion	Cheese wastes	[265]
4.6 and 8.6 $kgCOD/m^3/d$	The maximum methane productivity peaked to 2.78 L/L/d at OLR of 8.6 $kgCOD/m^3/d$, but the system was unstable	Effect of feeding with or without dilution	Food waste	[266]
2.0 to 6.0 gVS/L/d	Methane yield decreased as OLR increased for both two-stage and co-digestion reactors	Comparison between two-stage and co.digestion AD	Food waste and horticulture waste	[267]
1.53 to 5.04 gVS/L	0.44 LCH_4/gVS at OLR of 5.04 gVS/L	Determination of kinetics constant	Co-digestion of cattle manure and municipal food waste	[268]
Reactor ASBR: 0.93–25.0 gCOD/L.d Reactor AMBR: 1.04–19.65 gCOD/L.d	Maximum biogas yield at OLR of 10.08 gCOD/L.d, Biogas production decreased for OLR > 18.52 gCOD/L.d	Effect of OLR and series reactor AD	Composting leachate	[269]
0.5, 1.5 and 2.0 $VS/L_{sludge}/d$	H_2 uptake by homoacetogens increased at higher OLR resulting acetate accumulation	Acetate concentration during in situ methane upgradation	Sludge and H_2; fluromethane as inhibitor	[270]
1.12 to 3.88 $kgCOD/m^3/d$	Methane yield continued to increase up to OLR of 2 $kgCOD/m^3/d$. Methane production inhibited at OLR > 3.8 $kgCOD/m^3/d$	Co-digestion	Beverage waste and sewage sludge	[271]

3.9. Pressure

The effect of pressure on anaerobic digestion has not been studied extensively. Although anaerobic digestion typically occurs at atmospheric pressure, due to the accumulation and exchange of different gases into the reactor headspace, over or under pressure on a liquid surface can be developed. Based on the previous findings, it was reported that lower pressures on liquid surfaces resulted in higher biogas yields, as CH_4 solubility increases with pressure [272]. In an earlier study [273], the pressure change effect of various gas species was investigated using reaction models. The results showed that the partial pressure increase of CO_2 led to increasing the digester pH level, which consequently lowered the nonionized hydrogen sulfide concentration, reducing gas toxicity. Additionally, the hydrostatic pressure level within the digester can also affect the production of methane. Past experiments [274] observed that the methanogenesis activities maximized at digester depth of 4–5 m (400–500 mm H_2O) for a digester of over 10 m height. By reducing the height of this digester (i.e., making it horizontally oriented), a reduced hydrostatic pressure can be established, resulting in improved methanogenesis.

Anaerobic reactors working with high pressures (i.e., pressurized reactors), have been developed and investigated in lab-scale studies. As high pressure increases CO_2 solubility, the biogas released from the anaerobic digester partially strips of CO_2, giving net rise in CH_4 concentration. By using a high pressure within an anaerobic digester, a methane level of above 95% was achieved [275]. However, in some cases, elevated pressures in an anaerobic reactor did not produce a satisfactory improvement in biogas production. Technical challenges associated with leakages in the reactor system, pH decrement, and investment costs are identified as a few of the drawbacks of the pressurized reactor concept.

4. Conclusions

Optimizing operational parameters towards increasing methane yield has been and will be the most adopted technique within the application of the anaerobic digestion process in both industrial and lab-based studies. The role of individual and multiple parameters combined in different aspects of anaerobic digestion has been explored in the present review and the following stand out points, as conclusions, are expressed:

(1) Feedstock physical and chemical compositions substantially affect biogas production. Among the various types of feedstock materials, animal manure is still the dominant substrate or cosubstrate for biogas production because of its operational advantages of pH buffering and C:N ratio optimization. Lignin-rich substrates are found to be recalcitrant, while lipids are expected to have a high potential to boost methane. Nevertheless, LCFA inhibition from lignin remains to be of concern. Low or no lignin feedstock such as algae is an interesting biogas yield promoter. However, LCFA inhibition, increased pH level, and ammonia inhibition are highlighted as some of the barriers. For counteracting feedstock induced operational problems, codigestion, pretreatment, and use of additives are utilized in current R&D and real-life applications.

(2) Enhancing feedstock accessibility allows accelerated biological degradation and consequently high AD efficiency. Pretreatment and codigestion are broadly used options promoting feedstock accessibility. However, the choice of the pretreatment method is feedstock-dependent and often is a compromise between cost and energy. For pretreatment of lignocellulosic biomass (e.g., animal manure), by overcoming the recalcitrant lignin or crystalline cellulose barrier, the biogas production can be enhanced where approaches like steam explosion, enzyme addition, and sonication at present are widespread. For substrates with high-fat content, saponification is preferably used, while for algae-like substrates, thermal pretreatment is considered an option. Despite this, almost all the conventional pretreatment methods have both success and failure, as some pretreatment options are easily amenable, while others have side-effects that counteract their positive effects. Within the novel pretreatment approaches, a combination of various pretreatment methods has been examined, and the obtained results are reported to be promising. Additionally, as for future studies, exploiting genomic sequencing as a means of understanding feedstock degradation prior to anaerobic digestion is suggested.

(3) Manipulating reactor designs for achieving an optimum AD process performance has been emphasized, revealing many innovative approaches currently in practice. In terms of continuous operation, staged reactors give substantial increases in methane yields due to the establishment

of appropriate microbiological conditions at different anaerobic digestion phases. Consequently, three-staged reactor configurations have been developed and reported to be an attractive option in optimizing methane production. Moreover, an anaerobic membrane reactor, internal circulation reactor, and super-high-rate reactor are some of the novel configurations shown to facilitate efficient high solid substrate treatment by increasing bacterial cells containment and separating simultaneous removal of gases, solids, and liquids. Therefore, these reactor approaches are becoming increasingly popular.

(4) Among the various operational temperature regimes (psychrophilic, mesophilic, and thermophilic), the choice of an appropriate regime is largely investment and geographic specific. Thermophilic temperature enhances methane conversion rate and controls pathogens in the digestate liquids. However, the high heat requirement makes its application expensive. The recent investigations towards an optimum temperature system suggest that multitemperature, staged-digesters offer suitable conditions for diverse microbial activities, and hence give high biogas production efficiencies. By employing high-temperature post-treatment of digestate, further improvement in methane production has already been demonstrated.

(5) The pH affects the degree of conversion at different AD steps and the quality of the residual digestate. Optimal methanogenesis and biogas production occurs at around pH 7. A host of factors, such as ammonium formation, bicarbonate decomposition, mineralization and reduction of multivalent ions, and struvite formation result in digestate pH fluctuations. To enable pH controlling to the desired value, adding acid or basic solutions are traditionally the major options. Moreover, online pH monitoring and alert systems are also implemented in modern day applications.

(6) HRT is directly linked to the size of the anaerobic reactor, and a low HRT usually allows investment reduction. Among the all bacteria and archaea, methanogens grow slowly, and for these microorganisms, a higher HRT is required. Both feed rate and feed type influence the HRT. Consequently, the feedstock OLR regulation is a usual approach for HRT optimization. SRT, another term intertwined with HRT, represents the microbial culture and biomass retention in an operating digester. SRT is often optimized by incorporating OLR variation. A recent achievement suggests that a deliberate SRT variation can result in a shift in bacterial culture, causing a change in reaction pattern from one group to the other.

(7) Ammonia is produced via the degradation of proteins and nitrogen in the feedstock. Among the two forms of ammonia (free ammonia and ionized ammonia), free ammonia is very toxic to methanogens and is a strong function of combined pH and temperature. Lowering the generation of ammonia during AD has been targeted in a variety of approaches (see Section 3.6), among which, using ammonia tolerant microorganisms (or bio-augmentation) or combined thermal stripping and absorption process are some of the novel techniques receiving constant attention.

(8) VFAs are the intermediate products required for conversion to methane. Some of the VFA components are more sensitive than the others to methanogens, e.g., propionate. Unutilized VFAs accumulate, and in the worst case halt the production of biogas. To achieve increased VFA utilization and as such improved methane yield, regulation of AD process parameters such as temperature, OLR, pH, and H_2 partial pressure is critical. Moreover, many additives and trace metals of various origins have been suggested to have an improved utilization of VFAs.

(9) OLR refers to the amount of feedstock treated by a reactor on a daily basis. OLR variation allows the optimization of HRT, pH, VFA, ammonia, and methane production. High operational OLR enables reducing the size of the reactor and accordingly the investment cost. However, as a result of high OLR, implications such as bacteria wash out, VFA accumulation, or methane yield reduction can be experienced. In recent applications, OLR control has been used to suppress ammonia inhibition. Furthermore, a novel three-stage reactor configuration has been shown to successfully achieve high OLR treatment without compromising the production of biogas.

(10) Anaerobic digestion operates typically at atmospheric pressure, but recent investigations have identified that high-pressure systems are also possible. High pressure allows the increase of dissolved CO_2 in the liquid phase and consequently increased methane composition in the biogas. The other aspects of anaerobic digester pressure, such as partial pressure of headspace gas components and variation of hydrostatic pressure levels, were mentioned as potential causes of fluctuation in the production of methane.

Supplementary Materials: The following are available online at http://www.mdpi.com/2076-3417/9/9/1915/s1, Figure S1: Artwork Figures Revised, Table S1: Table Artwork Tables revised.

Author Contributions: Conceptualization, S.S.; Formal analysis, S.S., J.J.L.; Investigation, S.S.; Resources, D.R.H., K.M.L.; Writing—original draft preparation, S.S.; Writing—review and editing, S.S., J.J.L., K.L; Visualization, S.S.; Supervision, D.R.H., K.M.L.; Project administration, K.M.L., D.R.H.; Funding acquisition, K.M.L., D.R.H.

Funding: This research received no external funding. The APC was funded by the NTNU publication fund.

Acknowledgments: The authors gratefully acknowledge Norwegian University of Science and Technology (NTNU) for the internal funding of this study under the framework of 'ENERSENSE (Energy, sensor and storage technology)'.

Conflicts of Interest: The authors declare no conflict of interest.

References

1. Daioglou, V.; Doelman, J.C.; Wicke, B.; Faaij, A.; van Vuuren, D.P. Integrated assessment of biomass supply and demand in climate change mitigation scenarios. *Glob. Environ. Chang.* **2019**, *54*, 88–101. [CrossRef]
2. Batstone, D.J.; Virdis, B. The role of anaerobic digestion in the emerging energy economy. *Curr. Opin. Biotechnol.* **2014**, *27*, 142–149. [CrossRef]
3. Toop, T.A.; Ward, S.; Oldfield, T.; Hull, M.; Kirby, M.E.; Theodorou, M.K. AgroCycle–developing a circular economy in agriculture. *Energy Procedia* **2017**, *123*, 76–80. [CrossRef]
4. Simet, A. German biogas industry adds 150 plants in 2015. *Biomass Magazine*, 2016.
5. Zhang, T.; Yang, Y.; Xie, D. Insights into the production potential and trends of China's rural biogas. *Int. J. Energy Res.* **2015**, *39*, 1068–1082. [CrossRef]
6. Gu, L.; Zhang, Y.-X.; Wang, J.-Z.; Chen, G.; Battye, H. Where is the future of China's biogas? Review, forecast, and policy implications. *Pet. Sci.* **2016**, *13*, 604–624. [CrossRef]
7. Zhang, J.; Loh, K.-C.; Lee, J.; Wang, C.-H.; Dai, Y.; Tong, Y.W. Three-stage anaerobic co-digestion of food waste and horse manure. *Sci. Rep.* **2017**, *7*, 1269. [CrossRef] [PubMed]
8. Dollhofer, V.; Podmirseg, S.M.; Callaghan, T.M.; Griffith, G.W.; Fliegerová, K. Anaerobic fungi and their potential for biogas production. In *Biogas Science and Technology*; Springer: Berlin, Germany, 2015; pp. 41–61.
9. Sasidharan, K.; Martinez-Vernon, A.S.; Chen, J.; Fu, T.; Soyer, O. A low-cost DIY device for high resolution, continuous measurement of microbial growth dynamics. *bioRxiv* **2018**, 407742. [CrossRef]
10. Hublin, A.; Zokić, T.; Zelić, B. Optimization of biogas production from co-digestion of whey and cow manure. *Biotechnol. Bioprocess Eng.* **2012**, *17*, 1284–1293. [CrossRef]
11. Ward, A.J.; Hobbs, P.J.; Holliman, P.J.; Jones, D.L. Optimisation of the anaerobic digestion of agricultural resources. *Bioresour. Technol.* **2008**, *99*, 7928–7940. [CrossRef]
12. Mital, K. *Biogas Systems: Policies, Progress and Prospects*; Taylor & Francis: Abingdon, UK, 1997.
13. Kim, M.D.; Song, M.; Jo, M.; Shin, S.G.; Khim, J.H.; Hwang, S. Growth condition and bacterial community for maximum hydrolysis of suspended organic materials in anaerobic digestion of food waste-recycling wastewater. *Appl. Microbiol. Biotechnol.* **2010**, *85*, 1611–1618. [CrossRef]
14. Schnurer, A.; Jarvis, A. Microbiological handbook for biogas plants. *Swed. Waste Manag. U* **2010**, *2009*, 1–74.
15. Angelidaki, I.; Karakashev, D.; Batstone, D.J.; Plugge, C.M.; Stams, A.J. Biomethanation and its potential. *Methods Enzym.* **2011**, *494*, 327–351.
16. Nayono, S.E. *Anaerobic Digestion of Organic Solid Waste for Energy Production*; KIT Scientific Publishing: Karlsruhe, Germany, 2010; Volume 46.
17. EInstruments. Biomass to Biogas—Anaerobic Digestion. Available online: http://www.e-inst.com/biomass-to-biogas/ (accessed on 13 March 2019).

18. Felchner-Zwirello, M. *Propionic Acid Degradation by Syntrophic Bacteria during Anaerobic Biowaste Digestion*; KIT Scientific Publishing: Karlsruhe, Germany, 2014; Volume 49.
19. Gerardi, M.H. *The Microbiology of Anaerobic Digesters*; John Wiley & Sons: Hoboken, NJ, USA, 2003.
20. André, L.; Ndiaye, M.; Pernier, M.; Lespinard, O.; Pauss, A.; Lamy, E.; Ribeiro, T. Methane production improvement by modulation of solid phase immersion in dry batch anaerobic digestion process: Dynamic of methanogen populations. *Bioresour. Technol.* **2016**, *207*, 353–360. [CrossRef]
21. Steffen, R.; Szolar, O.; Braun, R. *Feedstocks for Anaerobic Digestion*; Institute for Agrobiotechnology Tulln, University of Agricultural Sciences: Vienna, Austria, 1998.
22. Achinas, S.; Euverink, G.J.W. Theoretical analysis of biogas potential prediction from agricultural waste. *Resour.-Effic. Technol.* **2016**, *2*, 143–147. [CrossRef]
23. Martinez, E.; Redondas, V.; Fierro, J.; Gómez, X.; Morán, A. Anaerobic digestion of high lipid content wastes: FOG co-digestion and milk processing fat digestion. *J. Residuals Sci. Technol.* **2011**, *8*, 53–60.
24. Pereira, M.; Sousa, D.; Mota, M.; Alves, M. Mineralization of LCFA associated with anaerobic sludge: Kinetics, enhancement of methanogenic activity, and effect of VFA. *Biotechnol. Bioeng.* **2004**, *88*, 502–511. [CrossRef]
25. Leung, D.Y.; Wang, J. An overview on biogas generation from anaerobic digestion of food waste. *Int. J. Green Energy* **2016**, *13*, 119–131. [CrossRef]
26. Rajagopal, R.; Massé, D.I.; Singh, G. A critical review on inhibition of anaerobic digestion process by excess ammonia. *Bioresour. Technol.* **2013**, *143*, 632–641. [CrossRef]
27. Long, J.H.; Aziz, T.N.; Francis, L.; Ducoste, J.J. Anaerobic co-digestion of fat, oil, and grease (FOG): A review of gas production and process limitations. *Process Saf. Environ. Prot.* **2012**, *90*, 231–245. [CrossRef]
28. Latha, K.; Velraj, R.; Shanmugam, P.; Sivanesan, S. Mixing strategies of high solids anaerobic co-digestion using food waste with sewage sludge for enhanced biogas production. *J. Clean. Prod.* **2019**, *210*, 388–400. [CrossRef]
29. Xiao, B.; Zhang, W.; Yi, H.; Qin, Y.; Wu, J.; Liu, J.; Li, Y.-Y. Biogas production by two-stage thermophilic anaerobic co-digestion of food waste and paper waste: Effect of paper waste ratio. *Renew. Energy* **2019**, *132*, 1301–1309. [CrossRef]
30. Sahoo, S.R.; Rao, P.V. Temperature-Phased Anaerobic Co-digestion of Food Waste and Rice Husk Using Response Surface Methodology. In *Water Resources and Environmental Engineering II*; Springer: Berlin, Germany, 2019; pp. 137–146.
31. Aldin, S.; Nakhla, G.; Ray, M.B. Modeling the influence of particulate protein size on hydrolysis in anaerobic digestion. *Ind. Eng. Chem. Res.* **2011**, *50*, 10843–10849. [CrossRef]
32. Moestedt, J.; Påledal, S.N.; Schnürer, A.; Nordell, E. Biogas production from thin stillage on an industrial scale—Experience and optimisation. *Energies* **2013**, *6*, 5642–5655. [CrossRef]
33. Zhang, L.; Jahng, D. Enhanced anaerobic digestion of piggery wastewater by ammonia stripping: Effects of alkali types. *J. Hazard. Mater.* **2010**, *182*, 536–543. [CrossRef]
34. Karlsson, A.; Ejlertsson, J. Addition of HCl as a means to improve biogas production from protein-rich food industry waste. *Biochem. Eng. J.* **2012**, *61*, 43–48. [CrossRef]
35. Feedstock. The Official Innovation Portal on Anaerobic Digestion. Available online: https://www.nnfcc.co.uk/publications/report-impact-biogas-sustainability-criteria (accessed on 13 March 2019).
36. Teghammar, A. *Biogas Production from Lignocelluloses: Pretreatment, Substrate Characterization, Co-Digestion and Economic Evaluation*; Chalmers Tekniska Högskola: Göteborg, Sweden, 2013.
37. Dębowski, M.; Zieliński, M.; Grala, A.; Dudek, M. Algae biomass as an alternative substrate in biogas production technologies—Review. *Renew. Sustain. Energy Rev.* **2013**, *27*, 596–604. [CrossRef]
38. Barbot, Y.N.; Al-Ghaili, H.; Benz, R. A Review on the Valorization of Macroalgal Wastes for Biomethane Production. *Mar. Drugs* **2016**, *14*, 120. [CrossRef]
39. Biller, P.; Ross, A.B.; Skill, S.C.; Lea-Langton, A.; Balasundaram, B.; Hall, C.; Riley, R.; Llewellyn, C.A. Nutrient recycling of aqueous phase for microalgae cultivation from the hydrothermal liquefaction process. *Algal Res.* **2012**, *1*, 70–76. [CrossRef]
40. Chinnasamy, S.; Bhatnagar, A.; Hunt, R.W.; Das, K.C. Microalgae cultivation in a wastewater dominated by carpet mill effluents for biofuel applications. *Bioresour. Technol.* **2010**, *101*, 3097–3105. [CrossRef]
41. Elliott, D.C.; Hart, T.R.; Schmidt, A.J.; Neuenschwander, G.G.; Rotness, L.J.; Olarte, M.V.; Zacher, A.H.; Albrecht, K.O.; Hallen, R.T.; Holladay, J.E. Process development for hydrothermal liquefaction of algae feedstocks in a continuous-flow reactor. *Algal Res.* **2013**, *2*, 445–454. [CrossRef]

42. Ma, J.; Zhao, Q.-B.; Laurens, L.L.; Jarvis, E.E.; Nagle, N.J.; Chen, S.; Frear, C.S. Mechanism, kinetics and microbiology of inhibition caused by long-chain fatty acids in anaerobic digestion of algal biomass. *Biotechnol. Biofuels* **2015**, *8*, 141. [CrossRef]
43. Peu, P.; Sassi, J.-F.; Girault, R.; Picard, S.; Saint-Cast, P.; Béline, F.; Dabert, P. Sulphur fate and anaerobic biodegradation potential during co-digestion of seaweed biomass (*Ulva* sp.) with pig slurry. *Bioresour. Technol.* **2011**, *102*, 10794–10802. [CrossRef]
44. Nielsen, H.B.; Heiske, S. Anaerobic digestion of macroalgae: Methane potentials, pre-treatment, inhibition and co-digestion. *Water Sci. Technol.* **2011**, *64*, 1723–1729. [CrossRef]
45. González-Fernández, C.; Sialve, B.; Bernet, N.; Steyer, J. Comparison of ultrasound and thermal pretreatment of Scenedesmus biomass on methane production. *Bioresour. Technol.* **2012**, *110*, 610–616. [CrossRef]
46. Tedesco, S.; Benyounis, K.; Olabi, A. Mechanical pretreatment effects on macroalgae-derived biogas production in co-digestion with sludge in Ireland. *Energy* **2013**, *61*, 27–33. [CrossRef]
47. Sarker, S.; Møller, H.B.; Bruhn, A. Influence of variable feeding on mesophilic and thermophilic co-digestion of Laminaria digitata and cattle manure. *Energy Convers. Manag.* **2014**, *87*, 513–520. [CrossRef]
48. Møller, H.B.; Sommer, S.G.; Ahring, B.K. Biological degradation and greenhouse gas emissions during pre-storage of liquid animal manure. *J. Environ. Qual.* **2004**, *33*, 27–36. [CrossRef] [PubMed]
49. Astals, S.; Nolla-Ardèvol, V.; Mata-Alvarez, J. Anaerobic co-digestion of pig manure and crude glycerol at mesophilic conditions: Biogas and digestate. *Bioresour. Technol.* **2012**, *110*, 63–70. [CrossRef] [PubMed]
50. Horváth, I.S.; Tabatabaei, M.; Karimi, K.; Kumar, R. Recent updates on biogas production—A review. *Biofuel Res. J.* **2016**, *10*, 394–402. [CrossRef]
51. Liu, T.; Sun, L.; Müller, B.; Schnürer, A. Importance of inoculum source and initial community structure for biogas production from agricultural substrates. *Bioresour. Technol.* **2017**, *245*, 768–777. [CrossRef]
52. Parra-Orobio, B.A.; Donoso-Bravo, A.; Ruiz-Sánchez, J.C.; Valencia-Molina, K.J.; Torres-Lozada, P. Effect of inoculum on the anaerobic digestion of food waste accounting for the concentration of trace elements. *Waste Manag.* **2018**, *71*, 342–349. [CrossRef] [PubMed]
53. Zhou, Y.; Li, C.; Nges, I.A.; Liu, J. The effects of pre-aeration and inoculation on solid-state anaerobic digestion of rice straw. *Bioresour. Technol.* **2017**, *224*, 78–86. [CrossRef]
54. Liu, G.; Zhang, R.; El-Mashad, H.M.; Dong, R. Effect of feed to inoculum ratios on biogas yields of food and green wastes. *Bioresour. Technol.* **2009**, *100*, 5103–5108. [CrossRef]
55. Sunarso, S.; Johari, S.; Widiasa, I.N.; Budiyono, B. The Effect of Feed to Inoculums Ratio on Biogas Production Rate from Cattle Manure Using Rumen Fluid as Inoculums. *Int. J. Sci. Eng.* **2010**, *1*, 41–45. [CrossRef]
56. González-Fernández, C.; García-Encina, P.A. Impact of substrate to inoculum ratio in anaerobic digestion of swine slurry. *Biomass Bioenergy* **2009**, *33*, 1065–1069. [CrossRef]
57. Fathya, S.; Assia, K.; Hamza, M. Influence of inoculums/substrate ratios (ISRs) on the mesophilic anaerobic digestion of slaughterhouse waste in batch mode: Process stability and biogas production. *Energy Procedia* **2014**, *50*, 57–63.
58. Gu, Y.; Chen, X.; Liu, Z.; Zhou, X.; Zhang, Y. Effect of inoculum sources on the anaerobic digestion of rice straw. *Bioresour. Technol.* **2014**, *158*, 149–155. [CrossRef] [PubMed]
59. Ferrer, I.; Ponsá, S.; Vázquez, F.; Font, X. Increasing biogas production by thermal (70 °C) sludge pre-treatment prior to thermophilic anaerobic digestion. *Biochem. Eng. J.* **2008**, *42*, 186–192. [CrossRef]
60. Bruni, E.; Jensen, A.P.; Angelidaki, I. Comparative study of mechanical, hydrothermal, chemical and enzymatic treatments of digested biofibers to improve biogas production. *Bioresour. Technol.* **2010**, *101*, 8713–8717. [CrossRef]
61. Montingelli, M.; Benyounis, K.; Quilty, B.; Stokes, J.; Olabi, A. Influence of mechanical pretreatment and organic concentration of Irish brown seaweed for methane production. *Energy* **2017**, *118*, 1079–1089. [CrossRef]
62. Tiehm, A.; Nickel, K.; Neis, U. The use of ultrasound to accelerate the anaerobic digestion of sewage sludge. *Water Sci. Technol.* **1997**, *36*, 121–128. [CrossRef]
63. Park, B.; Ahn, J.-H.; Kim, J.; Hwang, S. Use of microwave pretreatment for enhanced anaerobiosis of secondary sludge. *Water Sci. Technol.* **2004**, *50*, 17–23. [CrossRef]
64. Zhong, W.; Zhang, Z.; Luo, Y.; Sun, S.; Qiao, W.; Xiao, M. Effect of biological pretreatments in enhancing corn straw biogas production. *Bioresour. Technol.* **2011**, *102*, 11177–11182. [CrossRef]

65. Carrere, H.; Antonopoulou, G.; Affes, R.; Passos, F.; Battimelli, A.; Lyberatos, G.; Ferrer, I. Review of feedstock pretreatment strategies for improved anaerobic digestion: From lab-scale research to full-scale application. *Bioresour. Technol.* **2016**, *199*, 386–397. [CrossRef]
66. Bougrier, C.; Delgenes, J.-P.; Carrere, H. Combination of thermal treatments and anaerobic digestion to reduce sewage sludge quantity and improve biogas yield. *Process Saf. Environ. Prot.* **2006**, *84*, 280–284. [CrossRef]
67. Gavala, H.N.; Yenal, U.; Skiadas, I.V.; Westermann, P.; Ahring, B.K. Mesophilic and thermophilic anaerobic digestion of primary and secondary sludge. Effect of pre-treatment at elevated temperature. *Water Res.* **2003**, *37*, 4561–4572. [CrossRef]
68. Raju, C.S.; Sutaryo, S.; Ward, A.J.; Møller, H.B. Effects of high-temperature isochoric pre-treatment on the methane yields of cattle, pig and chicken manure. *Environ. Technol.* **2013**, *34*, 239–244. [CrossRef]
69. Marañón, E.; Castrillón, L.; Quiroga, G.; Fernández-Nava, Y.; Gómez, L.; García, M. Co-digestion of cattle manure with food waste and sludge to increase biogas production. *Waste Manag.* **2012**, *32*, 1821–1825. [CrossRef] [PubMed]
70. Ariunbaatar, J.; Panico, A.; Esposito, G.; Pirozzi, F.; Lens, P.N. Pretreatment methods to enhance anaerobic digestion of organic solid waste. *Appl. Energy* **2014**, *123*, 143–156. [CrossRef]
71. Menardo, S.; Airoldi, G.; Balsari, P. The effect of particle size and thermal pre-treatment on the methane yield of four agricultural by-products. *Bioresour. Technol.* **2012**, *104*, 708–714. [CrossRef]
72. Rodriguez, C.; Alaswad, A.; Benyounis, K.Y.; Olabi, A.G. Pretreatment techniques used in biogas production from grass. *Renew. Sustain. Energy Rev.* **2017**, *68*, 1193–1204. [CrossRef]
73. Schwede, S.; Rehman, Z.-U.; Gerber, M.; Theiss, C.; Span, R. Effects of thermal pretreatment on anaerobic digestion of Nannochloropsis salina biomass. *Bioresour. Technol.* **2013**, *143*, 505–511. [CrossRef] [PubMed]
74. Rafique, R.; Poulsen, T.G.; Nizami, A.-S.; Murphy, J.D.; Kiely, G. Effect of thermal, chemical and thermo-chemical pre-treatments to enhance methane production. *Energy* **2010**, *35*, 4556–4561. [CrossRef]
75. Carrère, H.; Sialve, B.; Bernet, N. Improving pig manure conversion into biogas by thermal and thermo-chemical pretreatments. *Bioresour. Technol.* **2009**, *100*, 3690–3694. [CrossRef] [PubMed]
76. Horn, S.J.; Estevez, M.M.; Nielsen, H.K.; Linjordet, R.; Eijsink, V.G. Biogas production and saccharification of Salix pretreated at different steam explosion conditions. *Bioresour. Technol.* **2011**, *102*, 7932–7936. [CrossRef] [PubMed]
77. Amin, F.R.; Khalid, H.; Zhang, H.; u Rahman, S.; Zhang, R.; Liu, G.; Chen, C. Pretreatment methods of lignocellulosic biomass for anaerobic digestion. *Amb Express* **2017**, *7*, 72. [CrossRef] [PubMed]
78. Vivekanand, V.; Olsen, E.F.; Eijsink, V.G.; Horn, S.J. Effect of different steam explosion conditions on methane potential and enzymatic saccharification of birch. *Bioresour. Technol.* **2013**, *127*, 343–349. [CrossRef]
79. Theuretzbacher, F.; Lizasoain, J.; Lefever, C.; Saylor, M.K.; Enguidanos, R.; Weran, N.; Gronauer, A.; Bauer, A. Steam explosion pretreatment of wheat straw to improve methane yields: Investigation of the degradation kinetics of structural compounds during anaerobic digestion. *Bioresour. Technol.* **2015**, *179*, 299–305. [CrossRef] [PubMed]
80. Valo, A.; Carrere, H.; Delgenes, J.P. Thermal, chemical and thermo-chemical pre-treatment of waste activated sludge for anaerobic digestion. *J. Chem. Technol. Biotechnol.* **2004**, *79*, 1197–1203. [CrossRef]
81. Izumi, K.; Okishio, Y.-K.; Nagao, N.; Niwa, C.; Yamamoto, S.; Toda, T. Effects of particle size on anaerobic digestion of food waste. *Int. Biodeterior. Biodegrad.* **2010**, *64*, 601–608. [CrossRef]
82. Mladenovska, Z.; Hartmann, H.; Kvist, T.; Sales-Cruz, M.; Gani, R.; Ahring, B.K. Thermal pretreatment of the solid fraction of manure: Impact on the biogas reactor performance and microbial community. *Water Sci. Technol.* **2006**, *53*, 59–67. [CrossRef]
83. Lin, Y.; Wang, D.; Wu, S.; Wang, C. Alkali pretreatment enhances biogas production in the anaerobic digestion of pulp and paper sludge. *J. Hazard. Mater.* **2009**, *170*, 366–373. [CrossRef]
84. Devlin, D.; Esteves, S.; Dinsdale, R.; Guwy, A. The effect of acid pretreatment on the anaerobic digestion and dewatering of waste activated sludge. *Bioresour. Technol.* **2011**, *102*, 4076–4082. [CrossRef] [PubMed]
85. Vervaeren, H.; Hostyn, K.; Ghekiere, G.; Willems, B. Biological ensilage additives as pretreatment for maize to increase the biogas production. *Renew. Energy* **2010**, *35*, 2089–2093. [CrossRef]
86. Bougrier, C.; Albasi, C.; Delgenès, J.-P.; Carrère, H. Effect of ultrasonic, thermal and ozone pre-treatments on waste activated sludge solubilisation and anaerobic biodegradability. *Chem. Eng. Process. Process Intensif.* **2006**, *45*, 711–718. [CrossRef]

87. Battimelli, A.; Torrijos, M.; Moletta, R.; Delgenès, J.P. Slaughterhouse fatty waste saponification to increase biogas yield. *Bioresour. Technol.* **2010**, *101*, 3388–3393. [CrossRef]
88. Ugwu, S.N.; Enweremadu, C.C. Effects of pre-treatments and co-digestion on biogas production from Okra waste. *J. Renew. Sustain. Energy* **2019**, *11*, 013101. [CrossRef]
89. Antonopoulou, G.; Gavala, H.N.; Skiadas, I.V.; Lyberatos, G. The effect of aqueous ammonia soaking pretreatment on methane generation using different lignocellulosic biomasses. *Waste Biomass Valorization* **2015**, *6*, 281–291. [CrossRef]
90. Romero-Güiza, M.; Vila, J.; Mata-Alvarez, J.; Chimenos, J.; Astals, S.J.R.; Reviews, S.E. The role of additives on anaerobic digestion: A review. *Renew. Sustain. Energy Rev.* **2016**, *58*, 1486–1499. [CrossRef]
91. Barakat, A.; Mayer-Laigle, C.; Solhy, A.; Arancon, R.A.; De Vries, H.; Luque, R. Mechanical pretreatments of lignocellulosic biomass: Towards facile and environmentally sound technologies for biofuels production. *RSC Adv.* **2014**, *4*, 48109–48127. [CrossRef]
92. Cheng, J.J.; Timilsina, G.R. Status and barriers of advanced biofuel technologies: A review. *Renew. Energy* **2011**, *36*, 3541–3549. [CrossRef]
93. Hjorth, M.; Gränitz, K.; Adamsen, A.P.; Møller, H.B. Extrusion as a pretreatment to increase biogas production. *Bioresour. Technol.* **2011**, *102*, 4989–4994. [CrossRef]
94. Mustafa, A.M.; Poulsen, T.G.; Xia, Y.; Sheng, K. Combinations of fungal and milling pretreatments for enhancing rice straw biogas production during solid-state anaerobic digestion. *Bioresour. Technol.* **2017**, *224*, 174–182. [CrossRef]
95. Peng, H.; Chen, H.; Qu, Y.; Li, H.; Xu, J. Bioconversion of different sizes of microcrystalline cellulose pretreated by microwave irradiation with/without NaOH. *Appl. Energy* **2014**, *117*, 142–148. [CrossRef]
96. Sridar, V. Microwave radiation as a catalyst for chemical reactions. *Curr. Sci.* **1998**, *74*, 446–450.
97. Diaz, A.B.; de Souza Moretti, M.M.; Bezerra-Bussoli, C.; Nunes, C.D.; Blandino, A.; da Silva, R.; Gomes, E. Evaluation of microwave-assisted pretreatment of lignocellulosic biomass immersed in alkaline glycerol for fermentable sugars production. *Bioresour. Technol.* **2015**, *185*, 316–323. [CrossRef] [PubMed]
98. Saha, B.C.; Biswas, A.; Cotta, M.A. Microwave pretreatment, enzymatic saccharification and fermentation of wheat straw to ethanol. *J. Biobased Mater. Bioenergy* **2008**, *2*, 210–217. [CrossRef]
99. Zhu, S.; Wu, Y.; Yu, Z.; Chen, Q.; Wu, G.; Yu, F.; Wang, C.; Jin, S. Microwave-assisted alkali pre-treatment of wheat straw and its enzymatic hydrolysis. *Biosyst. Eng.* **2006**, *94*, 437–442. [CrossRef]
100. Pilli, S.; Bhunia, P.; Yan, S.; LeBlanc, R.; Tyagi, R.; Surampalli, R. Ultrasonic pretreatment of sludge: A review. *Ultrason. Sonochemistry* **2011**, *18*, 1–18. [CrossRef]
101. Ormaechea, P.; Castrillón, L.; Marañón, E.; Fernández-Nava, Y.; Negral, L.; Megido, L. Influence of the ultrasound pretreatment on anaerobic digestion of cattle manure, food waste and crude glycerine. *Environ. Technol.* **2017**, *38*, 682–686. [CrossRef] [PubMed]
102. Carrère, H.; Dumas, C.; Battimelli, A.; Batstone, D.; Delgenès, J.; Steyer, J.; Ferrer, I. Pretreatment methods to improve sludge anaerobic degradability: A review. *J. Hazard. Mater.* **2010**, *183*, 1–15. [CrossRef]
103. Cesaro, A.; Naddeo, V.; Amodio, V.; Belgiorno, V. Enhanced biogas production from anaerobic codigestion of solid waste by sonolysis. *Ultrason. Sonochem.* **2012**, *19*, 596–600. [CrossRef]
104. Pérez-Elvira, S.; Fdz-Polanco, M.; Plaza, F.; Garralón, G.; Fdz-Polanco, F. Ultrasound pre-treatment for anaerobic digestion improvement. *Water Sci. Technol.* **2009**, *60*, 1525–1532. [CrossRef] [PubMed]
105. Tanjore, D.; Richard, T.L. A systems view of lignocellulose hydrolysis. In *Advances in Bioprocess Technology*; Springer: Berlin, Germany, 2015; pp. 387–419.
106. Schroyen, M.; Vervaeren, H.; Vandepitte, H.; Van Hulle, S.W.; Raes, K. Effect of enzymatic pretreatment of various lignocellulosic substrates on production of phenolic compounds and biomethane potential. *Bioresour. Technol.* **2015**, *192*, 696–702. [CrossRef] [PubMed]
107. Schimpf, U.; Hanreich, A.; Mähnert, P.; Unmack, T.; Junne, S.; Renpenning, J.; Lopez-Ulibarri, R. Improving the efficiency of large-scale biogas processes: Pectinolytic enzymes accelerate the lignocellulose degradation. *J. Sustain. Energy Environ.* **2013**, *4*, 53–60.
108. Jagadabhi, P.S.; Kaparaju, P.; Väisänen, A.; Rintala, J. Effect of macro-and micro-nutrients addition during anaerobic mono-digestion of grass silage in leach-bed reactors. *Environ. Technol.* **2019**, *40*, 418–429. [CrossRef]
109. Facchin, V.; Cavinato, C.; Fatone, F.; Pavan, P.; Cecchi, F.; Bolzonella, D. Effect of trace element supplementation on the mesophilic anaerobic digestion of foodwaste in batch trials: The influence of inoculum origin. *Biochem. Eng. J.* **2013**, *70*, 71–77. [CrossRef]

110. Jiang, Y.; Heaven, S.; Banks, C.J. Strategies for stable anaerobic digestion of vegetable waste. *Renew. Energy* **2012**, *44*, 206–214. [CrossRef]
111. Funk, T.; Gu, W.; Friedrich, S.; Wang, H.; Gencic, S.; Grahame, D.A.; Cramer, S.P. Chemically Distinct Ni Sites in the A-Cluster in Subunit β of the Acetyl-CoA Decarbonylase/Synthase Complex from Methanosarcina t hermophila: Ni L-Edge Absorption and X-ray Magnetic Circular Dichroism Analyses. *J. Am. Chem. Soc.* **2004**, *126*, 88–95. [CrossRef]
112. Chen, S.; Yakunin, A.F.; Kuznetsova, E.; Busso, D.; Pufan, R.; Proudfoot, M.; Kim, R.; Kim, S.-H. Structural and functional characterization of a novel phosphodiesterase from Methanococcus jannaschii. *J. Biol. Chem.* **2004**, *279*, 31854–31862. [CrossRef]
113. Kumar, A.; Miglani, P.; Gupta, R.; Bhattacharya, T. Impact of Ni (II), Zn (II) and Cd (II) on biogassification of potato waste. *J. Environ. Biol.* **2006**, *27*, 61–66. [PubMed]
114. Climenhaga, M.; Banks, C. Anaerobic digestion of catering wastes: Effect of micronutrients and retention time. *Water Sci. Technol.* **2008**, *57*, 687–692. [CrossRef] [PubMed]
115. Gonzalez-Gil, G.; Seghezzo, L.; Lettinga, G.; Kleerebezem, R. Kinetics and mass-transfer phenomena in anaerobic granular sludge. *Biotechnol. Bioeng.* **2001**, *73*, 125–134. [CrossRef]
116. Lehtomäki, A.; Huttunen, S.; Rintala, J. Laboratory investigations on co-digestion of energy crops and crop residues with cow manure for methane production: Effect of crop to manure ratio. *Resour. Conserv. Recycl.* **2007**, *51*, 591–609. [CrossRef]
117. Sarker, S.; Møller, H.B. Boosting biogas yield of anaerobic digesters by utilizing concentrated molasses from 2nd generation bioethanol plant. *Int. J. Energy Environ.* **2013**, *4*, 199–210.
118. Sarker, S.; Møller, H.B. Regulating feeding and increasing methane yield from co-digestion of C 5 molasses and cattle manure. *Energy Convers. Manag.* **2014**, *84*, 7–12. [CrossRef]
119. El-Mashad, H.M.; Zhang, R. Biogas production from co-digestion of dairy manure and food waste. *Bioresour. Technol.* **2010**, *101*, 4021–4028. [CrossRef] [PubMed]
120. Cavinato, C.; Fatone, F.; Bolzonella, D.; Pavan, P. Thermophilic anaerobic co-digestion of cattle manure with agro-wastes and energy crops: Comparison of pilot and full scale experiences. *Bioresour. Technol.* **2010**, *101*, 545–550. [CrossRef] [PubMed]
121. Callaghan, F.; Wase, D.; Thayanithy, K.; Forster, C. Continuous co-digestion of cattle slurry with fruit and vegetable wastes and chicken manure. *Biomass Bioenergy* **2002**, *22*, 71–77. [CrossRef]
122. Poulsen, T.G.; Adelard, L. Improving biogas quality and methane yield via co-digestion of agricultural and urban biomass wastes. *Waste Manag.* **2016**, *54*, 118–125. [CrossRef]
123. Mata-Alvarez, J. *Biomethanization of the Organic Fraction of Municipal Solid Wastes*; IWA Publishing: London, UK, 2003.
124. Bardiya, N.; Gaur, A. Effects of carbon and nitrogen ratio on rice straw biomethanation. *J. Rural Energy* **1997**, *4*, 1–16.
125. Muzenda, E. Bio-Methane Generation from Organic Waste: A Review. In Proceedings of the World Congress on Engineering and Computer Science, San Francisco, CA, USA, 22–24 October 2014.
126. Vintiloiu, A.; Lemmer, A.; Oechsner, H.; Jungbluth, T. Mineral substances and macronutrients in the anaerobic conversion of biomass: An impact evaluation. *Eng. Life Sci.* **2012**, *12*, 287–294. [CrossRef]
127. Gil, A.; Siles, A.J.; Serrano, A.; Chica, A.; Martín, M. Effect of variation in the C/[N+P] ratio on anaerobic digestion. *Environ. Prog. Sustain. Energy* **2018**, *38*, 228–236. [CrossRef]
128. Kothari, R.; Pandey, A.; Kumar, S.; Tyagi, V.; Tyagi, S. Different aspects of dry anaerobic digestion for bio-energy: An overview. *Renew. Sustain. Energy Rev.* **2014**, *39*, 174–195. [CrossRef]
129. Lay, C.-H.; Huang, C.-Y.; Chen, C.-C.; Lin, C.-Y. Biohydrogen production in an anaerobic baffled stacking reactor: Recirculation strategy and substrate concentration effects. *Biochem. Eng. J.* **2016**, *109*, 59–64. [CrossRef]
130. Hamdi, M.; Garcia, J.-L. Comparison between anaerobic filter and anaerobic contact process for fermented olive mill wastewaters. *Bioresour. Technol.* **1991**, *38*, 23–29. [CrossRef]
131. Elmitwalli, T.A.; Sklyar, V.; Zeeman, G.; Lettinga, G. Low temperature pre-treatment of domestic sewage in an anaerobic hybrid or an anaerobic filter reactor. *Bioresour. Technol.* **2002**, *82*, 233–239. [CrossRef]
132. Zhang, Z.-P.; Tay, J.-H.; Show, K.-Y.; Yan, R.; Liang, D.T.; Lee, D.-J.; Jiang, W.-J. Biohydrogen production in a granular activated carbon anaerobic fluidized bed reactor. *Int. J. Hydrogen Energy* **2007**, *32*, 185–191. [CrossRef]

133. Hung, C.-H.; Lee, K.-S.; Cheng, L.-H.; Huang, Y.-H.; Lin, P.-J.; Chang, J.-S. Quantitative analysis of a high-rate hydrogen-producing microbial community in anaerobic agitated granular sludge bed bioreactors using glucose as substrate. *Appl. Microbiol. Biotechnol.* **2007**, *75*, 693–701. [CrossRef]
134. Saddoud, A.; Hassaïri, I.; Sayadi, S. Anaerobic membrane reactor with phase separation for the treatment of cheese whey. *Bioresour. Technol.* **2007**, *98*, 2102–2108. [CrossRef]
135. Zhang, R.; Yin, Y.; Sung, S.; Dague, R. Anaerobic treatment of swine waste by the anaerobic sequencing batch reactor. *Trans. ASAE* **1997**, *40*, 767. [CrossRef]
136. Angenent, L.T.; Sung, S. Development of anaerobic migrating blanket reactor (AMBR), a novel anaerobic treatment system. *Water Res.* **2001**, *35*, 1739–1747. [CrossRef]
137. Xie, S.; Lawlor, P.G.; Frost, J.P.; Hu, Z.; Zhan, X. Effect of pig manure to grass silage ratio on methane production in batch anaerobic co-digestion of concentrated pig manure and grass silage. *Bioresour. Technol.* **2011**, *102*, 5728–5733. [CrossRef] [PubMed]
138. Lee, K.S.; Wu, J.F.; Lo, Y.S.; Lo, Y.C.; Lin, P.J.; Chang, J.S. Anaerobic hydrogen production with an efficient carrier-induced granular sludge bed bioreactor. *Biotechnol. Bioeng.* **2004**, *87*, 648–657. [CrossRef]
139. Wu, S.Y.; Hung, C.H.; Lin, C.N.; Chen, H.W.; Lee, A.S.; Chang, J.S. Fermentative hydrogen production and bacterial community structure in high-rate anaerobic bioreactors containing silicone-immobilized and self-flocculated sludge. *Biotechnol. Bioeng.* **2006**, *93*, 934–946. [CrossRef]
140. Fang, C.; Boe, K.; Angelidaki, I. Biogas production from potato-juice, a by-product from potato-starch processing, in upflow anaerobic sludge blanket (UASB) and expanded granular sludge bed (EGSB) reactors. *Bioresour. Technol.* **2011**, *102*, 5734–5741. [CrossRef]
141. Alves, M.; Vieira, J.M.; Pereira, R.A.; Pereira, M.; Mota, M. Effects of lipids and oleic acid on biomass development in anaerobic fixed-bed reactors. Part II: Oleic acid toxicity and biodegradability. *Water Res.* **2001**, *35*, 264–270. [CrossRef]
142. Jagadish, K.; Chanakya, H.; Rajabapaiah, P.; Anand, V. Plug flow digestors for biogas generation from leaf biomass. *Biomass Bioenergy* **1998**, *14*, 415–423. [CrossRef]
143. Huang, Z.; Ong, S.L.; Ng, H.Y. Submerged anaerobic membrane bioreactor for low-strength wastewater treatment: Effect of HRT and SRT on treatment performance and membrane fouling. *Water Res.* **2011**, *45*, 705–713. [CrossRef]
144. Chen, J.; Tang, C.; Zheng, P.; Zhang, L. Performance of lab-scale SPAC anaerobic bioreactor with high loading rate. *Chin. J. Biotechnol.* **2008**, *24*, 1413–1419. [CrossRef]
145. Sung, S.; Santha, H. Performance of temperature-phased anaerobic digestion (TPAD) system treating dairy cattle wastes. *Water Res.* **2003**, *37*, 1628–1636. [CrossRef]
146. Bouallagui, H.; Cheikh, R.B.; Marouani, L.; Hamdi, M. Mesophilic biogas production from fruit and vegetable waste in a tubular digester. *Bioresour. Technol.* **2003**, *86*, 85–89. [CrossRef]
147. Chiang, C.-F. Effects of Reactor Configuration on the Performance Of Static-Bed Submerged Media Anaerobic Reactors. Ph.D. Thesis, Iowa State University, Ames, IA, USA, 1988.
148. Intanoo, P.; Chaimongkol, P.; Chavadej, S. Hydrogen and methane production from cassava wastewater using two-stage upflow anaerobic sludge blanket reactors (UASB) with an emphasis on maximum hydrogen production. *Int. J. Hydrogen Energy* **2016**, *41*, 6107–6114. [CrossRef]
149. Chiumenti, A.; da Borso, F.; Limina, S. Dry anaerobic digestion of cow manure and agricultural products in a full-scale plant: Efficiency and comparison with wet fermentation. *Waste Manag.* **2018**, *71*, 704–710. [CrossRef]
150. Lissens, G.; Vandevivere, P.; De Baere, L.; Biey, E.; Verstraete, W. Solid waste digestors: Process performance and practice for municipal solid waste digestion. *Water Sci. Technol.* **2001**, *44*, 91–102. [CrossRef]
151. Heo, N.H.; Park, S.C.; Kang, H. Effects of mixture ratio and hydraulic retention time on single-stage anaerobic co-digestion of food waste and waste activated sludge. *J. Environ. Sci. Health Part A* **2004**, *39*, 1739–1756. [CrossRef]
152. Mata-Alvarez, J.; Cecchi, F.; Llabrés, P.; Pavan, P. Anaerobic digestion of the Barcelona central food market organic wastes. Plant design and feasibility study. *Bioresour. Technol.* **1992**, *42*, 33–42. [CrossRef]
153. Nasir, I.M.; Mohd Ghazi, T.I.; Omar, R. Anaerobic digestion technology in livestock manure treatment for biogas production: A review. *Eng. Life Sci.* **2012**, *12*, 258–269. [CrossRef]

154. Estevez, M.M.; Sapci, Z.; Linjordet, R.; Schnürer, A.; Morken, J. Semi-continuous anaerobic co-digestion of cow manure and steam-exploded Salix with recirculation of liquid digestate. *J. Environ. Manag.* **2014**, *136*, 9–15. [CrossRef]
155. Nordberg, Å.; Jarvis, Å.; Stenberg, B.; Mathisen, B.; Svensson, B.H. Anaerobic digestion of alfalfa silage with recirculation of process liquid. *Bioresour. Technol.* **2007**, *98*, 104–111. [CrossRef]
156. Verrier, D.; Roy, F.; Albagnac, G. Two-phase methanization of solid vegetable wastes. *Biol. Wastes* **1987**, *22*, 163–177. [CrossRef]
157. Bouallagui, H.; Touhami, Y.; Cheikh, R.B.; Hamdi, M. Bioreactor performance in anaerobic digestion of fruit and vegetable wastes. *Process Biochem.* **2005**, *40*, 989–995. [CrossRef]
158. Kübler, H.; Schertler, C. Three-phase anaerobic digestion of organic wastes. *Water Sci. Technol.* **1994**, *30*, 367–374. [CrossRef]
159. Kim, S.W.; Park, J.Y.; Kim, J.K.; Cho, J.H.; Chun, Y.N.; Lee, I.H.; Lee, J.S.; Park, J.S.; Park, D.-H. Development of a modified three-stage methane production process using food wastes. *Appl. Biochem. Biotechnol.* **2000**, *84*, 731–741. [CrossRef]
160. Rodriguez, R. *Upflow Anaerobic Sludge Blanket Reactor: Modelling*; The Royal Institute of Technology: Stockholm, Sweden, 2011.
161. Ding, J.N.; Wang, D.Z. Influence of external circulation on sludge characteristics during start-up of internal circulation reactor. *J. Cent. South Univ. Technol.* **2005**, *12*, 425–429. [CrossRef]
162. Visvanathan, C.; Abeynayaka, A. Developments and future potentials of anaerobic membrane bioreactors (AnMBRs). *Membr. Water Treat* **2012**, *3*, 1–23. [CrossRef]
163. Gaida, D.; Wolf, C.; Bongards, M. Feed control of anaerobic digestion processes for renewable energy production: A review. *Renew. Sustain. Energy Rev.* **2017**, *68*, 869–875. [CrossRef]
164. Barjenbruch, M.; Hoffmann, H.; Kopplow, O.; Tränckner, J. Minimizing of foaming in digesters by pre-treatment of the surplus-sludge. *Water Sci. Technol.* **2000**, *42*, 235–241. [CrossRef]
165. Karim, K.; Hoffmann, R.; Klasson, K.T.; Al-Dahhan, M. Anaerobic digestion of animal waste: Effect of mode of mixing. *Water Res.* **2005**, *39*, 3597–3606. [CrossRef] [PubMed]
166. Chen, T.; Chynoweth, P.; Biljetina, R. Anaerobic digestion of municipal solid waste in a nonmixed solids concentrating digestor. *Appl. Biochem. Biotechnol.* **1990**, *24*, 533–544. [CrossRef]
167. Özdemir, G. *Optimering av Biogasproduktion Från Gödsel*; Linnaeus University: Växjö, Sweden, 2009.
168. Wang, B.; Wu, D.; Ekama, G.A.; Huang, H.; Lu, H.; Chen, G.-H. Optimizing mixing mode and intensity to prevent sludge flotation in sulfidogenic anaerobic sludge bed reactors. *Water Res.* **2017**, *122*, 481–491. [CrossRef]
169. Ong, H.; Greenfield, P.; Pullammanappallil, P. Effect of mixing on biomethanation of cattle-manure slurry. *Environ. Technol.* **2002**, *23*, 1081–1090. [CrossRef]
170. Kaparaju, P.; Buendia, I.; Ellegaard, L.; Angelidakia, I. Effects of mixing on methane production during thermophilic anaerobic digestion of manure: Lab-scale and pilot-scale studies. *Bioresour. Technol.* **2008**, *99*, 4919–4928. [CrossRef]
171. Dachs, G.; Rehm, W. *Der Eigenstromverbrauch von Biogasanlagen und Potenziale zu Dessen Reduzierung*; Solarenergieförderverein Bayern: München, Germany, 2006.
172. Kowalczyk, A.; Harnisch, E.; Schwede, S.; Gerber, M.; Span, R. Different mixing modes for biogas plants using energy crops. *Appl. Energy* **2013**, *112*, 465–472. [CrossRef]
173. Kowalczyk, A.A. Untersuchung der Übertragbarkeit und Reproduzierbarkeit von Versuchen zur Biogasbildung in kontinuierlichen Prozessen sowie erste Studien zu deren Optimierung. Ph.D. Thesis, Ruhr-Universität Bochum, Bochum, Germany, 2012.
174. Teng, Z.; Hua, J.; Wang, C.; Lu, X. Chapter 4—Design and optimization principles of biogas reactors in large scale applications A2—Shi, Fan. In *Reactor and Process Design in Sustainable Energy Technology*; Elsevier: Amsterdam, The Netherlands, 2014; pp. 99–134.
175. Prieto, A.L.; Futselaar, H.; Lens, P.N.; Bair, R.; Yeh, D.H. Development and start up of a gas-lift anaerobic membrane bioreactor (Gl-AnMBR) for conversion of sewage to energy, water and nutrients. *J. Membr. Sci.* **2013**, *111*, 158–167. [CrossRef]
176. Xu, F.; Huang, Z.; Miao, H.; Ren, H.; Zhao, M.; Ruan, W. Identical full-scale biogas-lift reactors (BLRs) with anaerobic granular sludge and residual activated sludge for brewery wastewater treatment and kinetic modeling. *J. Environ. Sci.* **2013**, *25*, 2031–2040. [CrossRef]

177. Schnürer, A.; Jarvis, Å. *Microbiological Handbook for Biogas Plants-Swedish Waste Management U2009: 03, Swedish Gas Centre Report 207*; Avfall Sverige, Svenskt Gastekniskt Center AB: Malmo, Sweden, 2009.
178. Bramstedt, S. *Temperature Optimization of Anaerobic Digestion at the Käppala Waste Water Treatment Plant*; The Royal Institute of Technology: Stockholm, Sweden, 2016.
179. Bouallagui, H.; Torrijos, M.; Godon, J.; Moletta, R.; Cheikh, R.B.; Touhami, Y.; Delgenes, J.; Hamdi, M. Two-phases anaerobic digestion of fruit and vegetable wastes: Bioreactors performance. *Biochem. Eng. J.* **2004**, *21*, 193–197. [CrossRef]
180. Cysneiros, D.; Thuillier, A.; Villemont, R.; Littlestone, A.; Mahony, T.; O'Flaherty, V. Temperature effects on the trophic stages of perennial rye grass anaerobic digestion. *Water Sci. Technol.* **2011**, *64*, 70–76. [CrossRef]
181. Neshat, S.A.; Mohammadi, M.; Najafpour, G.D.; Lahijani, P. Anaerobic co-digestion of animal manures and lignocellulosic residues as a potent approach for sustainable biogas production. *Renew. Sustain. Energy Rev.* **2017**, *79*, 308–322. [CrossRef]
182. Angelidaki, I.; Sanders, W. Assessment of the anaerobic biodegradability of macropollutants. *Rev. Environ. Sci. Biotechnol.* **2004**, *3*, 117–129. [CrossRef]
183. Nelson, R. *Methane Generation from Anaerobic Digesters: Considering Different Substrates*; Environmental Biotechnology, Lowa State University: Ames, IA, USA, 2010.
184. Angelidaki, I.; Ellegaard, L.; Ahring, B.K. Applications of the anaerobic digestion process. In *Biomethanation II*; Springer: Berlin, Germany, 2003; pp. 1–33.
185. Labatut, R.A.; Angenent, L.T.; Scott, N.R. Conventional mesophilic vs. thermophilic anaerobic digestion: A trade-off between performance and stability? *Water Res.* **2014**, *53*, 249–258. [CrossRef]
186. Nielsen, M.; Holst-Fischer, C.; Malmgren-Hansen, B.; Bjerg-Nielsen, M.; Kragelund, C.; Møller, H.B.; Ottosen, L.D. Small temperature differences can improve the performance of mesophilic sludge-based digesters. *Biotechnol. Lett.* **2017**, *39*, 1689–1698. [CrossRef] [PubMed]
187. David, B.; Francesco, F.; Paolo, P.; Franco, C. Mesophilic, thermophilic and temperature phased anaerobic digestion of waste activated sludge. *Chem. Eng. Trans.* **2008**, *17*, 385–390.
188. Voelklein, M.A.; O'Shea, R.; Jacob, A.; Murphy, J.D. Role of trace elements in single and two-stage digestion of food waste at high organic loading rates. *Energy* **2017**, *121*, 185–192. [CrossRef]
189. Puchajda, B.; Oleszkiewicz, J. Impact of sludge thickening on energy recovery from anaerobic digestion. *Water Sci. Technol.* **2008**, *57*, 395–401. [CrossRef]
190. Hartmann, H.; Ahring, B.K. A novel process configuration for anaerobic digestion of source-sorted household waste using hyper-thermophilic post-treatment. *Biotechnol. Bioeng.* **2005**, *90*, 830–837. [CrossRef]
191. Alepu, O.; Li, Z.; Ikhumhen, H.; Kalakodio, L.; Wang, K. Effect of Hydraulic Retention Time on Anaerobic Digestion of Xiao Jiahe Municipal Sludge. *Int. J. Waste Resour.* **2016**, *6*, 2.
192. Passos, F.; García, J.; Ferrer, I. Impact of low temperature pretreatment on the anaerobic digestion of microalgal biomass. *Bioresour. Technol.* **2013**, *138*, 79–86. [CrossRef]
193. Lohani, S.P.; Wang, S.; Bergland, W.H.; Khanal, S.N.; Bakke, R. Modeling temperature effects in anaerobic digestion of domestic wastewater. *Water-Energy Nexus* **2018**, *1*, 56–60. [CrossRef]
194. Verma, S. Anaerobic Digestion of Biodegradable Organics in Municipal Solid Wastes. Master's Thesis, Columbia University, New York, NY, USA, 2002.
195. Liu, D.; Zeng, R.J.; Angelidaki, I. Effects of pH and hydraulic retention time on hydrogen production versus methanogenesis during anaerobic fermentation of organic household solid waste under extreme-thermophilic temperature (70 °C). *Biotechnol. Bioeng.* **2008**, *100*, 1108–1114. [CrossRef]
196. Sibiya, N.T.; Muzenda, E.; Tesfagiorgis, H.B. Effect of Temperature and pH on the Anaerobic Digestion of Grass Silage. In Proceedings of the 6th International Conference on Green Technology, Renewable Energy and Environmental Engineering, Cape Town, South Africa, 27–28 November 2014.
197. Liu, Y.; Zhang, Y.; Quan, X.; Li, Y.; Zhao, Z.; Meng, X.; Chen, S. Optimization of anaerobic acidogenesis by adding Fe 0 powder to enhance anaerobic wastewater treatment. *Chem. Eng. J.* **2012**, *192*, 179–185. [CrossRef]
198. Georgacakis, D.; Sievers, D.; Iannotti, E.J. Buffer stability in manure digesters. *Agric. Wastes* **1982**, *4*, 427–441. [CrossRef]
199. Sommer, S.G.; Husted, S. A simple model of pH in slurry. *J. Agric. Sci.* **1995**, *124*, 447–453. [CrossRef]
200. Kovács, E.; Wirth, R.; Maróti, G.; Bagi, Z.; Nagy, K.; Minárovits, J.; Rákhely, G.; Kovács, K.L. Augmented biogas production from protein-rich substrates and associated metagenomic changes. *Bioresour. Technol.* **2015**, *178*, 254–261. [CrossRef]

201. Yu, D.; Liu, J.; Sui, Q.; Wei, Y. Biogas-pH automation control strategy for optimizing organic loading rate of anaerobic membrane bioreactor treating high COD wastewater. *Bioresour. Technol.* **2016**, *203*, 62–70. [CrossRef] [PubMed]
202. Wu, C.; Huang, Q.; Yu, M.; Ren, Y.; Wang, Q.; Sakai, K. Effects of digestate recirculation on a two-stage anaerobic digestion system, particularly focusing on metabolite correlation analysis. *Bioresour. Technol.* **2018**, *251*, 40–48. [CrossRef]
203. Demirel, B.; Yenigun, O. Anaerobic acidogenesis of dairy wastewater: The effects of variations in hydraulic retention time with no pH control. *J. Chem. Technol. Biotechnol. Int. Res. Process Environ. Clean Technol.* **2004**, *79*, 755–760. [CrossRef]
204. Nayono, S.E.; Gallert, C.; Winter, J. Co-digestion of press water and food waste in a biowaste digester for improvement of biogas production. *Bioresour. Technol.* **2010**, *101*, 6987–6993. [CrossRef]
205. Yoon, S.-H.; Kim, H.-S.; Yeom, I.-T. The optimum operational condition of membrane bioreactor (MBR): Cost estimation of aeration and sludge treatment. *Water Res.* **2004**, *38*, 37–46. [CrossRef]
206. Buysman, E. Anaerobic Digestion for Developing Countries with Cold Climates. Master's Thesis, Wageningen University, Wageningen, The Netherlands, 2009.
207. Ueno, Y.; Tatara, M.; Fukui, H.; Makiuchi, T.; Goto, M.; Sode, K. Production of hydrogen and methane from organic solid wastes by phase-separation of anaerobic process. *Bioresour. Technol.* **2007**, *98*, 1861–1865. [CrossRef]
208. Yang, F.; Chen, R.; Yue, Z.; Liao, W.; Marsh, T.L. Phylogenetic analysis of anaerobic co-digestion of animal manure and corn stover reveals linkages between bacterial communities and digestion performance. *AiM* **2016**, *6*, 879. [CrossRef]
209. Nges, I.A.; Liu, J. Effects of solid retention time on anaerobic digestion of dewatered-sewage sludge in mesophilic and thermophilic conditions. *Renew. Energy* **2010**, *35*, 2200–2206. [CrossRef]
210. Akram, A.; Stuckey, D. Biomass acclimatisation and adaptation during start-up of a submerged anaerobic membrane bioreactor (SAMBR). *Environ. Technol.* **2008**, *29*, 1053–1065. [CrossRef]
211. Vanwonterghem, I.; Jensen, P.D.; Rabaey, K.; Tyson, G.W. Temperature and solids retention time control microbial population dynamics and volatile fatty acid production in replicated anaerobic digesters. *Sci. Rep.* **2015**, *5*, 8496. [CrossRef]
212. Gagliano, M.; Braguglia, C.; Gallipoli, A.; Gianico, A.; Rossetti, S. Microbial diversity in innovative mesophilic/thermophilic temperature-phased anaerobic digestion of sludge. *Environ. Sci. Pollut. Res.* **2015**, *22*, 7339–7348. [CrossRef] [PubMed]
213. Krakat, N.; Demirel, B.; Anjum, R.; Dietz, D. Methods of ammonia removal in anaerobic digestion: A review. *Water Sci. Technol.* **2017**, *76*, 1925–1938. [CrossRef] [PubMed]
214. Yabu, H.; Sakai, C.; Fujiwara, T.; Nishio, N.; Nakashimada, Y. Thermophilic two-stage dry anaerobic digestion of model garbage with ammonia stripping. *J. Biosci. Bioeng.* **2011**, *111*, 312–319. [CrossRef] [PubMed]
215. Gallert, C.; Winter, J. Mesophilic and thermophilic anaerobic digestion of source-sorted organic wastes: Effect of ammonia on glucose degradation and methane production. *Appl. Microbiol. Biotechnol.* **1997**, *48*, 405–410. [CrossRef]
216. Bujoczek, G.; Oleszkiewicz, J.; Sparling, R.; Cenkowski, S. High solid anaerobic digestion of chicken manure. *J. Agric. Eng. Res.* **2000**, *76*, 51–60. [CrossRef]
217. Sahu, N.; Deshmukh, S.; Chandrashekhar, B.; Sharma, G.; Kapley, A.; Pandey, R. Optimization of hydrolysis conditions for minimizing ammonia accumulation in two-stage biogas production process using kitchen waste for sustainable process development. *J. Environ. Chem. Eng.* **2017**, *5*, 2378–2387. [CrossRef]
218. Emerson, K.; Russo, R.C.; Lund, R.E.; Thurston, R.V. Aqueous ammonia equilibrium calculations: Effect of pH and temperature. *J. Fish. Board Can.* **1975**, *32*, 2379–2383. [CrossRef]
219. Hansen, K.H.; Angelidaki, I.; Ahring, B.K. Anaerobic digestion of swine manure: Inhibition by ammonia. *Water Res.* **1998**, *32*, 5–12. [CrossRef]
220. De Baere, L.; Devocht, M.; Van Assche, P.; Verstraete, W. Influence of high NaCl and NH_4Cl salt levels on methanogenic associations. *Water Res.* **1984**, *18*, 543–548. [CrossRef]
221. Farrow, C. *Anaerobic Digestion of Poultry Manure: Implementation of Ammonia Control to Optimize Biogas Yield*; University of Guelph: Guelph, ON, Canada, 2016.
222. Kroeker, E.; Schulte, D.; Sparling, A.; Lapp, H. Anaerobic treatment process stability. *J. Water Pollut. Control Fed.* **1979**, *51*, 718–727.

223. Niu, Q.; Qiao, W.; Qiang, H.; Li, Y.-Y. Microbial community shifts and biogas conversion computation during steady, inhibited and recovered stages of thermophilic methane fermentation on chicken manure with a wide variation of ammonia. *Bioresour. Technol.* **2013**, *146*, 223–233. [CrossRef] [PubMed]
224. Song, M.; Shin, S.G.; Hwang, S. Methanogenic population dynamics assessed by real-time quantitative PCR in sludge granule in upflow anaerobic sludge blanket treating swine wastewater. *Bioresour. Technol.* **2010**, *101*, S23–S28. [CrossRef] [PubMed]
225. Jarrell, K.F.; Saulnier, M.; Ley, A. Inhibition of methanogenesis in pure cultures by ammonia, fatty acids, and heavy metals, and protection against heavy metal toxicity by sewage sludge. *Can. J. Microbiol.* **1987**, *33*, 551–554. [CrossRef]
226. Schnürer, A.; Nordberg, Å. Ammonia, a selective agent for methane production by syntrophic acetate oxidation at mesophilic temperature. *Water Sci. Technol.* **2008**, *57*, 735–740. [CrossRef]
227. Gallert, C.; Bauer, S.; Winter, J. Effect of ammonia on the anaerobic degradation of protein by a mesophilic and thermophilic biowaste population. *Appl. Microbiol. Biotechnol.* **1998**, *50*, 495–501. [CrossRef]
228. Angelidaki, I.; Ahring, B. Anaerobic thermophilic digestion of manure at different ammonia loads: Effect of temperature. *Water Res.* **1994**, *28*, 727–731. [CrossRef]
229. Chen, Y.; Cheng, J.J.; Creamer, K.S. Inhibition of anaerobic digestion process: A review. *Bioresour. Technol.* **2008**, *99*, 4044–4064. [CrossRef]
230. Bonmatı, A.; Flotats, X. Air stripping of ammonia from pig slurry: Characterisation and feasibility as a pre-or post-treatment to mesophilic anaerobic digestion. *Waste Manag.* **2003**, *23*, 261–272. [CrossRef]
231. Zhang, Y.; Angelidaki, I. Submersible microbial desalination cell for simultaneous ammonia recovery and electricity production from anaerobic reactors containing high levels of ammonia. *Bioresour. Technol.* **2015**, *177*, 233–239. [CrossRef]
232. Bodalo, A.; Gomez, J.-L.; Gomez, E.; Leon, G.; Tejera, M. Ammonium removal from aqueous solutions by reverse osmosis using cellulose acetate membranes. *Desalination* **2005**, *184*, 149–155. [CrossRef]
233. Serna-Maza, A.; Heaven, S.; Banks, C. Biogas stripping of ammonia from fresh digestate from a food waste digester. *Bioresour. Technol.* **2015**, *190*, 66–75. [CrossRef]
234. Nzila, A. Mini review: Update on bioaugmentation in anaerobic processes for biogas production. *Anaerobe* **2016**, *46*, 3–12. [CrossRef]
235. Kimura, Y.; Isaka, K.; Kazama, F.; Sumino, T. Effects of nitrite inhibition on anaerobic ammonium oxidation. *Appl. Microbiol. Biotechnol.* **2010**, *86*, 359–365. [CrossRef]
236. Lin, L.; Chen, J.; Xu, Z.; Yuan, S.; Cao, M.; Liu, H.; Lu, X. Removal of ammonia nitrogen in wastewater by microwave radiation: A pilot-scale study. *J. Hazard. Mater.* **2009**, *168*, 862–867. [CrossRef]
237. Çelen, I.; Türker, M. Recovery of ammonia as struvite from anaerobic digester effluents. *Environ. Technol.* **2001**, *22*, 1263–1272. [CrossRef]
238. Zeshan; Karthikeyan, O.P.; Visvanathan, C. Effect of C/N ratio and ammonia-N accumulation in a pilot-scale thermophilic dry anaerobic digester. *Bioresour. Technol.* **2012**, *113*, 294–302. [CrossRef]
239. Ukwuani, A.T.; Tao, W. Developing a vacuum thermal stripping–acid absorption process for ammonia recovery from anaerobic digester effluent. *Water Res.* **2016**, *106*, 108–115. [CrossRef] [PubMed]
240. Labatut, R.A.; Gooch, C.A. Monitoring of Anaerobic Digestion Process to Optimize Performance and Prevent System Failure. In Proceedings of the Got Manure? Enhancing Environmental and Economic Sustainability Conference, Liverpool, NY, USA, 28–29 March 2012.
241. Inanc, B.; Matsui, S.; Ide, S. Propionic acid accumulation in anaerobic digestion of carbohydrates: An investigation on the role of hydrogen gas. *Water Sci. Technol.* **1999**, *40*, 93–100. [CrossRef]
242. Khan, M.; Ngo, H.H.; Guo, W.; Liu, Y.; Nghiem, L.D.; Hai, F.I.; Deng, L.; Wang, J.; Wu, Y. Optimization of process parameters for production of volatile fatty acid, biohydrogen and methane from anaerobic digestion. *Bioresour. Technol.* **2016**, *219*, 738–748. [CrossRef]
243. Hao, T.-W.; Luo, J.-H.; Wei, L.; Mackey, H.R.; Chi, K.; Chen, G.-H. Example study for granular bioreactor stratification: Three-dimensional evaluation of a sulfate-reducing granular bioreactor. *Sci. Rep.* **2016**, *6*, 31718. [CrossRef] [PubMed]
244. Wijekoon, K.C.; Visvanathan, C.; Abeynayaka, A. Effect of organic loading rate on VFA production, organic matter removal and microbial activity of a two-stage thermophilic anaerobic membrane bioreactor. *Bioresour. Technol.* **2011**, *102*, 5353–5360. [CrossRef]

245. Yuan, Q.; Sparling, R.; Oleszkiewicz, J.A. VFA generation from waste activated sludge: Effect of temperature and mixing. *Chemosphere* **2011**, *82*, 603–607. [CrossRef]
246. Öztürk, M. Degradation of acetate, propionate, and butyrate under shock temperature. *J. Environ. Eng.* **1993**, *119*, 321–331. [CrossRef]
247. Yu, L.; Bule, M.; Ma, J.; Zhao, Q.; Frear, C.; Chen, S. Enhancing volatile fatty acid (VFA) and bio-methane production from lawn grass with pretreatment. *Bioresour. Technol.* **2014**, *162*, 243–249. [CrossRef]
248. Wu, H.; Gao, J.; Yang, D.; Zhou, Q.; Liu, W. Alkaline fermentation of primary sludge for short-chain fatty acids accumulation and mechanism. *Chem. Eng. J.* **2010**, *160*, 1–7. [CrossRef]
249. Sarker, S.; Lamb, J.J.; Hjelme, D.R.; Lien, K.M. Overview of recent progress towards in-situ biogas upgradation techniques. *Fuel* **2018**, *226*, 686–697. [CrossRef]
250. Speece, R.; Boonyakitsombut, S.; Kim, M.; Azbar, N.; Ursillo, P. Overview of Anaerobic Treatment: Thermophilic and Propionate Implications-Keynote Address—Association of Environmental Engineering and Science Professors—78th Annual Water Environment Federation Technical Exposition and Conference, Washington, DC, Oct. 29–Nov. 2, 2005. *Water Environ. Res.* **2006**, *78*, 460–473.
251. Fukuzaki, S.; Nishio, N.; Shobayashi, M.; Nagai, S. Inhibition of the fermentation of propionate to methane by hydrogen, acetate, and propionate. *Appl. Environ. Microbiol.* **1990**, *56*, 719–723.
252. Liu, X.; Wang, W.; Shi, Y.; Zheng, L.; Gao, X.; Qiao, W.; Zhou, Y. Pilot-scale anaerobic co-digestion of municipal biomass waste and waste activated sludge in China: Effect of organic loading rate. *Waste Manag.* **2012**, *32*, 2056–2060. [CrossRef]
253. Liu, C.; Wang, W.; Anwar, N.; Ma, Z.; Liu, G.; Zhang, R. Effect of organic loading rate on anaerobic digestion of food waste under mesophilic and thermophilic conditions. *Energy Fuels* **2017**, *31*, 2976–2984. [CrossRef]
254. Zhang, W.; Lang, Q.; Pan, Z.; Jiang, Y.; Liebetrau, J.; Nelles, M.; Dong, H.; Dong, R. Performance evaluation of a novel anaerobic digestion operation process for treating high-solids content chicken manure: Effect of reduction of the hydraulic retention time at a constant organic loading rate. *Waste Manag.* **2017**, *64*, 340–347. [CrossRef]
255. Zhang, C.; Su, H.; Wang, Z.; Tan, T.; Qin, P. Biogas by semi-continuous anaerobic digestion of food waste. *Appl. Biochem. Biotechnol.* **2015**, *175*, 3901–3914. [CrossRef]
256. Ferguson, R.M.; Coulon, F.; Villa, R. Organic loading rate: A promising microbial management tool in anaerobic digestion. *Water Res.* **2016**, *100*, 348–356. [CrossRef]
257. Shen, F.; Li, H.; Wu, X.; Wang, Y.; Zhang, Q. Effect of organic loading rate on anaerobic co-digestion of rice straw and pig manure with or without biological pretreatment. *Bioresour. Technol.* **2018**, *250*, 155–162. [CrossRef]
258. Ding, L.; Gutierrez, E.C.; Cheng, J.; Xia, A.; O'Shea, R.; Guneratnam, A.J.; Murphy, J.D. Assessment of continuous fermentative hydrogen and methane co-production using macro-and micro-algae with increasing organic loading rate. *Energy* **2018**, *151*, 760–770. [CrossRef]
259. Zhang, E.; Li, J.; Zhang, K.; Wang, F.; Yang, H.; Zhi, S.; Liu, G. Anaerobic digestion performance of sweet potato vine and animal manure under wet, semi-dry, and dry conditions. *AMB Express* **2018**, *8*, 45. [CrossRef]
260. Polizzi, C.; Alatriste-Mondragón, F.; Munz, G. The role of organic load and ammonia inhibition in anaerobic digestion of tannery fleshing. *Water Resour. Ind.* **2018**, *19*, 25–34. [CrossRef]
261. Wang, S.; Hawkins, G.L.; Kiepper, B.H.; Das, K.C. Treatment of slaughterhouse blood waste using pilot scale two-stage anaerobic digesters for biogas production. *Renew. Energy* **2018**, *126*, 552–562. [CrossRef]
262. Solli, L.; Schnürer, A.; Horn, S.J. Process performance and population dynamics of ammonium tolerant microorganisms during co-digestion of fish waste and manure. *Renew. Energy* **2018**, *125*, 529–536. [CrossRef]
263. Muñoz, P.J.W.; Valorization, B. Assessment of Batch and Semi-continuous Anaerobic Digestion of Food Waste at Psychrophilic Range at Different Food Waste to Inoculum Ratios and Organic Loading Rates. *Waste Biomass Valoriz.* **2018**, 1–10. [CrossRef]
264. Micolucci, F.; Gottardo, M.; Pavan, P.; Cavinato, C.; Bolzonella, D. Pilot scale comparison of single and double-stage thermophilic anaerobic digestion of food waste. *J. Clean. Prod.* **2018**, *171*, 1376–1385. [CrossRef]
265. Fontana, A.; Campanaro, S.; Treu, L.; Kougias, P.G.; Cappa, F.; Morelli, L.; Angelidaki, I. Performance and genome-centric metagenomics of thermophilic single and two-stage anaerobic digesters treating cheese wastes. *Water Res.* **2018**, *134*, 181–191. [CrossRef] [PubMed]

266. Park, J.-H.; Kumar, G.; Yun, Y.-M.; Kwon, J.-C.; Kim, S.-H. Effect of feeding mode and dilution on the performance and microbial community population in anaerobic digestion of food waste. *Bioresour. Technol.* **2018**, *248*, 134–140. [CrossRef] [PubMed]
267. Li, W.; Loh, K.-C.; Zhang, J.; Tong, Y.W.; Dai, Y. Two-stage anaerobic digestion of food waste and horticultural waste in high-solid system. *Appl. Energy* **2018**, *209*, 400–408. [CrossRef]
268. Morken, J.; Gjetmundsen, M.; Fjørtoft, K. Determination of kinetic constants from the co-digestion of dairy cow slurry and municipal food waste at increasing organic loading rates. *Renew. Energy* **2018**, *117*, 46–51. [CrossRef]
269. Eslami, H.; Hashemi, H.; Fallahzadeh, R.A.; Khosravi, R.; Fard, R.F.; Ebrahimi, A.A. Effect of organic loading rates on biogas production and anaerobic biodegradation of composting leachate in the anaerobic series bioreactors. *Ecol. Eng.* **2018**, *110*, 165–171. [CrossRef]
270. Agneessens, L.M.; Ottosen, L.D.M.; Andersen, M.; Olesen, C.B.; Feilberg, A.; Kofoed, M.V. Parameters affecting acetate concentrations during in-situ biological hydrogen methanation. *Bioresour. Technol.* **2018**, *258*, 33–40. [CrossRef]
271. Wickham, R.; Xie, S.; Galway, B.; Bustamante, H.; Nghiem, L.D. Anaerobic digestion of soft drink beverage waste and sewage sludge. *Bioresour. Technol.* **2018**, *262*, 141–147. [CrossRef]
272. Petersson, A.; Wellinger, A. Biogas upgrading technologies—Developments and innovations. *IEA Bioenergy* **2009**, *20*, 1–19.
273. Vavilin, V.; Vasiliev, V.; Rytov, S. Modelling of gas pressure effects on anaerobic digestion. *Bioresour. Technol.* **1995**, *52*, 25–32. [CrossRef]
274. Singh, G.; Jain, V.K.; Singh, A. Effect of Temperature and other factors on Anaerobic Digestion Process, responsible for Bio Gas Production. *Int. J. Theor. Appl. Mech.* **2017**, *12*, 637–657.
275. Lindeboom, R.; Fermoso, F.; Weijma, J.; Zagt, K.; Van Lier, J. Autogenerative high pressure digestion: Anaerobic digestion and biogas upgrading in a single step reactor system. *Water Sci. Technol.* **2011**, *64*, 647–653. [CrossRef] [PubMed]

© 2019 by the authors. Licensee MDPI, Basel, Switzerland. This article is an open access article distributed under the terms and conditions of the Creative Commons Attribution (CC BY) license (http://creativecommons.org/licenses/by/4.0/).

MDPI
St. Alban-Anlage 66
4052 Basel
Switzerland
Tel. +41 61 683 77 34
Fax +41 61 302 89 18
www.mdpi.com

Applied Sciences Editorial Office
E-mail: applsci@mdpi.com
www.mdpi.com/journal/applsci

www.ingramcontent.com/pod-product-compliance
Lightning Source LLC
Chambersburg PA
CBHW040315270926
51850CB00007BC/1487

9783036511429